Progress of Fiber-Reinforced Composites: Design and Applications

Progress of Fiber-Reinforced Composites: Design and Applications

Editor

Ioannis Kartsonakis

MDPI • Basel • Beijing • Wuhan • Barcelona • Belgrade • Manchester • Tokyo • Cluj • Tianjin

Editor
Ioannis Kartsonakis
School of Chemical Engineering
National Technical University of Athens
Zographos
Greece

Editorial Office
MDPI
St. Alban-Anlage 66
4052 Basel, Switzerland

This is a reprint of articles from the Special Issue published online in the open access journal *Applied Sciences* (ISSN 2076-3417) (available at: www.mdpi.com/journal/applsci/special_issues/fiber_reinforced_composites).

For citation purposes, cite each article independently as indicated on the article page online and as indicated below:

LastName, A.A.; LastName, B.B.; LastName, C.C. Article Title. *Journal Name* **Year**, *Volume Number*, Page Range.

ISBN 978-3-0365-5182-1 (Hbk)
ISBN 978-3-0365-5181-4 (PDF)

Cover image courtesy of Ioannis Kartsonakis

© 2022 by the authors. Articles in this book are Open Access and distributed under the Creative Commons Attribution (CC BY) license, which allows users to download, copy and build upon published articles, as long as the author and publisher are properly credited, which ensures maximum dissemination and a wider impact of our publications.
The book as a whole is distributed by MDPI under the terms and conditions of the Creative Commons license CC BY-NC-ND.

Contents

About the Editor vii

Preface to "Progress of Fiber-Reinforced Composites: Design and Applications" ix

Ioannis A. Kartsonakis
Special Issue on "Progress of Fiber-Reinforced Composites: Design and Applications"
Reprinted from: *Appl. Sci.* **2022**, *12*, 8030, doi:10.3390/app12168030 1

Juliane Troschitz, Julian Vorderbrüggen, Robert Kupfer, Maik Gude and Gerson Meschut
Joining of Thermoplastic Composites with Metals Using Resistance Element Welding
Reprinted from: *Appl. Sci.* **2020**, *10*, 7251, doi:10.3390/app10207251 5

Yanlei Wang, Wanxin Zhu, Xue Zhang, Gaochuang Cai and Baolin Wan
Influence of Thickness on Water Absorption and Tensile Strength of BFRP Laminates in Water or Alkaline Solution and a Thickness-Dependent Accelerated Ageing Method for BFRP Laminates
Reprinted from: *Appl. Sci.* **2020**, *10*, 3618, doi:10.3390/app10103618 17

Yang Peng, Wei Chen, Zhe Wu, Jun Zhao and Jun Dong
Experimental Study on the Performance of GFRP–GFRP Slip-Critical Connections with and without Stainless-Steel Cover Plates
Reprinted from: *Appl. Sci.* **2020**, *10*, 4393, doi:10.3390/app10124393 39

Donato Di Vito, Mikko Kanerva, Jan Järveläinen, Alpo Laitinen, Tuomas Pärnänen and Kari Saari et al.
Safe and Sustainable Design of Composite Smart Poles for Wireless Technologies
Reprinted from: *Appl. Sci.* **2020**, *10*, 7594, doi:10.3390/app10217594 55

A. Riccio, S. Saputo, A. Sellitto and F. Di Caprio
A Numerical Assessment on the Influences of Material Toughness on the Crashworthiness of a Composite Fuselage Barrel
Reprinted from: *Appl. Sci.* **2020**, *10*, 2019, doi:10.3390/app10062019 75

Andrea Sellitto, Salvatore Saputo, Angela Russo, Vincenzo Innaro, Aniello Riccio and Francesco Acerra et al.
Numerical-Experimental Investigation into the Tensile Behavior of a Hybrid Metallic–CFRP Stiffened Aeronautical Panel
Reprinted from: *Appl. Sci.* **2020**, *10*, 1880, doi:10.3390/app10051880 91

Mohamad Zaki Hassan, S. M. Sapuan, Zainudin A. Rasid, Ariff Farhan Mohd Nor, Rozzeta Dolah and Mohd Yusof Md Daud
Impact Damage Resistance and Post-Impact Tolerance of Optimum Banana-Pseudo-Stem-Fiber-Reinforced Epoxy Sandwich Structures
Reprinted from: *Appl. Sci.* **2020**, *10*, 684, doi:10.3390/app10020684 109

Wei Wang, Chen Zhang, Jia Guo, Na Li, Yuan Li and Hang Zhou et al.
Investigation on the Triaxial Mechanical Characteristics of Cement-Treated Subgrade Soil Admixed with Polypropylene Fiber
Reprinted from: *Appl. Sci.* **2019**, *9*, 4557, doi:10.3390/app9214557 131

Jong-Han Lee, Eunsoo Choi and Baik-Soon Cho
Shear Failure Mode and Concrete Edge Breakout Resistance of Cast-In-Place Anchors in Steel Fiber-Reinforced Normal Strength Concrete
Reprinted from: *Appl. Sci.* **2020**, *10*, 6883, doi:10.3390/app10196883 147

Kailong Xu, Wei Chen, Lulu Liu, Gang Luo and Zhenhua Zhao
Longitudinal Compressive Property of Three-Dimensional Four-Step Braided Composites after Cyclic Hygrothermal Aging under High Strain Rates
Reprinted from: *Appl. Sci.* **2020**, *10*, 2061, doi:10.3390/app10062061 . . . . **167**

Yanfeng Zhang, Zhengong Zhou and Zhiyong Tan
Compression Shear Properties of Bonded–Bolted Hybrid Single-Lap Joints of C/C Composites at High Temperature
Reprinted from: *Appl. Sci.* **2020**, *10*, 1054, doi:10.3390/app10031054 . . . . **185**

Panagiotis Goulis, Ioannis A. Kartsonakis, George Konstantopoulos and Costas A. Charitidis
Synthesis and Processing of Melt Spun Materials from Esterified Lignin with Lactic Acid
Reprinted from: *Appl. Sci.* **2019**, *9*, 5361, doi:10.3390/app9245361 . . . . **199**

About the Editor

Ioannis Kartsonakis

Ioannis Kartsonakis is a chemist with a PhD in Chemistry and a MSc in Polymer science. His PhD subject was on electrochemical deposition of multifunctional conductive polymers for corrosion protection of metals (doi: 10.12681/eadd/25933), using sol-gel technique, electro-polymerization and radical polymerization. His work is focused on: Management of R&D projects; implementation of research and development tasks in the field of corrosion and corrosion protection; studies on physicochemical properties of nanomaterials; and synthesis and characterization of hybrid inorganic/organic nanomaterials, coatings, fibers, and cement/polymer composites. He has experience in many European and National projects related to: recycling of textile and plastic materials; concrete reinforcement; self-healing, antifouling, anticorrosion, coatings and materials.

Preface to "Progress of Fiber-Reinforced Composites: Design and Applications"

Fiber-reinforced composites (FRC) are hybrid materials consisting of metallic, polymeric, or ceramic matrices in which reinforcing fibers of high modulus and strength are incorporated in or bonded to. Therefore, FRC materials consist of three components that are the fibers as the discontinuous or dispersed phase, the matrix as the continuous phase, and the fine interphase region. The combination of each phase performance enables these materials to achieve an assembly of properties that cannot be obtained with the constituent acting alone and subsequently to get either a synergistic or antagonistic effect of the materials.

Materials of FRC are widely used in advanced structures and are often applied in order to replace traditional materials such as metal components, especially those used in corrosive environments. They have become essential materials for maintaining and strengthening existing infrastructure due to the fact that they combine low weight and density with high strength, corrosion resistance, and high durability, providing many benefits in performance and durability. Modified fiber-based composites exhibit better mechanical properties, impact resistance, wear resistance, and fire resistance. Therefore, the FRC materials have reached a significant level of applications ranging from aerospace, aviation, and automotive systems to industrial, civil engineering, military, biomedical, marine facilities, and renewable energy.

In order to update the field of design and development of composites with the use of organic or inorganic fibers, a Special Issue entitled "Progress of Fiber-Reinforced Composites: Design and Applications"has been introduced. This book gathers and reviews the collection of twelve article contributions, with authors from Europe, Asia and America accepted for publication in the aforementioned Special Issue of Applied Sciences. This book aims to attract all researchers working in this research field, and collects new findings and recent advances on the development, synthesis, structure–activity relationships, and future applications of composites including fibers.

Ioannis Kartsonakis
Editor

Editorial

Special Issue on "Progress of Fiber-Reinforced Composites: Design and Applications"

Ioannis A. Kartsonakis

Laboratory of Advanced, Composite, Nanomaterials and Nanotechnology (R-Nano Lab), School of Chemical Engineering, National Technical University of Athens, Zografou Campus, 9 Heroon Polytechniou Str., 15773 Athens, Greece; ikartso@chemeng.ntua.gr

1. Introduction

Fiber-reinforced composite (FRC) materials are widely used in advanced structures and are often used to replace traditional materials such as metal components, especially those used in corrosive environments. They have become essential materials for maintaining and strengthening existing infrastructure due to the fact that they combine low weight and density with high strength, corrosion resistance, and high durability, providing many benefits in performance and durability. Modified fiber-based composites exhibit better mechanical properties, impact resistance, wear resistance, and fire resistance. Therefore, the FRC materials have reached a significant level of applications ranging from aerospace, aviation, and automotive systems to industrial, civil engineering, military, biomedical, marine facilities, and renewable energy.

In order to update the field of design and development of composites with the use of organic or inorganic fibers, a Special Issue entitled "Progress of Fiber-Reinforced Composites: Design and Applications" has been introduced. This editorial manuscript gathers and reviews the collection of twelve article contributions, with authors from Europe, Asia and America accepted for publication in the aforementioned Special Issue of *Applied Sciences*.

2. Fiber-Reinforced Plastic Composites

In recent years, the uses of composite fiber-reinforced plastic (CFRP) for several manufacturing structures have increased. Seven articles have been published in this Special Issue related to the synthesis and application of CFRP. The work of Troschitz and his team [1] reports the procedure of embedding metal weld inserts as an interface in glass-fiber-reinforced polypropylene sheet thermoplastic composites via compression molding in high quality and without fiber damage. The composite component was joined with steel sheets using conventional spot-welding guns. Numeric welding simulations were used for the determination of welding process parameters. Using to this innovative technology, high-quality joints were obtained.

A very interesting concept is developed by Yanlei Wang and his team [2] based on an thickness-dependent accelerated ageing method for tensile strength retention and water absorption of basalt fiber-reinforced polymer (BFRP) laminates. The evaluation of the tensile properties, water absorption, and degradation mechanism of BFRP laminates was conducted after their immersion on either alkaline solution or deionized water. The BFRP laminates were synthesized by a wet-layup method, and the influence of thickness on their tensile properties and water absorption were investigated under hygrothermal environment together with degradation mechanism. The obtained results revealed that the tensile properties and the water absorption of BFRP laminates were affected by specimen thickness.

Vito and his team [3] performed studies focuses on the physical structure of a 5G smart light pole and its multidisciplinary design process. Experiments were conducted for the determination of the design drivers of a composite 5G smart pole and the connecting design between signal penetration, finite element modeling, and computational fluid dynamics

Citation: Kartsonakis, I.A. Special Issue on "Progress of Fiber-Reinforced Composites: Design and Applications". *Appl. Sci.* **2022**, *12*, 8030. https://doi.org/10.3390/app12168030

Received: 8 August 2022
Accepted: 10 August 2022
Published: 11 August 2022

Publisher's Note: MDPI stays neutral with regard to jurisdictional claims in published maps and institutional affiliations.

Copyright: © 2022 by the author. Licensee MDPI, Basel, Switzerland. This article is an open access article distributed under the terms and conditions of the Creative Commons Attribution (CC BY) license (https://creativecommons.org/licenses/by/4.0/).

for thermal analysis. The results indicate the significant effects of thermal loading on the material selection.

In the work of Peng et al. [4], the slip load and slip factor of the glass-fiber-reinforced plastic (GFRP) GFRP–GFRP slip-critical connections were investigated. The impact on the long-term effects of the creep property in composite elements under the pressure of high-strength bolts was also evaluated via pre-tension force relaxation tests. It was proved that a high-efficiency fastener connection can be obtained using stainless steel cover plates with a grit-blasting surface treatment, while the effects of the creep property are negligible.

In the work of Riccio et al. [5] the influence of the material fracture toughness on the capability of a composite fuselage barrel to tolerate an impact on a rigid surface was investigated. The effects of intralaminar fracture energy variations on the impact deformation of the barrel were evaluated comparing the numerical results in terms of displacements and damage evolution for the three analyzed material configurations. According to the obtained results, a relevant influence of the in-plane toughness on the global dynamic response of the fuselage barrel was observed.

In another study, Sellitto et al. [6] reported the investigation of tensile behavior of a hybrid metallic–composite stiffened panel. A numerical–experimental investigation into the mechanical behavior of a fastened omega-reinforced composite fiber-reinforced plastic (CFRP) panel joined with a Z-reinforced aluminum plate was discussed. It was proved that the investigated hybrid panel does not experience any damage in composite sub-components up to a tensile load of 300 kN, while, at this loading level, some metallic sub-components experience extensive plastic deformation. Moreover, it was revealed that the omega stringer joints are able to arrest the skin-stringer debonding growth avoiding a drastic reduction of the load-carrying capability of the panel.

In another study, Hassan et al. [7] investigated the optimal conditions for a banana/epoxy composite in order to synthesize a sandwich structure where carbon/Kevlar twill plies acted as the skins. The structure was examined based on low-velocity impact and compression after impact tests. The results depicted a low peak load and larger damage area in the optimal banana/epoxy structures. The impact damage area increased with increasing impact energy. The optimal banana composite and synthetic fiber systems were proven to offer a similar residual strength and normalized strength when higher impact energies were applied.

3. Fiber-Reinforced Concrete Composites

Concrete and cement-based materials are commonly used in structural members and equipment in civil and industrial infrastructure. Two articles have been published in this Special Issue related to the fabrication, characterization, and application of these composites.

The improvement effect of fiber on the brittle failure of cement-treated subgrade soil was investigated by Wei Wang et al. [8]. A series of triaxial unconsolidated undrained (UU) tests were performed on samples of polypropylene fiber-cement-treated subgrade soil (PCS) in several polypropylene fiber mass contents. The outcome of this study denotes that it is feasible to modify cement subgrade soil with an appropriate amount of polypropylene fiber to mitigate its brittle failure.

The study of Jong-Han Lee et al. [9] examined the effect of steel fibers on the shear failure mode and breakout resistance of anchors embedded in steel fiber-reinforced concrete (SFRC). The relationship between the tensile performance of SFRC beams and the shear resistance of SFRC anchors was also assessed. The obtained results revealed that the calculated shear resistance of anchors in both SFRC and the plain concrete were in good agreement with the measurements. Moreover, the energy absorption capacity depicted a linear increase with that of the SFRC beam.

4. Fiber-Reinforced Carbonaceous Composites

Carbon-fiber-reinforced carbon, or carbon-carbon, is a composite material based on carbon fibers incorporated into a carbonaceous matrix. Three articles have been published in this Special Issue related to the production, characterization and application of these composites. In the work of Kailong Xu et al. [10], 3D four-step braided composites were aged in a cyclic hygrothermal environment. An accelerated hygrothermal aging spectrum for a military aircraft was applied. The microscopic damage morphologies of the composites were examined in order for the damage evolution determination with cyclic hygrothermal aging days to be determined. A split Hopkinson pressure bar was employed to evaluate the dynamic compressive mechanical property along the longitudinal direction of the 3D four-step braided composites at various cyclic hygrothermal aging days.

In the paper of Yanfeng Zhang et al. [11], the mechanical properties and compressive shear failure behavior of bonded–bolted hybrid single-lap joints of C/C composites at high temperature were studied. The interrelationship of the compression shear loading mechanism and the variations in stress distribution between bonded joints and bonded–bolted hybrid joints at high temperature were explored. The progressive damage of hybrid joints and the variations in the ratio of the bolt load to the total load with displacement were obtained. The variations in the ratio of the load shared by the bolt to the total load with displacement were obtained.

Finally, in the study of Kartsonakis et al. [12], the fabrication of a carbon fiber process was investigated, using high-density polyethylene (HDPE) and lignin esterified with either lactic acid or poly(lactic acid). The modified compounds were blended with HDPE thermoplastic polymer to synthesize composite precursors suitable for carbon fiber production. Stabilization and carbonization were performed, and the corresponding carbon fibers were obtained. This bottom-up approach was evaluated as a viable route considering large scale production for the transformation of lignin in a value-added product.

5. Future Strategies

Although the Special Issue has been closed, more in-depth research in the field of fabrication and characterization of fiber-reinforced composites is expected. It can be anticipated that more friendly production methods of fibers and their corresponding composites will be demanded in the future for several applications. In this case, suitable strategies should be ready for consolidation and utilization.

Funding: This research received no external funding.

Acknowledgments: The Guest Editor would like to thank all the authors and peer reviewers for their fruitful and valuable contributions to this Special Issue. The confluence of the editorial team of Applied Sciences is highly appreciated.

Conflicts of Interest: The author declares no conflict of interest.

References

1. Troschitz, J.; Vorderbrüggen, J.; Kupfer, R.; Gude, M.; Meschut, G. Joining of Thermoplastic Composites with Metals Using Resistance Element Welding. *Appl. Sci.* **2020**, *10*, 7251. [CrossRef]
2. Wang, Y.; Zhu, W.; Zhang, X.; Cai, G.; Wan, B. Influence of Thickness on Water Absorption and Tensile Strength of BFRP Laminates in Water or Alkaline Solution and a Thickness-Dependent Accelerated Ageing Method for BFRP Laminates. *Appl. Sci.* **2020**, *10*, 3618. [CrossRef]
3. Di Vito, D.; Kanerva, M.; Järveläinen, J.; Laitinen, A.; Pärnänen, T.; Saari, K.; Kukko, K.; Hämmäinen, H.; Vuorinen, V. Safe and Sustainable Design of Composite Smart Poles for Wireless Technologies. *Appl. Sci.* **2020**, *10*, 7594. [CrossRef]
4. Peng, Y.; Chen, W.; Wu, Z.; Zhao, J.; Dong, J. Experimental Study on the Performance of GFRP–GFRP Slip-Critical Connections with and without Stainless-Steel Cover Plates. *Appl. Sci.* **2020**, *10*, 4393. [CrossRef]
5. Riccio, A.; Saputo, S.; Sellitto, A.; Di Caprio, F. A Numerical Assessment on the Influences of Material Toughness on the Crashworthiness of a Composite Fuselage Barrel. *Appl. Sci.* **2020**, *10*, 2019. [CrossRef]
6. Sellitto, A.; Saputo, S.; Russo, A.; Innaro, V.; Riccio, A.; Acerra, F.; Russo, S. Numerical-Experimental Investigation into the Tensile Behavior of a Hybrid Metallic–CFRP Stiffened Aeronautical Panel. *Appl. Sci.* **2020**, *10*, 1880. [CrossRef]
7. Hassan, M.Z.; Sapuan, S.M.; Rasid, Z.A.; Nor, A.F.M.; Dolah, R.; Md Daud, M.Y. Impact Damage Resistance and Post-Impact Tolerance of Optimum Banana-Pseudo-Stem-Fiber-Reinforced Epoxy Sandwich Structures. *Appl. Sci.* **2020**, *10*, 684. [CrossRef]
8. Wang, W.; Zhang, C.; Guo, J.; Li, N.; Li, Y.; Zhou, H.; Liu, Y. Investigation on the Triaxial Mechanical Characteristics of Cement-Treated Subgrade Soil Admixed with Polypropylene Fiber. *Appl. Sci.* **2019**, *9*, 4557. [CrossRef]

9. Lee, J.-H.; Choi, E.; Cho, B.-S. Shear Failure Mode and Concrete Edge Breakout Resistance of Cast-In-Place Anchors in Steel Fiber-Reinforced Normal Strength Concrete. *Appl. Sci.* **2020**, *10*, 6883. [CrossRef]
10. Xu, K.; Chen, W.; Liu, L.; Luo, G.; Zhao, Z. Longitudinal Compressive Property of Three-Dimensional Four-Step Braided Composites after Cyclic Hygrothermal Aging under High Strain Rates. *Appl. Sci.* **2020**, *10*, 2061. [CrossRef]
11. Zhang, Y.; Zhou, Z.; Tan, Z. Compression Shear Properties of Bonded–Bolted Hybrid Single-Lap Joints of C/C Composites at High Temperature. *Appl. Sci.* **2020**, *10*, 1054. [CrossRef]
12. Goulis, P.; Kartsonakis, I.; Konstantopoulos, G.; Charitidis, C. Synthesis and Processing of Melt Spun Materials from Esterified Lignin with Lactic Acid. *Appl. Sci.* **2019**, *9*, 5361. [CrossRef]

Article

Joining of Thermoplastic Composites with Metals Using Resistance Element Welding

Juliane Troschitz [1,*], Julian Vorderbrüggen [2], Robert Kupfer [1], Maik Gude [1] and Gerson Meschut [2]

1. Institute of Lightweight Engineering and Polymer Technology, Technische Universität Dresden, Holbeinstraße 3, 01307 Dresden, Germany; robert.kupfer@tu-dresden.de (R.K.); maik.gude@tu-dresden.de (M.G.)
2. Laboratory for Material and Joining Technology (LWF), Paderborn University, Pohlweg 47-49, 33098 Paderborn, Germany; julian.vorderbrueggen@lwf.upb.de (J.V.); gerson.meschut@lwf.upb.de (G.M.)
* Correspondence: juliane.troschitz@tu-dresden.de; Tel.: +49-351-463-38480

Received: 30 September 2020; Accepted: 15 October 2020; Published: 16 October 2020

Abstract: Joining is a key enabler for a successful application of thermoplastic composites (TPC) in future multi-material systems. To use joining technologies, such as resistance welding for composite-metal joints, auxiliary joining elements (weld inserts) can be integrated into the composite and used as an interface. The authors pursue the approach of embedding metal weld inserts in TPC during compression moulding without fibre damage. The technology is based on the concept of moulding holes by a pin and simultaneously placing the weld insert in the moulded hole. Subsequently, the composite component can be joined with metal structures using conventional spot welding guns. For this purpose, two different types of weld inserts were embedded in glass fibre reinforced polypropylene sheets and then welded to steel sheets. A simulation of the welding process determined suitable welding parameters. The quality of the joints was analysed by microsections before and after the welding process. In addition, the joint strength was evaluated by chisel tests as well as single-lap shear tests for the different weld insert designs. It could be shown that high-quality joints can be achieved by using the innovative technology and that the load-bearing capacity is significantly influenced by the weld inserts head design.

Keywords: multi-material design; thermoplastic composites; joining; resistance spot welding; metal inserts

1. Introduction

Thermoplastic composites (TPC) have an important contribution to modern lightweight construction due to their excellent density-related mechanical properties and established efficient production processes [1]. However, one of the obstacles for the use of TPC in multi-material systems is the availability of suitable joining technologies. To increase the efficiency and reduce the complexity, automobile manufactures intend to keep the number of different joining technologies as low as possible [2]. Resistance spot welding is a standardized joining technology, which is widely used for sheet metal structures [2]. Nevertheless, the use of resistance spot welding for multi-material joints is not directly possible due to different physical properties of dissimilar materials, such as melting temperatures and micro-structural incompatibilities [3]. For this reason, various studies address research regarding the welding of dissimilar materials, such as composites or aluminium to steel. The objective is to use conventional spot-welding guns in established metal-oriented processes for composites as well. One approach is to integrate additional metallic elements as joining interfaces. The usage of an auxiliary joining element allows to avoid the welding incompatibility of dissimilar materials [4]. These elements can be integrated into composites during or after the part manufacturing process. Joesbury et al. [5] presented a process for thermoset matrix composites based on the integration of a metallic intermediate plate in the joining zone prior to the infiltration process. Subsequently, the weld was created, and finally, the whole assembly was infused with resin. Roth et al. applied a flat weld insert into the composite during preforming, whereby the fibres were displaced by a pin in the welding area [6]. Subsequently, the component was infiltrated in a resin transfer moulding (RTM) process, excluding the welding zone.

An example for the application of additional metallic elements after the composite part manufacturing process is described by Shah et al. [7]. The composite joining part was pre-punched in the flange area and then additional steel doubler strips were positioned on the composite and bonded with adhesive. In the hole area, the additional steel doubler strips were then welded to the metal component creating a form fit between composite and steel. Other studies have investigated the principle of resistance element welding, which was developed for joining high-strength (e.g., ultrahigh-strength steel) and low-ductile materials (e.g., aluminium) in multi-material joints [8]. The additional steel element can be positioned in a pre-hole [9] of the composite component or directly inserted by a punching process and joined to the steel component by resistance welding [10]. Weykenat et al. presented an approach to insert multiple metal layers (multilayer insert) locally into a composite using automated fibre placement [11]. These multilayer inserts lead to a reinforcement of the hole, into which a Flexweld® insert was inserted to join the composite to a steel component by resistance welding [12].

These joining techniques allow the use of conventional spot-welding guns and the integration of composite components in production lines established for metals. Besides, they cause local damage to the reinforcing fibres, which results in reduced load bearing capacity as shown for moulded holes compared to drilled ones [13]. For this reason, Obruch et al. [14] developed a new type of weld insert with a head plate and pins, which is suitable for low-damage integration into thermoplastic composites. Despite the reduction of fibre damage, as no pre-hole is necessary, it has the disadvantage that the embedding of the weld insert is an additional process step. Therefore, the authors pursue a new approach, in which the weld insert is integrated during composite component manufacture in the compression mould without fibre damage.

2. Materials and Methods

2.1. Process-Integrated Embedding of Weld Inserts into TPC

The technology of process-integrated embedding of weld inserts into TPC uses the principle of moulding holes by a pin tool and simultaneously places weld inserts in the moulded hole [15]. Thereby, the reinforcing fibres are not cut by punching or drilling but shifted aside by a tapered pin tool while the TPC is hot and formable [16]. The integration of the embedding process of the weld inserts into the composite component manufacturing process provides the possibility of reducing the number of process steps and manufacturing costs.

The process is schematically illustrated in Figure 1. At first, a pre-consolidated TPC sheet is warmed up above melting temperature of the matrix polymer by an infrared heating device (Figure 1a). Subsequently, the TPC sheet is transferred into the open compression mould. Immediately after mould closing (Figure 1b), a pin tool is shifted forward, forming a hole by displacing the reinforcing fibres and the still molten thermoplastic matrix (Figure 1c). The pin tool consist of a pin retainer containing a magnet to fix the steel weld insert and the tapered pin. After the pin movement, a ring shaped counterpunch recompresses the squeezed-out material whereby the undercut of the weld insert is filled with fibres and matrix material (Figure 1d). The embedding of the weld insert (steps c and d) takes less than one second. After cooling and solidification of the TPC specimen, the pin retainer is retracted and the tapered pin is separated. Finally, the TPC specimen with integrated weld insert is demoulded (Figure 1e).

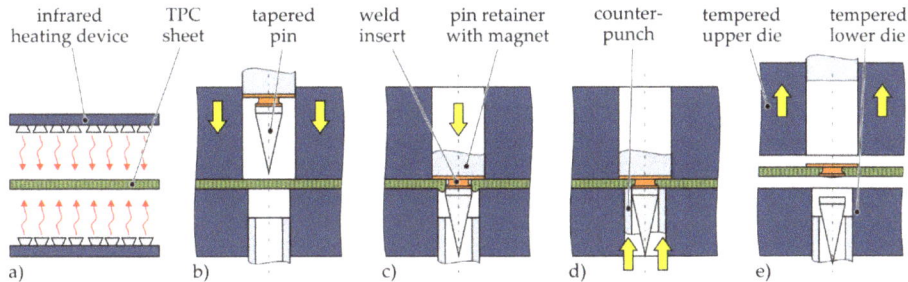

Figure 1. Schematic illustration of process-integrated embedding of weld inserts in thermoplastic composites (TPC): (**a**) warming up the TPC sheet, (**b**) closing the compression mould, (**c**) shifting forward the pin tool, (**d**) recompressing the squeezed-out material by the counterpunch, (**e**) demoulding of the TPC specimen.

The process parameters for the embedding of weld inserts in TPC were identified based on the experience of moulding holes [17] and process-integrated embedding of metal inserts into TPC [16]. A tempered steel mould (in this case 40 °C, cf. Table 1) on laboratory scale with vertical flash face is applied to produce plane test specimens with embedded weld inserts. The pin tool is pneumatically actuated, such as the counterpunch. The feed forces have been set at approximately 2.5 kN and 5 kN, respectively (cf. Table 1). A schematic illustration of the temperature curve during process-integrated embedding of weld inserts in TPC with the heating temperature of the TPC set to 210 °C and photos of TPC test specimens with integrated weld inserts are shown in Figure 2.

Table 1. Manufacturing parameters of the process-integrated embedding of weld inserts in TPC.

Heating temperature TPC	210 °C
Mould temperature	40 °C
Moulding pressure	8 bar
Feed force of the pin tool	2.5 kN
Feed force of the counterpunch	5 kN

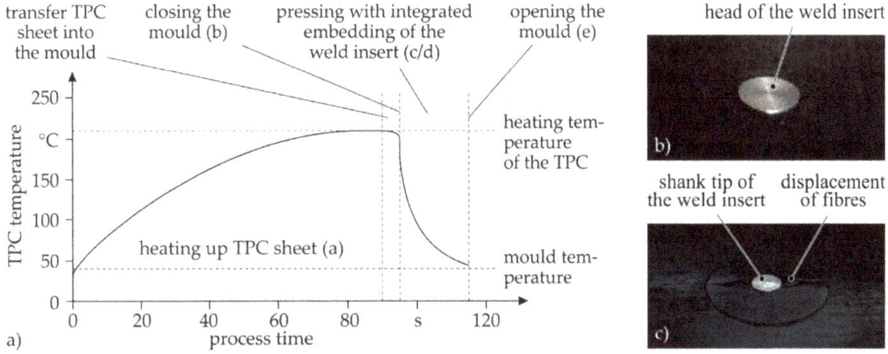

Figure 2. Schematic illustration of the temperature curve during process-integrated embedding of weld inserts in TPC with associated process steps from Figure 1. (**a**), TPC test specimen with integrated weld insert type A: top view (**b**) and bottom view (**c**).

2.2. Material Specification and Geometry of the Weld Inserts

The relevant investigations were performed on TPC sheets made of unidirectional glass fibre reinforced polypropylene (GF/PP) tapes, which were produced in an autoclave. GF/PP is a typical material for TPC applications with moderate thermal and mechanical requirements. For example, it is used for the large-series production of thermoplastic door module carriers [18]. A low-alloyed steel with high yield strength and a good weldability was chosen for the steel sheet as a typical representative of reinforcement parts e.g., in the automotive industry [19]. A high availability and good weldability are the requirements for the weld insert. An unalloyed structural steel was selected for this purpose. The properties of the TPC as well as the metal sheets and weld inserts used are shown in Table 2. Two different types of weld inserts were embedded in TPC specimens, both rotationally symmetric, see Figure 3. Type A has a head geometry that protrudes from the composite surface as a disturbing contour, whereas type B has a countersunk head and therefore is flush with the composite surface. Both weld inserts have similar shank tips, in order to ensure that the volumes in the welding area are comparable during the welding process.

Table 2. Material specification.

TPC	Unidirectional semi-finished product	Celstran® CFR-TP PP-GF70
	Fibre volume content	45 vol.-% E-glass
	Matrix	Polypropylene (PP)
	Laminate thickness	2 mm
	Laminate structure	$[(0°/90°)_2]_s$
Steel sheet	Material	HC340LA ([19])
	Sheet thickness	1.5 mm
Weld insert	Material	S235JR ([20])

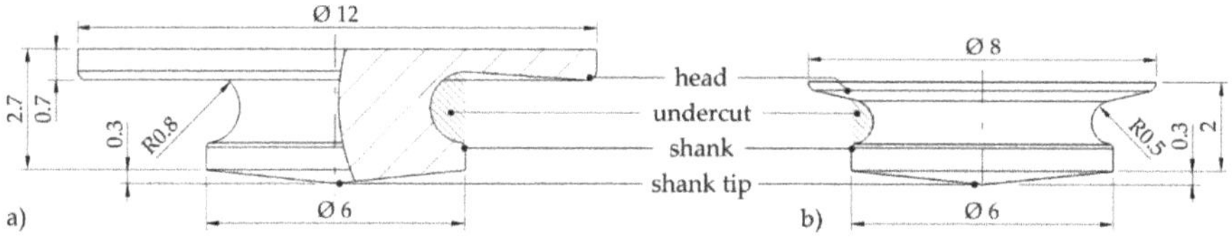

Figure 3. Prototypic weld inserts for process-integrated embedding: (a) type A, (b) type B.

2.3. Resistance Element Welding (REW)

After embedding the weld inserts in TPC, experimental investigations regarding resistance element welding were conducted using a powerGUN 2-C type C-frame welding gun by NIMAK International GmbH (Wissen, Germany). This welding gun has a servomotor drive with a projection of 700 mm and a maximum electrode force of 8.0 kN. The medium-frequency direct current welding device is equipped with a welding case of the type SK-Genius HWI436WA by Harms & Wende GmbH & Co. KG (Hamburg, Germany), which has a constant current control and provides a maximum weld current of 65 kA. Standardised Electrode caps of type ISO 5821-A0-20-20-100 made from CuCrZr material were used.

The identification of welding areas and the selection of process reliable welding parameters for the qualification of the developed weld inserts were carried out using the numeric simulation software SORPAS® 2D Welding V13.83 Enterprise Edition by SWANTEC Software and Engineering ApS (Kongens Lyngby, Denmark, 2020). In this study the welded joint is created between the isotropic metallic weld insert and a steel sheet. In this regard a two-dimensional, axisymmetric simulation model was built, as shown in Figure 4. The model was built according to the experimental setup. The thermal and the electrical properties of the metallic joining partners and the electrode caps were taken from the SORPAS database. The TPC was modelled in a simplified form as a thermal isotropic material. The electrical resistivity $\left(16*10^{15}\ (\Omega*cm)\right)$ and the thermal conductivity (0.69 (W/m∗K)) of the TPC were adjusted according to the data sheet values. All parts were modelled as deformable bodies with heat conduction. For the mesh, quadrilateral elements (954 elements; 1108 nodes) were used. The associated results are presented in Chapter 3.2.

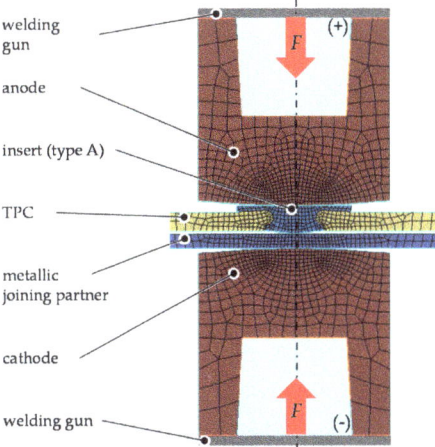

Figure 4. Two-dimensional (2D) axisymmetric model used for the simulation of the resistance element welding process.

2.4. Mechanical Testing

Both prototypic weld inserts have a shank with an undercut to prevent the weld insert from falling out when handling the composite component before joining by REW. Push-out tests were carried out on TPC specimens with embedded weld inserts before welding in order to investigate whether the handling strength is guaranteed. For this purpose, the weld inserts were loaded against the embedding direction. The push-out tests were performed on a universal testing machine (see Figure 5a). The size of the TPC test specimens was 100×100 mm^2 and the weld inserts were positioned in the centre. The test specimens were clamped between steel plates with a clearance hole diameter of 18 mm. The load was applied to the weld insert by a pressure pin at a constant crosshead velocity of 1 mm/min.

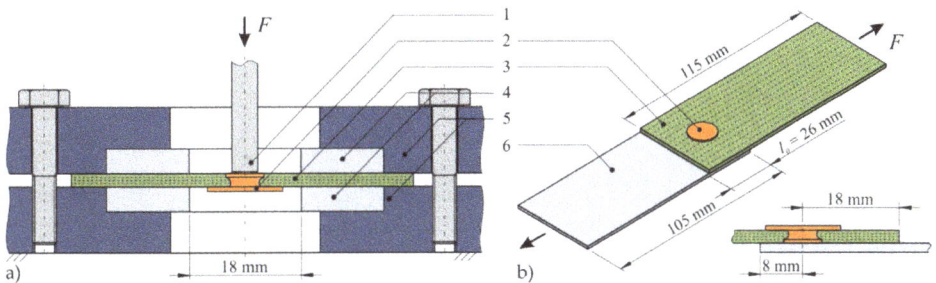

Figure 5. (**a**) Schematic illustration of the push-out test setup, (**b**) geometry of single-lap shear specimen (l_o: overlap length); 1: pressure punch, 2: weld insert, 3: TPC specimen, 4: exchangeable steel clamping, 5: base steel plates, 6: steel specimen.

After welding the TPC sheets to steel sheets, the welded joints were investigated by chisel tests in accordance with [21] to verify the chosen welding parameters. Chisel test as well as microsections were conducted on quadratic specimen (45×45 mm^2) with the weld inserts placed in the centre. Joint strength tests of welded TPC-steel-joints were conducted on single-lap shear specimen in accordance with [22] at quasi-static load. Before welding, the TPC specimen with embedded weld inserts were cut from size 160×100 mm^2 to 115×45 mm^2. The geometry of the specimen is shown in Figure 5b. The tests were performed on a high-rigid universal testing machine (Zwick Z100, by ZwickRoell AG, Ulm, Germany) at a testing velocity of 10 mm/min. The displacement was measured locally in the area of the joint using long-stroke extensometers. The extensometers were positioned each at a distance of 22 mm from the joint centre. The tests ended in the complete separation of the specimen.

3. Results

3.1. Process-Integrated Embedding of Weld Inserts in TPC

Microscopic examinations of cross sections before REW were carried out to evaluate whether the selected process parameters are suitable for manufacturing high-quality joining zones. As is evident from Figure 6, a complete filling of the undercut with reinforcing fibres and matrix can be achieved for both types of weld inserts. In addition, it is evident that the weld inserts are well positioned with the shank tip protruding from the laminate, thus enabling the welding process. Thus, the selection of the process parameters could be confirmed as suitable for further investigations.

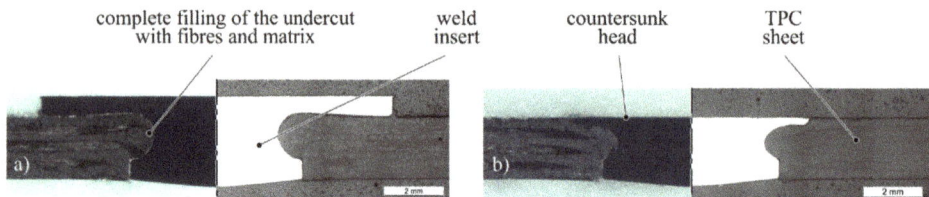

Figure 6. Analysis of the joining zone after embedding: (**a**) type A, (**b**) type B.

After qualifying the embedding process by means of microscopic examinations, push-out tests were performed on weld inserts embedded with the selected parameters before REW. The findings are summarised in Figure 7. All load-displacement curves show a progressive initial increase, then an almost linear section before the gradient of the curve declines. From then on, the load increases almost linear up to the maximum push-out force (cf. Figure 7a). The achievable maximum push-out force depends on the weld insert's geometry. In [16] it was demonstrated that the undercut volume has significant influence on the achievable out-of-plane loads for embedded inserts. Weld insert type A has an undercut volume of 16.8 mm^3 compared to 6.9 mm^3 of type B (cf. Figure 3). Due to the larger undercut, type A with 1.7 kN (arithmetic mean) reaches a higher ultimate push-out load than type B with 0.8 kN (arithmetic mean). For both types of weld inserts, the push-out loads are considered high enough for the further handling process.

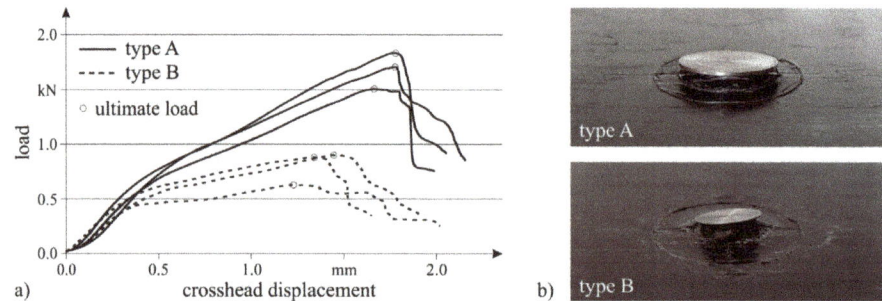

Figure 7. Results of the push-out tests: (**a**) load-displacement curves, (**b**) typical failure behaviour.

3.2. Welding Process Simulation

After qualifying the embedding of weld inserts in TPC, the aim was to determine suitable welding parameters for resistance element welding of the TPC-steel-joints. To ensure a resource-efficient determination of welding areas and reliable welding parameters for the joints, numeric parameter studies were carried out prior the experimental investigations. In order to validate the numeric simulation model, the results of a process simulation were compared to the results of corresponding experimentally welded joints with respect to both the geometry (height and diameter) of the weld nugget and the shape of the heat-affected zone.

Böddeker et al. [23] have reported that for REW of a TPC-steel combination acceptable joints can be provided using electrode caps of the type ISO 58211-A0-20-16-100 and an axial electrode force of 3.0 kN. For this reason, similar boundary conditions were selected for this study. For the validation of the simulation model, a weld current of 9.0 kA and a weld time of 60 ms were chosen. The geometric comparison of the experimental and simulative results as well as the evaluation of the weld nuggets are shown in Figure 8.

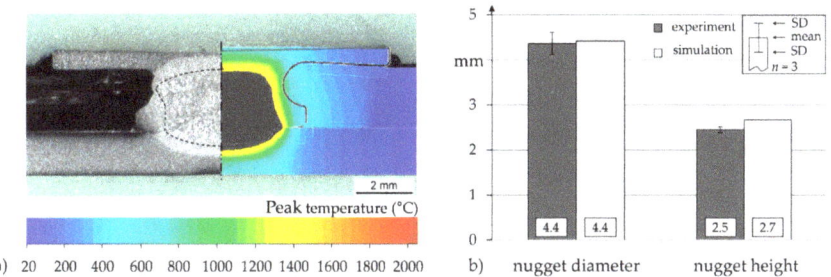

Figure 8. Validation of the welding simulation model via comparison: (**a**) of the shapes, (**b**) of the sizes of experimentally and simulatively generated weld nuggets.

By comparing the geometry of the welded area and the heat-affected zone, it can be seen that the process simulation has a good correlation with the experiment. The deviations of the nugget diameter and the nugget height are less than 10%. Therefore, it can be stated that the simulation model ensures a satisfying agreement with the experiments. It can be seen in Figure 8a that the chosen welding parameters lead to a slight deformation of the weld insert in the undercut area. A reason for this phenomena may be a too high thermal energy input during the welding process, which causes a softening of the material in the area of the undercut, where the cross sectional area is minimal. Due to the reduced cross-sectional area, the yield strength of the material is exceeded as a result of the contact pressure of the electrodes, causing the deformation of the weld insert. The thermal energy induced in the weld insert depends on several factors, such as the geometry of the insert, the weld time and weld current, or the electrode force, which has an influence on the contact resistance.

Afterwards, using the optimisation tool available in SORPAS, weld growth curves were generated for discrete weld times in order to define suitable weld current ranges in which appropriate joints can be generated. For this purpose, the weld current (I) was increased in 0.5 kA steps for a given axial electrode force of 3.0 kN. Weld times (t_w) were increased in 10 ms steps in between 40 ms and 70 ms. For the welding areas, a nugget diameter (d_n) of 3.25 mm was defined as the lower limit and the occurrence of splashes as the upper limit. Figure 9 shows the weldability lobes for weld insert types A and B determined by this approach.

According to the results determined by numeric simulation, weldability lobes extending over a range of approximately 2.5 kA (type A) to 3.5 kA (type B) can be expected. With the decreasing weld time, higher currents are required to produce sufficient weld nuggets. At the same time, an extension of the weldability lobes following a reduction of the weld time can be observed, which is consistent with the investigations from [23]. Based on the determined weldability lobes, suitable welding parameters were selected for the subsequent mechanical tests. As Roth et al. [24] showed that for REW short welding times at higher welding currents result in a lower heat input and, consequently, lead to a significantly smaller heat-affected zone, a short weld time is expected to minimise the deformation of the weld insert compared to Figure 8 and the thermal impact of the thermoplastic matrix of the

TPC. Furthermore, high weld times [23], as well as too high weld currents [25] may lead to thermal damage of resistance element welded polymer-steel-joints. Therefore, a weld time of 40 ms and a weld current of 11 kA was selected for both types of weld inserts, as this current is located centrally in the determined weldability lobes, thus, ensuring a high level of process reliability.

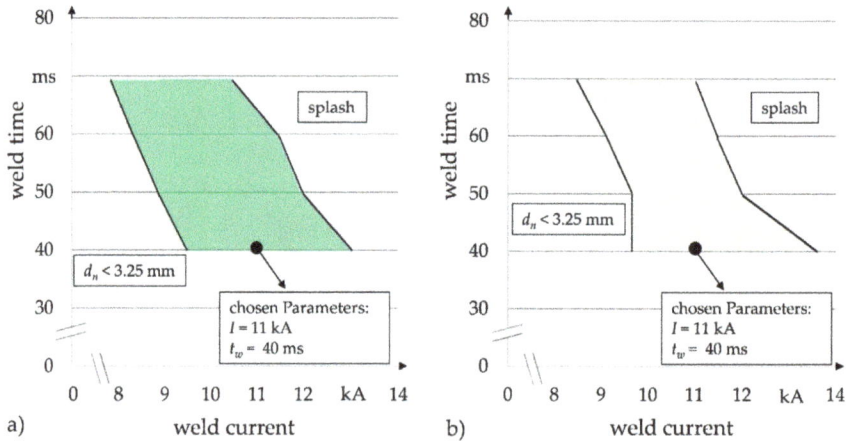

Figure 9. Simulative generated weldability lobes for weld insert type A (a) and type B (b).

3.3. Mechanical Testing of Welded Joints

The verification of the selected welding parameters determined by numeric simulation was carried out based on etched microsections for both weld insert types. In addition, chisel tests were conducted, in order to verify a sufficient strength of the weld nugget. In Figure 10, the mentioned microsections and photographs of samples after the chisel tests are shown.

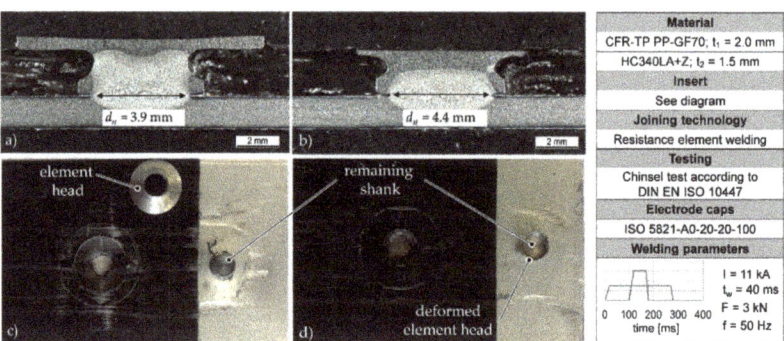

Figure 10. Analysis of the joining zone after resistance element welding (REW) (top: micrographs of welded joints using weld insert type A (a), and type B (b), bottom: pictures of welded specimen using weld insert type A (c) and type B (d) after chisel tests).

For the weld nuggets, diameters of 3.9 mm (type A) and 4.4 mm (type B) as well as heights of 1.2 mm (type A) and 1.3 mm (type B) were determined. The diameters of the weld nuggets as well as the shape of the heat affected zone are within the range predicted by numeric simulation. With regard to the chisel test, the element heads were peeled off (type A) or deformed (type B). Subsequently, the TPC was unbuttoned, with the weld nugget withstanding. Accordingly, a sufficient strength of the welded joints can be assumed, as the failure mode matches the aspired failure for REW, which is unbuttoning of the weld nugget from the base sheet or a head failure, respectively an unbuttoning of the TPC. As the pictures of the chiselled specimen show, no residues of molten plastic or smoke can be seen on the metallic joining partner. Furthermore, as can be seen from the pictures, the plastic matrix of the TPC does not show any thermal damage.

For further investigation of the load-bearing capacity of the joints, shear tests on single-lap joints were carried out under quasi-static load application on five specimens each for both types of weld inserts. The corresponding load-displacement curves, evaluated arithmetic means regarding the ultimate shear loads, the corresponding standard deviations (SD) and exemplary failure pictures of the specimens are shown in Figure 11.

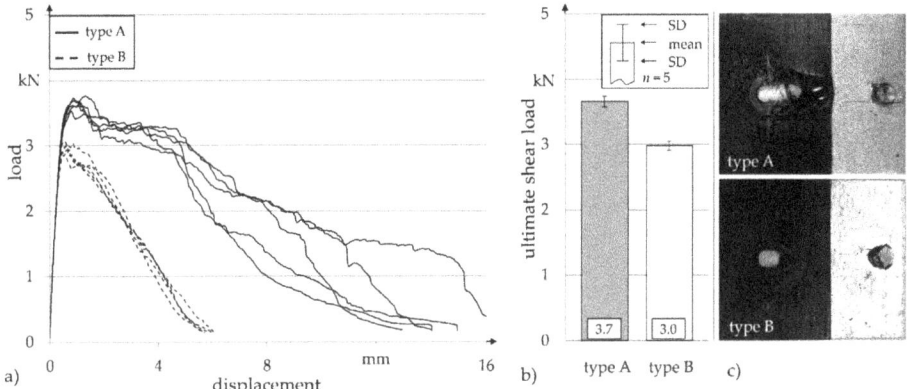

Figure 11. Results of single-lap shear tests under quasi-static load for both types of weld inserts: (**a**) load-displacement curves, (**b**) ultimate shear load, (**c**) characteristic failure behaviour.

It can be seen that the ultimate shear load of the welded joints using weld insert type A is 3.7 kN on average with a maximum displacement of 13 mm to 15 mm. The average ultimate shear load of joints using type B weld inserts is significantly lower (3.0 kN) with a maximum displacement of approximately 5 mm to 6 mm. This phenomenon can be explained by the corresponding failure characteristics. For type A insert, the failure behaviour of the specimens is characterised by the deformation with subsequent peeling off of the element head. Due to the fact that the specimen bend during the shear tests, the joint is loaded with an additional bending moment, which causes the flat head of the weld insert to be deformed before breaking off. Thus a bearing failure in the TPC is induced before the remaining element shank is unbuttoned, which explains the higher load-bearing capacity and in particular the increased displacement at failure. In contrast to that the smaller element head of the type B insert merely deforms during the shear test. Due to the smaller head diameter, the weld insert unbuttons from the TPC at lower load and lower displacement.

Table 3 summarises the mechanical characteristics of embedded weld inserts and welded joints as well as geometric sizes of the weld nuggets for both weld insert types.

Table 3. Mechanical properties of joints and geometric sizes of weld nuggets.

Parameter	Type A	Type B
push-out load before welding (mean)	1.7 kN	0.8 kN
nugget height (mean)	1.2 mm	1.3 mm
nugget diameter (mean)	3.9 mm	4.4 mm
ultimate shear load (mean)	3.7 kN	3.0 kN

4. Discussion

Microscopic examinations of cross sections of embedded weld inserts before REW verified a complete filling of the undercut with reinforcing fibres and matrix for both types of weld inserts. Furthermore, the weld inserts are well positioned with the shank tip protruding from the laminate to enable the subsequent welding process. Thus, the selected process parameters for process-integrated embedding of weld inserts into TPC (cf. Table 1) could be confirmed as suitable for further investigations. The developed technology can be applied to other types of TPC in terms of matrix material as well as architecture and material of the reinforcement. For this purpose, the process parameters such as the heating temperature of the TPC sheet and the mould temperature need to be adjusted accordingly.

For an application of the developed technology, especially in series production, the joint strength of the embedded weld inserts and the TPC component must be high enough to provide reliable handling of the component up to the final part assembly by REW. Due to the larger undercut, the measured ultimate push-out load of type A with 1.7 kN is higher than type B with 0.8 kN. For the further handling process, the push-out loads are considered high enough for both types of weld inserts.

For the subsequent welding process, suitable process parameters were determined by numeric welding simulations, which were verified by the analysis of etched microsections. The shape and diameter of the weld nuggets were within the range predicted by the numeric simulation. With these types of welding simulations, a prediction of suitable process parameters can be obtained rapidly for applications with alternative materials or sheet thicknesses. In this context, the short welding times of 40 ms are particularly notable, which promise efficient processing.

In chisel tests, a sufficient strength of the welded joints could be determined, as the failure mode matched the aspired failure for REW, which is unbuttoning of the weld nugget from the steel sheet or unbuttoning of the TPC. Tensile test using a weld stud welded on the head of the insert to transfer the tensile load will be carried out in order to quantify the ultimate strength of the weld nugget between the insert and the steel sheet.

The ultimate shear load of the welded joints is higher for weld insert type A (3.7 kN) as compared to type B (3.0 kN). This is due to the significantly smaller countersunk head of type B compared to the head geometry of type A that protrudes from the composite surface as a disturbing contour. The load bearing capacity could be further increased by adjusting the head geometry, for example by increasing the head thickness. This could change the failure behaviour under shear load to a laminate bearing before element pull-out of the TPC. In addition, cross-tension tests will be carried out for further evaluation of the load bearing capability.

It could be shown that high-quality joints can be achieved with the developed technology. This provides the opportunity to use conventional spot-welding guns in established metal-oriented processes for TPC as well. Compared to other technologies using weld inserts (such as [14]), the proposed approach allows the embedding of the weld insert during composite component manufacture without an additional process. Thereby the embedding of the weld insert does not require the preparation of a pre-hole in the TPC. For these reasons, the developed technology offers an excellent opportunity to integrate TPC into steel dominated multi-material systems. Of particular interest are fields of application where resistance spot welding is already implemented in series production, such as in the automotive industry.

5. Conclusions

TPC components can be welded to steel sheets by using embedded weld inserts as an interface. It could be shown, that weld inserts can be embedded in TPC during compression moulding in high quality and without fibre damage. Push-out tests before REW confirmed that the joint strength is high enough for further handling processes.

For the subsequent welding process, suitable process parameters were determined by numeric welding simulations. The ultimate shear load of the welded joints is significantly dependent on the head design of the embedded weld inserts. By analysing the quality and strength of the joints after the different process steps, it could be shown that high-quality joints can be achieved with this innovative technology.

Author Contributions: Process-integrated embedding of weld inserts in TPC, J.T.; analysis of the joint before welding, J.T.; simulation of the welding process and welding, J.V.; mechanical testing of welded joints, J.V.; writing and visualization, J.T. and J.V.; writing—review, R.K., M.G., and G.M.; funding acquisition J.T., J.V., R.K., M.G., and G.M.; project administration, M.G. and G.M. All authors have read and agreed to the published version of the manuscript.

Funding: The research project "Entwicklung multifunktionaler Schnittstellen zum Verbinden von FKV mit Metallen unter Nutzung etablierter Fügeverfahren" of the European Research Association for Sheet Metal Working (EFB) is carried out in the framework of the industrial collective research programme (IGF No. 20870 BG/EFB No. 08/119). It is supported by the Federal Ministry for Economic Affairs and Energy (BMWi) through the AiF (German Federation of Industrial Research Associations eV) based on a decision taken by the German Bundestag.

Conflicts of Interest: The authors declare no conflict of interest.

References

1. Behrens, B.-A.; Raatz, A.; Hübner, S.; Bonk, C.; Bohne, F.; Bruns, C.; Micke-Camuz, M. Automated Stamp Forming of Continuous Fiber Reinforced Thermoplastics for Complex Shell Geometries. *Procedia CIRP* **2017**, *66*, 113–118. [CrossRef]
2. Modi, S.; Stevens, M.; Chess, M. *Mixed Material Joining—Advancements and Challenges*; Center for Automotive Research: Ann Arbor, MI, USA, 2017.
3. Schmid, D.; Neudel, C.; Zäh, M.F.; Merklein, M. Pressschweißen von Aluminium-Stahl-Mischverbindungen. *Lightweight Des.* **2012**, *5*, 14–19. [CrossRef]
4. Meschut, G.; Janzen, V.; Olfermann, T. Innovative and highly productive joining technologies for multi-material lightweight car body structures. *J. Mater. Eng. Perform.* **2014**, *23*, 1515–1523. [CrossRef]
5. Joesbury, A.M.; Colegrove, P.A.; Rymenant, P.V.; Ayre, D.S.; Ganguly, S.; Williams, S. Weld-bonded stainless steel to carbon fibre-reinforced plastic joints. *J. Mater. Process. Technol.* **2018**, *251*, 241–250. [CrossRef]
6. Roth, S.; Warnck, M.; Coutandin, S.; Fleischer, J. RTM Process Manufacturing of Spot-weldable CFRP-metal Components. *Lightweight Des. Worldw.* **2019**, *5*, 18–23. [CrossRef]
7. Shah, B.; Frame, B.; Dove, C.; Fuchs, H. Structural Performance Evaluation of Composite-to-Steel Weld Bonded Joint. In Proceedings of the 10th Annual Automotive Composites Conference and Exhibition, Troy, MI, USA, 15–16 September 2010; pp. 545–561.
8. Holtschke, N.; Jüttner, S. Joining lightweight components by short-time resistance spot welding. *Weld. World* **2017**, *61*, 413–421. [CrossRef]
9. Meschut, G.; Hahn, O.; Janzen, V.; Olfermann, T. Innovative joining technologies for multi-material structures. *Weld. World* **2014**, *58*, 65–75. [CrossRef]
10. Meschut, G.; Matzke, M.; Hoerhold, R.; Olfermann, T. Hybrid technologies for joining ultra-high-strength boron steels with aluminum alloys for lightweight car body structures. *Procedia CIRP* **2014**, *23*, 19–23. [CrossRef]
11. Herwig, A.; Horst, P.; Schmidt, C.; Pottmeyer, F.; Weidenmann, K.A. Design and mechanical characterisation of a layer wise build AFP insert in comparison to a conventional solution. *Prod. Eng.* **2018**, *12*, 121–130. [CrossRef]
12. Weykenat, J.; Denkena, B.; Horst, P.; Meiners, D.; Schmidt, C.; Groß, L.; Herwig, A.; Nagel, L.; Serna, J. Local Fiber-Metal-Laminate used as Load Introduction Element for Thin-Walled CFRP Structures. In Proceedings of the 4th International Conference Hybrid 2020 Materials and Structures, Web Conference, Germany, 28–29 April 2020.
13. Hufenbach, W.; Adam, F.; Kupfer, R. A novel textile-adapted notching technology for bolted joints in textile-reinforced thermoplastic composites. In Proceedings of the ECCM, Budapest, Hungary, 6–7 June 2010; pp. 1–9.
14. Obruch, O.; Jüttner, S.; Ballschmiter, G.; Kühn, M.; Dröder, K. Production of hybrid FRP/steel structures with a new sheet metal connecting element. *Biul. Inst. Spaw.* **2016**, *5*, 60–66.

15. Troschitz, J.; Kupfer, R.; Gude, M. Experimental investigation of the load bearing capacity of inserts em-bedded in thermoplastic composites. In Proceedings of the 4th International Conference Hybrid 2020 Materials and Structures, Web Conference, Germany, 28–29 April 2020; pp. 249–254.
16. Troschitz, J.; Kupfer, R.; Gude, M. Process-integrated embedding of metal inserts in continuous fibre reinforced thermoplastics. *Procedia CIRP* **2019**, *85*, 84–89. [CrossRef]
17. Kupfer, R. Zur Warmlochformung in Textil-Thermoplast-Strukturen—Technologie, Phänomenologie, Modellierung. Ph.D. Thesis, Technische Universität Dresden, Dresden, Germany, 2016.
18. Cetin, M.; Thienel, M. Large-series Production of Thermoplastic Door Module Carriers. *Lightweight Des. Worldw.* **2019**, *12*, 12–17. [CrossRef]
19. *DIN EN 10268 Cold Rolled Steel Flat Products with High Yield Strength for Cold Forming—Technical Delivery Conditions*; Deutsches Institut für Normung E.V.: Berlin, Germany, 2013.
20. *DIN EN 10025-2 Hot Rolled Products of Structural Steels—Part 2: Technical Delivery Conditions for Non-Alloy Structural Steels*; Deutsches Institut für Normung E.V.: Berlin, Germany, 2019.
21. *DIN EN ISO 10447 Resistance Welding—Testing of Welds—Peel and Chisel Testing of Resistance Spot and Projection Welds*; Deutsches Institut für Normung E.V.: Berlin, Germany, 2015.
22. *DVS/EFB 3480-1 Testing of Properties of Joints—Testing of Properties of Mechanical and Hybrid (Mechanical/Bonded) Joints*; DVS: Düsseldorf, Germany, 2007.
23. Böddeker, T.; Chergui, A.; Ivanjko, M.; Gili, F.; Behrens, S.; Runkel, D.; Folgar, H. Joining TWIP—TWIP-Steels for Multi-Material Design in Automotive Industry Using Low heat Joining Technologies. In Proceedings of the 6th Fügetechnische Gemeinschaftskolloquium, Munich, Germany, 7–8 December 2016; pp. 77–82. (In German).
24. Roth, S.; Hezler, A.; Pampus, O.; Coutandin, S.; Fleischer, J. Influence of the process parameter of resistance spot welding and the geometry of weldable load introducing elements for FRP/metal joints on the heat input. *J. Adv. Join. Process.* **2020**, *2*, 100032. [CrossRef]
25. Schmal, C.; Meschut, G. Process characteristics and influences of production-related disturbances in resistance element welding of hybrid materials with steel cover sheets and polymer core. *Weld. World* **2020**, *64*, 437–448. [CrossRef]

Publisher's Note: MDPI stays neutral with regard to jurisdictional claims in published maps and institutional affiliations.

© 2020 by the authors. Licensee MDPI, Basel, Switzerland. This article is an open access article distributed under the terms and conditions of the Creative Commons Attribution (CC BY) license (http://creativecommons.org/licenses/by/4.0/).

Article

Influence of Thickness on Water Absorption and Tensile Strength of BFRP Laminates in Water or Alkaline Solution and a Thickness-Dependent Accelerated Ageing Method for BFRP Laminates

Yanlei Wang [1], Wanxin Zhu [1], Xue Zhang [1,*], Gaochuang Cai [2] and Baolin Wan [3]

[1] State Key Laboratory of Coastal and Offshore Engineering, School of Civil Engineering, Dalian University of Technology, Dalian 116024, China; wangyanlei@dlut.edu.cn (Y.W.); zhuwanxinpg@foxmail.com (W.Z.)
[2] Laboratoire de Tribologie et de Dynamique des Systèmes, Ecole Nationale d'Ingénieurs de Saint-Etienne (ENISE), University of Lyon, UMR 5513, 58 Rue Jean Parot, 42023 Saint-Etienne CEDEX 2, France; gaochuang.cai@enise.fr
[3] Department of Civil, Construction and Environmental Engineering, Marquette University, Milwaukee, WI 53201, USA; baolin.wan@marquette.edu
* Correspondence: xuezhang@dlut.edu.cn; Tel.: +86-138-0408-9214

Received: 12 April 2020; Accepted: 20 May 2020; Published: 23 May 2020

Abstract: This paper first presented an experimental study on water absorption and tensile properties of basalt fiber-reinforced polymer (BFRP) laminates with different specimen thicknesses (i.e., 1, 2, and 4 mm) subjected to 60 °C deionized water or alkaline solution for an ageing time up to 180 days. The degradation mechanism of BFRP laminates in solution immersion was also explored combined with micro-morphology analysis by scanning electronic microscopy (SEM). The test results indicated that the water absorption and tensile properties of BFRP laminates were dramatically influenced by specimen thickness. When the BFRP laminates with different thicknesses were immersed in the solution for the same ageing time, the water absorption of the specimens decreased firstly before reaching their peak water absorption and then increased in the later stage with the increase of specimen thickness, while the tensile strength retention sustaining increased as specimen thickness increased. The reason is that the thinner the specimen, the more severe the degradation. In this study, a new accelerated ageing method was proposed to predict the long-term water absorption and tensile strength of BFRP laminates. The accelerated factor of the proposed method was determined based on the specimen thickness. The proposed method was verified by test results with a good accuracy, indicating that the method could be used to predict long-term water absorption and tensile strength retention of BFRP laminates by considering specimen thickness in accelerating tests.

Keywords: basalt fiber-reinforced polymer (BFRP); thickness; durability; hygrothermal ageing; accelerated ageing method

1. Introduction

Fiber-reinforced polymer (FRP) composites such as Carbon FRP (CFRP), Glass FRP (GFRP) and Aramid FRP (AFRP) have been widely used in infrastructure construction and other fields [1–11]. However, large-scale applications of FRP composites in infrastructure construction are still limited for some reasons, e.g., high cost of CFRP and AFRP composites, poor chemical stability of GFRP composites [12–16]. Over the past few years, basalt fiber has gradually received more attention as a new inorganic green fiber for its environment-friendly features. The basalt fiber is made of basalt stone after melting at 1450–1500 °C. There is no pollution during its production process [17,18]. The basalt fiber has a greater elongation at break than carbon fiber, higher elastic modulus, and greater chemical stability than glass fiber, and lower cost than carbon fiber and aramid fiber [18–21]. Moreover, the basalt fiber is also a good flame retardant [21,22]. Therefore, basalt FRP (BFRP) composites are increasingly used in civil engineering, such as externally strengthening sheets of reinforced concrete (RC) structure. The literature [23,24] revealed that BFRP laminates would be degraded due to the strong alkaline environment inside concrete (pH = 12–13), and such degradation would be further accelerated when the BFRP laminates were exposed to hygrothermal environments. The reason is that the resin matrix of BFRP laminates is highly sensitive to the change of the temperature and the moisture [25–27], which is seriously harmful to BFRP laminates. Therefore, the hygrothermal ageing properties of BFRP laminates in alkaline solution need to be studied.

At present, some researches [28–32] have been done on the hygrothermal ageing properties of BFRP laminates subjected to alkaline solution. Lu et al. [28,29] investigated the long-term mechanical properties of pultruded BFRP laminates after a long-term immersion. The test results showed that as the increase of immersion time, the water absorption increased, and the tensile strength and interlaminar shear strength decreased dramatically due to the severe interfacial debonding between the basalt fiber and resin matrix after ageing. Ma et al. [30] tested the tensile strength of BFRP laminates fabricated by the vacuum assistant resin infusion (VARI) method. The tensile strength of BFRP specimens decreased by 37% and 34% at 20 °C and 40 °C for distilled water and decreased by 67% and 90% for alkaline solution after 180 ageing days, respectively. Xiao et al. [31] studied the tensile properties of wet-layup BFRP laminates. After immersion of 180 ageing days, it was observed that a higher temperature (e.g., 60 or 80 °C) would cause a greater reduction of tensile strength during the same ageing time. Wu et al. [32] studied the degradation of basalt fiber and BFRP laminates and concluded that the tensile properties of BFRP laminates were superior to that of basalt fiber in alkaline solution. The tensile properties reduction of BFRP laminates might be attributed to the severe degradation of interfacial adhesion between the basalt fiber and epoxy resin, which was observed by scanning electron microscopy (SEM) images, rather than the ageing of basalt fibers. It can be concluded from the existing studies that the long-term mechanical properties of BFRP laminates were greatly affected by the hygrothermal environment. A higher temperature and alkaline solution would aggravate the ageing process, leading to a severe interfacial debonding between the basalt fiber and resin matrix.

Although the above researches showed that the hygrothermal ageing properties of BFRP laminates have been studied by many test methods, the long-term durability of BFRP laminates still cannot be evaluated comprehensively. One reason is that different specimen thicknesses (ranging from 0.45 to 4.5 mm) were adopted for test specimens in the existing researches [21,28,30–33]. For example, Lu et al. [28] measured the water absorption of BFRP specimens with a thickness of 1.4 mm, which were immersed in 60 °C distilled water, and found that water absorption was 0.32% after 90 days of ageing. Xiao et al. [31] also immersed the BFRP specimens with a thickness of 1.0 mm in the 60 °C distilled water for the same ageing time (i.e., 90 days), while the water absorption of the specimens was 2.8%. Lu et al. [29] studied the tensile properties of the BFRP laminates with a thickness of 1.4 mm in alkaline solution. After ageing for 90 days, the tensile strength retention of the BFRP laminates decreased to 71.9%, 61.1%, and 56.8%, respectively, at 20, 40, and 60 °C. As reported by Xiao et al. [31], the tensile strength retention of 1.0 mm thickness BFRP laminates decreased to 77.3%, 33.3%, and 40.8% after the immersion in alkaline solution for 90 days. It can be seen from the above studies that although the same hygrothermal ageing tests were conducted for BFRP laminates, the test results cannot be compared directly due to the difference of specimen thickness. It has been reported that the durability test results of FRP composites were probably dependent on the dimensions of the adopted specimen. For example, the influence of diameter on the long-term durability of FRP bars was previously investigated [34,35]. According to the existing test results [34,35], for the same ageing time, the deterioration degree of BFRP bars was inversely related to the diameter of the specimen, which means that the specimen dimension should not be neglected on exploring the durability of BFRP composites. Therefore, the investigation of the influence of specimen dimension (especially specimen thickness) on long-term durability of BFRP laminates was very necessary.

Additionally, the ageing time adopted in above hygrothermal ageing tests was much shorter than the actual service duration of FRP composites, meaning that these test results were unsuitable to evaluate the long-term performances. Therefore, some efforts have been made to explore test methods to accelerate the ageing process of FRP composites. So far, the temperature-dependent accelerated ageing method was widely adopted in hygrothermal ageing tests [36,37]. The efficiency of the accelerated test is highly dependent on the accelerated factor, which is determined by experimental parameters. For example, the accelerated factor of temperature-dependent acceleration is calculated by considering at least three ageing temperatures in the test. In theory, the greater accelerated effect could be achieved by using higher test temperatures. However, the accelerated efficiency of temperature-dependent acceleration for FRP composites was not improved significantly because the adopted ageing temperature must be lower than the glass transition temperature (T_g) of FRP composites [36]. The excessive ageing temperature would change the resin matrix of FRP composites from glassy to viscous fluid status, deteriorating the properties severely and changing the ageing mechanism. To this end, it is significant to explore alternative accelerated ageing methods for which the aforementioned limitation should be solved.

In this paper, the influence of specimen thickness on the water absorption behavior and tensile properties of BFRP laminates in 60 °C deionized water and alkaline solution were first experimentally investigated. Through analyzing and summarizing the rules between the specimen thickness and the test results, an accelerated ageing method dependent on specimen thickness was proposed for the water absorption and tensile strength of BFRP laminates. Two accelerated ageing factors were theoretically established considering specimen thicknesses for water absorption and tensile strength of BFRP laminates, respectively. The test results were also adopted to verify the accuracy of the proposed method. The proposed accelerated ageing method in this study provides a new way for the prediction of the long-term ageing properties of BFRP laminates, which is simple and easy to apply.

2. Materials and Methods

2.1. Materials and Laminates Fabrication

Commercial unidirectional basalt fiber fabric (purchased from Aerospace Tuoxin Basalt Industrial Co., Ltd., Chengdu, China) with an areal density of 300 g/m^2 and a nominal thickness of 0.16 mm was adopted in the current study. Epoxy main agent and curing agent (Model JGN-T, produced by Kaihua New Technology Engineering Co., Ltd., Dalian, China) with a weight ratio of 3:1 were used as the matrix of BFRP laminates. It is one kind of commercial epoxy resin commonly used in FRP strengthening engineering in China [38]. No other fillers were added in the matrix except those added by the manufacturer. The mechanical parameters of the basalt fiber fabric and epoxy resin were provided by the corresponding manufacturers as shown in Table 1.

Table 1. Mechanical properties of materials.

Material	Tensile Strength (MPa)	Elastic Modulus (GPa)	Elongation at Break (%)
Basalt fiber fabric	2100	91	2.6
Epoxy resin	>40	>2.5	>1.8

BFRP laminates used in the current study were fabricated using wet lay-up method. The fiber direction of basalt fabrics was properly stacked to ensure that the unidirectional BFRP laminates were fabricated. The laminates with specific number of fabric layers (i.e., 2, 4, and 8 layers) were fabricated. Specially designed molds with the thickness $h = 1$, $h = 2$, and $h = 4$ mm to fabricate the laminates need to be used to ensure the required thickness of the laminates. The fiber volume fraction of the BFRP laminates was 0.32. In this paper, BFRP laminates with the thickness $h = 1$, $h = 2$, and $h = 4$ mm were represented with B_1, B_2, and B_4, respectively.

2.2. Water Absorption Test

According to ASTM D5229 [39], the water absorption specimens were subjected to deionized water and alkaline solution (pH = 13) at a temperature of 60 °C for a duration of 180 days. The setting pH of the alkaline solution is 13, which was used to simulate the pore water of the concrete [40]. The square specimens of water absorption (shown in Figure 1a) were cut from the fabricated BFRP laminates, and the dimensions of the specimens were 60 × 60 × h mm, where h represents the thickness of the BFRP laminates. The initial dry weight (W_0) of the specimen was weighed using a precision electronics balance (Model FA2004N, Shanghai Precision Scientific Instruments Co., Ltd., Shanghai, China) with an accuracy of 0.1 mg before immersion. After specific immersion time, the wet specimen was re-weighed. Before re-weighing, the water on the surface of the specimen was wiped using absorbent paper. The water absorption $M(t)$ at specific ageing time is calculated by Equation (1). Ten specimens for each thickness were tested and the average values and standard deviation of water absorption for each thickness at specific immersion time were calculated.

$$M(t) = \left(\frac{W_t - W_0}{W_0}\right) \times 100\% \tag{1}$$

where W_t is the weight of wet specimen at specific ageing time t.

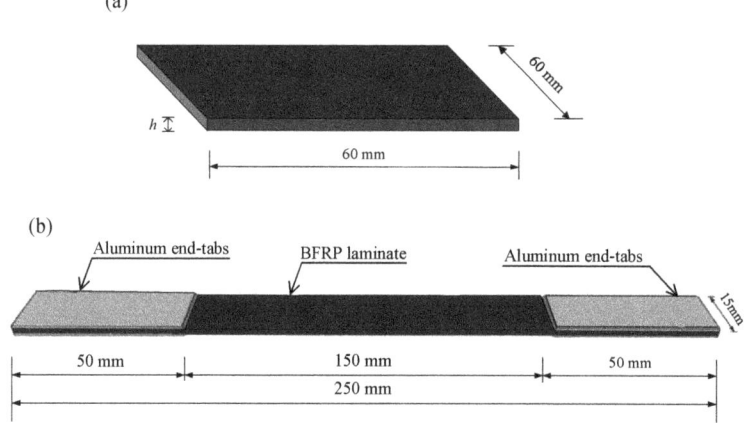

Figure 1. Test specimens of BFRP laminates. (a) water absorption test and (b) tensile test.

2.3. Tensile Test

The requirements for the tensile test of BFRP laminates were conducted in accordance with ASTM D3039 [41]. The rectangular tensile specimens with the dimensions of 250 × 15 × h mm (shown in Figure 1b) were cut from the fabricated BFRP laminates, which were soaked in deionized water and alkaline solution (pH = 13) at 60 °C last up to 180 days. As shown in Figure 1b, aluminum end-tabs were used in two ends to avoid the premature damage of the specimen during the tensile test. According to ASTM D3039 [41], the monotonic tensile test was conducted using a universal machine (Model WDE-200E, Jinan Gold Testing Machines Inc., Jinan, China) through displacement control with a displacement speed of 2 mm/min. The tensile stress σ could be calculated according to the Equation (2) and the strain was measured using an extensometer with a gauge length (Model Y50/10, Changchun Sanjing Test Instrument Co., Ltd., Changchun, China) of 50 mm in the test, as shown in

Figure 2. It should be noted that the acquired tensile stress was nominal tensile stress, which is more convenient to apply in practice engineering. Five specimens of each thickness were tested at specific duration, the average values and standard deviation of tensile strength retention and tensile elastic modulus were respectively calculated.

$$\sigma = \frac{F}{A} \quad (2)$$

where F represents the measured tensile force. A represents the nominal cross-sectional area of the specimens.

Figure 2. Tensile test of BFRP laminates.

2.4. Scanning Electron Microscopy

Fractured surfaces of tensile failure specimens were observed using scanning electron microscopy (SEM) (Model Nova NanoSEM450, FEI Inc., Oregon, USA) at an accelerated voltage of 15 kV. All the observed specimens were sprayed with gold powder to increase their conductivity for easy observation [22,29].

3. Test Results

3.1. Water Absorption

Figure 3 shows the average values of the water absorption of BFRP laminates soaked in 60 °C deionized water and alkaline solution. It can be seen that the values of error bars were generally small, indicating that the dispersions of results were acceptable. It is noted that the water absorption of all specimens increased first before reaching their peak water absorption and then decreased as the ageing time increased. When soaked in deionized water and alkaline solution at 60 °C, the peak water absorption of the specimens B_1, B_2 and B_4 were 1.31%, 1.16%, 0.96% and 1.60%, 1.52%, 1.50%, respectively. After a 180-day immersion, the water absorption of specimens B_1, B_2 and B_4 decreased to 0.18%, −0.43%, −1.16% for deionized water immersion and to −0.27%, −0.73% and −2.15% for alkaline solution immersion, respectively. Generally, the water absorption of FRP composites increased gradually in the initial ageing duration. After that, the water absorption reached the dynamic moisture equilibrium or kept increasing. However, the water absorption curve of this paper, as shown in Figure 3, was quite different from the general cases reported in the existing literature [42,43]. The decline in water absorption after reaching its peak water absorption was probably caused by the serious degradation of the resin matrix in hygrothermal environment. The water absorption changes of BFFRP laminates were mainly affected by water immersion and hydrolysis of epoxy resin. When the mass gain of epoxy matrix from water immersion was less than the mass loss of epoxy resin due to the hydrolysis, the water absorption of BFRP laminates began to decline. As the degree of hydrolysis of epoxy resin increased, the water absorption of aged BFRP laminates were lower than that without ageing. This abnormal water absorption behavior (i.e., the decline in water absorption after reaching its peak water absorption) was also reported in the other literature [44–46].

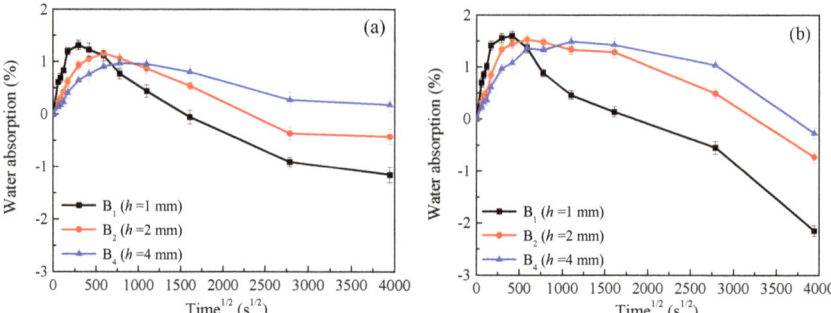

Figure 3. Water absorption of BFRP specimens soaked in 60 °C (**a**) deionized water and (**b**) alkaline solution.

Figure 4 shows the degradation mechanism of epoxy resin soaked in deionized water or alkaline solution. It should be noted that the proportion of each part of the schematic is not drawn strictly according to the actual size, but just to better explain what needs to be expressed. The complete epoxy molecular chains of epoxy resin are presented in Figure 4a. When the BFRP laminates were immersed in high-temperature (i.e., 60 °C) deionized water or alkaline solution for a period of immersion time, lots of water molecules and/or OH^- accelerate the erosion to epoxy resin, leading to the hydrolysis reaction of epoxy resin. The hydrolysis reaction is that the ether bonds (C-O) of epoxy molecular chains are broken as shown in Figure 4a. At this time, the epoxy molecular chains were broken down and formed some short molecular chains, dissolved in the solution finally. Macroscopically, the hydrolysis process of epoxy resin could be characterized by the appearance of some voids and micro-cracks after the water molecules and/or OH^- deteriorated to epoxy resin as shown in Figure 4b. In order to verify the degradation mechanism, SEM observations were also conducted. The representative SEM images were selected from lots of SEM images; similar phenomena were observed in different regions of different samples. Figure 5a shows the surface of unaged specimen that was intact without ageing defects. Figure 5b,c represent the aged specimen in 60 °C deionized water and alkaline solution for 180 days, respectively. It can be seen that some obvious voids and micro-cracks appeared on the surface of the aged specimens, which supported the degradation mechanism of epoxy resin shown in Figure 4. Moreover, it can also be seen that, compared with the micro-cracks soaked in deionized water in Figure 5b, the extended micro-cracks appeared in Figure 5c due to the erosion of OH^- in alkaline solution. A more severe hydrolysis reaction occurred on the specimens soaked in alkaline solution compared with the specimens soaked in deionized water. Therefore, the change range of water absorption was greater for alkaline solution immersion. In addition, it should be noted that it is significant to quantify the micro-cracks for a better understanding the hygrothermal ageing on epoxy resin. The related work will be included in a future study.

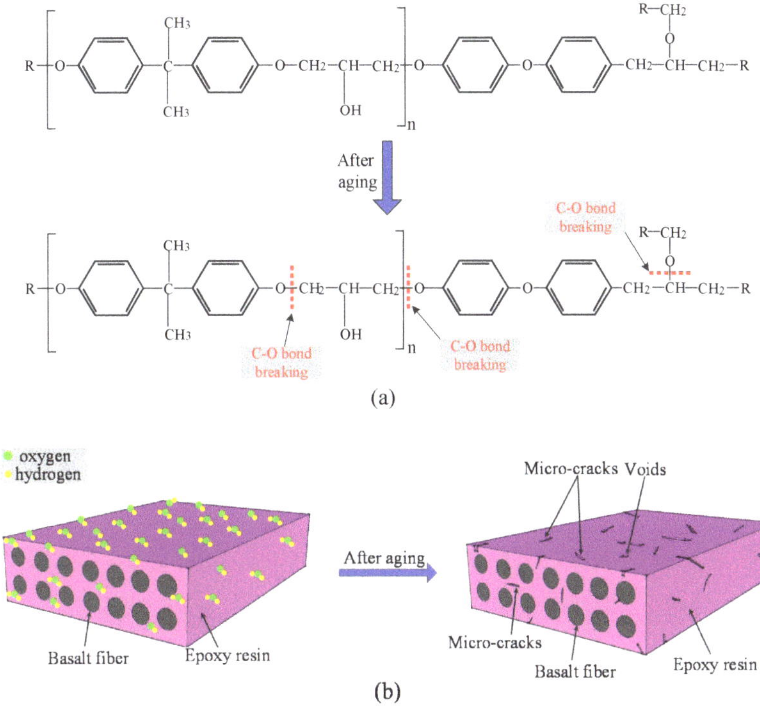

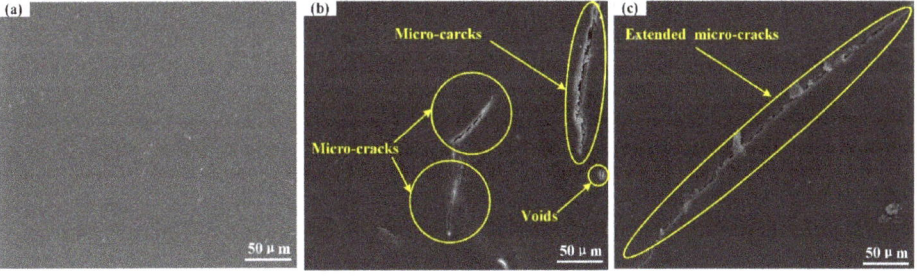

Figure 4. Degradation mechanism of epoxy resin in 60 °C deionized water or alkaline solution. (**a**) breaking down of epoxy molecular chains (**b**) ageing damage of epoxy resin in macroscopy.

Figure 5. SEM images of (**a**) unaged specimen surface; Aged surfaces of the specimens soaked (**b**) deionized water and (**c**) alkaline solution at 60 °C after 180 ageing days.

3.2. Tensile Properties

Table 2 gives the average values with standard deviations of the tensile strength (f) and elastic modulus (E) of the specimens at specific ageing time. The average values of the tensile strength retention and elastic modulus retention are plotted against ageing time in Figures 6 and 7, respectively. It can be seen that the variation of error values was small overall, which can be considered that the results of the tensile test were accurate and reasonable. It can be found from Figure 6, in the first 14 days, the tensile strength decreased rapidly by 20%–40%. Afterward, the tensile strength showed a slower decline with the increase of ageing time. After an ageing of 180 days, the tensile strength retention of specimens B_1, B_2, and B_4 in deionized water and alkaline solution were 34.0%, 37.7%, 40.6% and 20.1%, 26.3%, 33.1%, respectively. It can be seen that the thicker specimens had higher tensile strength retention than the thinner specimens. Figure 7 shows elastic modulus retention in deionized water and alkaline solution. The change trends of elastic modulus retention were similar to those of tensile strength retention. However, the degradation rate of elastic modulus is far less than that of tensile strength. After the ageing of 180 days, the maximum decrease of elastic modulus retention of all specimens was around 5%–20%.

Table 2. Degradation of tensile strength (f) and elastic modulus (E) of BFRP laminates with different thicknesses.

Specimen Thickness	Ageing Days	Deionized Water		Alkaline Solution	
		f (MPa)	E (GPa)	f (MPa)	E (GPa)
1 mm (B_1)	0	1763 ± 41	81.8 ± 0.8	1763 ± 41	81.8 ± 0.8
	7	1345 ± 34	79.1 ± 0.7	1227 ± 34	78.6 ± 0.7
	14	1059 ± 28	75.3 ± 0.7	1056 ± 29	78.0 ± 0.6
	30	1042 ± 28	74.6 ± 0.7	1012 ± 27	79.7 ± 0.6
	90	851 ± 22	73.5 ± 0.7	599 ± 16	78.2 ± 0.6
	180	600 ± 14	70.3 ± 0.6	354 ± 10	65.5 ± 0.7
2 mm (B_2)	0	1830 ± 44	85.1 ± 0.9	1830 ± 44	85.1 ± 0.8
	7	1390 ± 38	82.5 ± 0.7	1252 ± 34	81.5 ± 0.9
	14	1148 ± 31	80.2 ± 0.7	1159 ± 30	79.3 ± 0.7
	30	1121± 26	79.9 ± 0.6	1057 ± 28	77.0 ± 0.6
	90	885± 22	77.3 ± 0.6	759 ± 21	75.8 ± 0.5
	180	689 ± 17	75.7 ± 0.7	481 ± 13	71.7 ± 0.6
4 mm (B_4)	0	1782 ± 42	83.5 ± 0.8	1782 ± 42	83.5 ± 0.8
	7	1420 ± 39	80.4 ± 0.7	1354 ± 38	82.9 ± 0.6
	14	1279 ± 34	78.5 ± 0.7	1230 ± 34	79.5 ± 0.6
	30	1174 ± 32	78.2 ± 0.6	1099 ± 30	79.1 ± 0.6
	90	993 ± 26	78.8 ± 0.6	946 ± 25.4	77.5 ± 0.7
	180	723 ± 16.2	77.5 ± 0.7	590 ± 15.2	75.6 ± 0.6

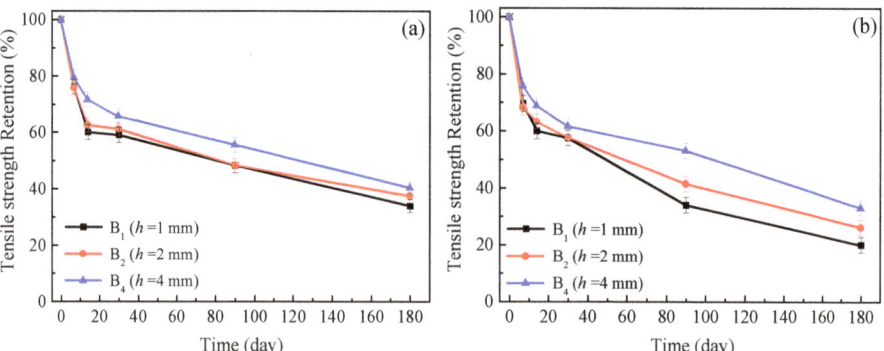

Figure 6. Variations of tensile strength retention of BFRP laminates vs. ageing time soaked in (**a**) deionized water and (**b**) alkaline solution at 60 °C.

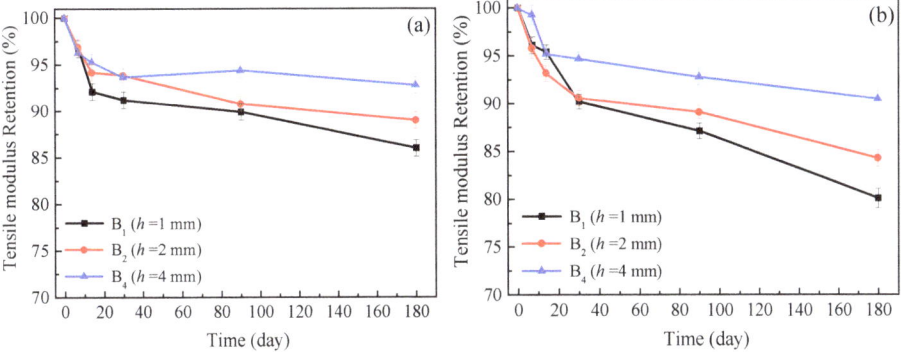

Figure 7. Variations of elastic modulus retention of BFRP laminates vs. ageing time soaked in (**a**) deionized water and (**b**) alkaline solution at 60 °C.

According to the SEM images of tensile failure surfaces of unaged and aged specimens shown in Figure 8, the degradation of tensile properties of BFRP laminates mainly comes from two parts. One is the degradation of epoxy resin and the corresponding interfacial debonding between the basalt fiber and the epoxy resin. Figure 8a exhibits the tensile failure surface of unaged specimen, where the basalt fiber was perfectly bonded to epoxy resin. It means that the basalt fiber and the epoxy resin play a synergistic role in the process of stress transmission during tension for unaged specimens. At the beginning of ageing, as shown in Figure 8b, the interfacial bond between basalt fiber and epoxy resin was gradually weakened due to the hydrolysis reaction of epoxy resin, leading to partial interfacial debonding. As the increase of ageing time, the hydrolysis of epoxy resin would be more severe as shown in Figure 8c (180 days). The other is the ageing of basalt fiber. SEM images of basalt fiber before and after ageing were shown in Figure 8d–f, respectively. It can be seen that the basalt fiber was also deteriorated due to the chemical reactions. Basalt fibers are composed of SiO_2 with tetrahedral network structure, in which silicon and oxygen account for 27% and 44% of the total elements, respectively [47]. When basalt fiber was exposed to the solution, H_2O and OH^- played a significant role in the deterioration of basalt fiber. The chemical reaction formula of basalt fiber is shown as Equation (3). OH^- attacks the molecular chains of SiO_2, causing a part of SiO_2 chains to relax. After that, the produced SiO^- further react with H_2O to produce OH^- to intensify the reaction of Equation (3) [47,48], as follows in Equation (4). Besides, SiOH produced from SiO^- is a kind of white colloid, which can transfer H_2O and OH^- to approach SiO_2 bone chains, and further deteriorate basalt fiber. Since the reaction rate of SiO_2 bone chains is relatively slow, the degradation of tensile strength and elastic modulus decreased slowly in the late stage, consistent with the experimental results (shown in Figures 6 and 7).

$$\left[-\text{Si}-\text{O}-\text{Si}- \right]_n + OH^- \rightarrow -\text{Si}-\text{OH} + -\text{Si}-\text{O}^- \tag{3}$$

$$\left[-\text{Si}-\text{O}^- \right]_n + H_2O \rightarrow -\text{Si}-\text{OH} + OH^- \tag{4}$$

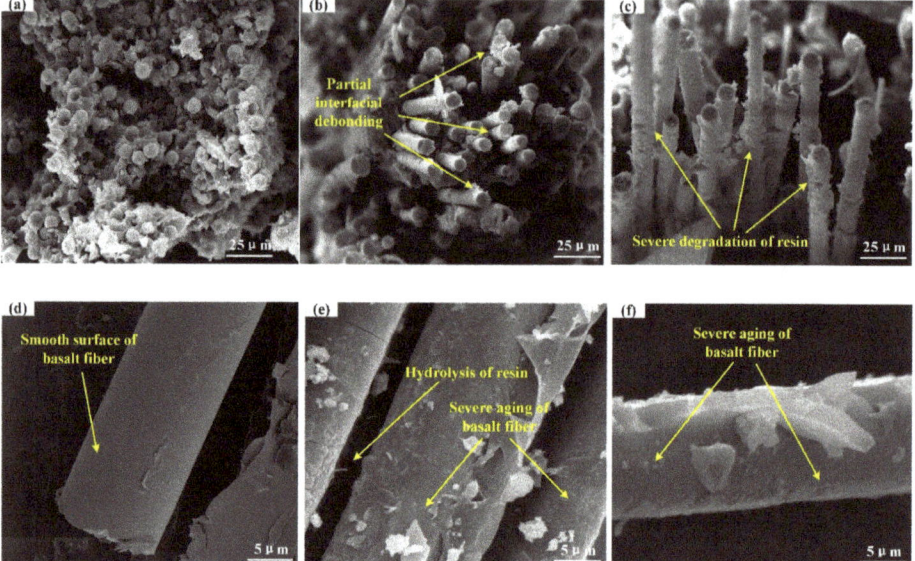

Figure 8. SEM images of tensile failure surfaces: (**a**) unaged specimen; aged specimen soaked in 60 °C alkaline solution last (**b**) 14 days and (**c**) 180 days; (**d**) unaged basalt fiber; aged basalt fiber soaked in 60 °C (**e**) deionized water and (**f**) alkaline solution last 180 days.

4. Discussions

4.1. Influence of Thickness on Water Absorption

The measured water absorption was analyzed against specimen thickness at specific ageing days (Figures 9 and 10). Figure 9a,b show that water absorption of the specimens decreased by increasing specimen thickness in the early stage (less than two days) of the immersion in deionized water and alkaline solution, respectively. In other words, the thinner the specimen thickness, the faster the water absorption saturation was achieved. As shown in Figure 9c,d, when the ageing time increased to four and seven days, the water absorption rose and then declined as the increasing specimen thickness. This is because in this ageing stage the change in the weight of specimen B_1 began to be dominated by the hydrolysis of the epoxy resin (reducing the weight of the specimen) rather than water absorption (increasing the weight of the specimen). While the changes in the weight of specimen B_2 and B_3 were still dominated by water absorption. At this time, the mass gain of specimen B_1 from water absorption was less than the mass loss of epoxy resin due to the hydrolysis, leading to a decrease of measured water absorption of specimen B_1. As shown in Figure 10, the water absorption trend in late stage of the immersion (14–180 days) was opposite to that in early stage, indicating that the mass loss of all specimens due to the hydrolysis of epoxy resin was more than the mass gain of water absorption in this stage. The mass loss of thinner specimen (B_1) was still greater than that of thicker specimens (B_2, B_4) in late stage, meaning that a thinner specimen had a greater hydrolysis degree at the given ageing time.

It is noteworthy that the water absorption of BFRP laminates was studied by considering the epoxy resin and the basalt fiber as a whole in this study. It is significant to determine water absorptions of the resin and basalt fiber respectively for a better understanding of the water absorption mechanism. In a future study, the related work (i.e., separately studying the water absorption of the resin and the basalt fiber) will be incorporated in the authors' study target.

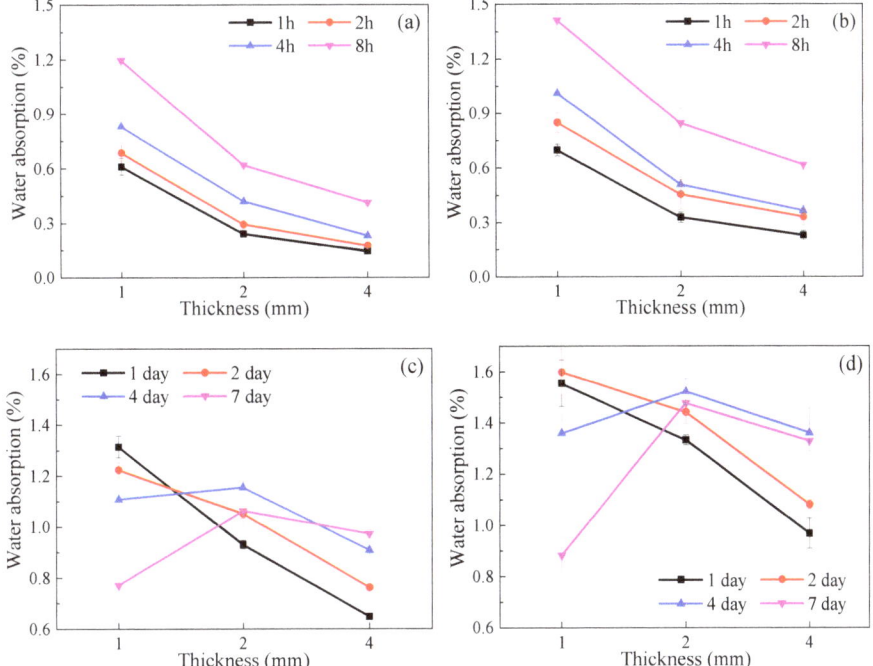

Figure 9. Variation of water absorption vs. specimen thickness soaked in (**a**,**c**) deionized water and (**b**,**d**) alkaline solution in early stage.

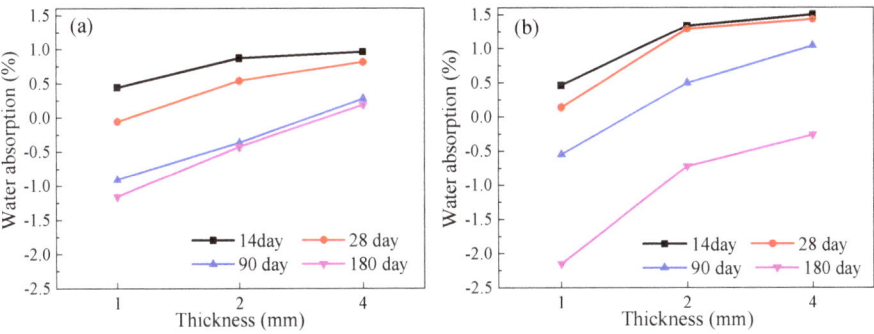

Figure 10. Variation of water absorption vs. specimen thickness soaked in (**a**) deionized water and (**b**) alkaline solution in late stage.

4.2. Influence of Thickness on Tensile Strength

The tensile strength retention vs. specimen thickness under different ageing durations was plotted in Figure 11. It can be observed that the specimens with thinner thickness had smaller retention than that of specimens with thicker thickness at given ageing time. For example, the tensile strength retention of specimen B_1 was 5%–20% lower than that of specimen B_4. Combining the test results of tensile test and water absorption, it can be concluded that at the beginning of ageing (around 30 days), the specimen with a thinner thickness had a faster deterioration rate than that of the specimen with a thicker thickness. After that, the deterioration rate was not sensitive to the thickness, while the ageing degree of the specimen with a thinner thickness was more severe than that of the specimen with a thicker thickness during a long period of ageing. Figure 12 shows the ageing of the specimens with different thicknesses through the cross section. It should be noted that the aged areas shown in Figure 12 only represent the aged areas of the specimens at a certain ageing time and the aged areas will gradually increase with the increasing ageing time. Additionally, note that the proportion of each part of the schematic is not drawn strictly according to the actual size, but just to better explain what needs to be expressed. Since the thickness of the specimen is much smaller than its length and width, it can be assumed that the deterioration of the specimen is carried out along the direction of the thickness. The aged area gradually expands to the interior of the specimen with the increase of ageing time. Although the thicknesses of B_1, B_2, and B_4 are different, the diffusion coefficient of the immersion solution is identical due to the same fabrication materials [49,50]. Therefore, the aged areas of B_1, B_2 and B_4 are the same after a same certain ageing time. It can be seen clearly that the thicker specimen has a larger unaged area, i.e., $S_1 < S_2 < S_4$. As the ageing time increases, the relationship between the unaged area S of BFRP laminates with different thicknesses was still established, i.e., $S_1 < S_2 < S_4$. Therefore, a thinner specimen is inferior to a thicker one on long-term hygrothermal properties due to a faster degradation rate.

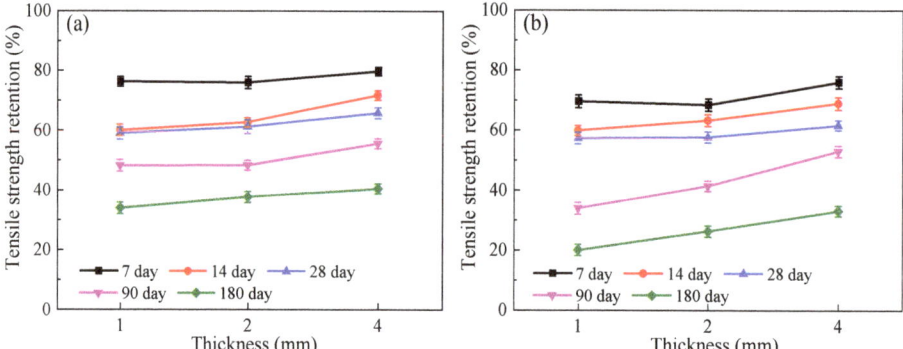

Figure 11. Variation of tensile strength retention vs. specimen thickness soaked in (**a**) deionized water and (**b**) alkaline solution at different ageing days.

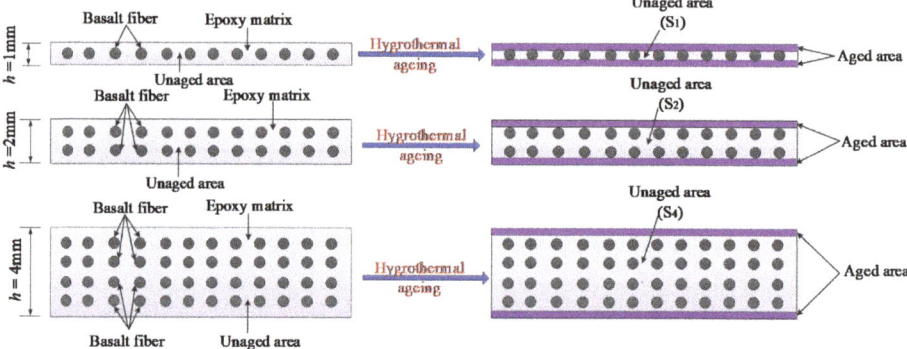

Figure 12. Schematic diagram of the ageing of the specimens with different thicknesses through cross section at a same certain ageing time.

5. Theoretical Model of Accelerated Factor

The above experimental results indicated clearly that the water absorption and tensile strength retention of BFRP laminate were dramatically influenced by specimen thickness. As illustrated in

Figures 3 and 6, it could be observed that the ageing degradation (i.e., the changing trends of water absorption and reduction of tensile strength retention) of BFRP laminates with thinner specimens was more severe than that of thicker specimens when the BFRP laminates were soaked in the solution for a same long time. Therefore, there is a possibility to deduce an accelerated ageing model based on specimen thickness for water absorption and tensile strength retention of BFRP laminates under hygrothermal ageing environment. In this section, a theoretical accelerated ageing model based on specimen thickness was developed and the accelerated factors (AFs) related to specimen thickness were theoretically deduced.

5.1. Accelerated Factor of Water Absorption

Considerable researches show that the water absorption of FRP composites can be described by two models as shown in Figure 13, i.e., the Fick's model and Two-stage model [39,42,51]. The Fick's model assumes that the initial phase of water absorption increases linearly with $t^{1/2}$ in the initial phase, and then increases non-linearly until the water absorption reaches a dynamic equilibrium without obvious changes. For the Two-stage model, the water absorption is identical to Fick's model in the initial phase. But the water absorption cannot reach the equilibrium stage due to the water immersion constantly and the degradation of resin matrix. In the current study, the test results of water absorption conform to the Two-stage model.

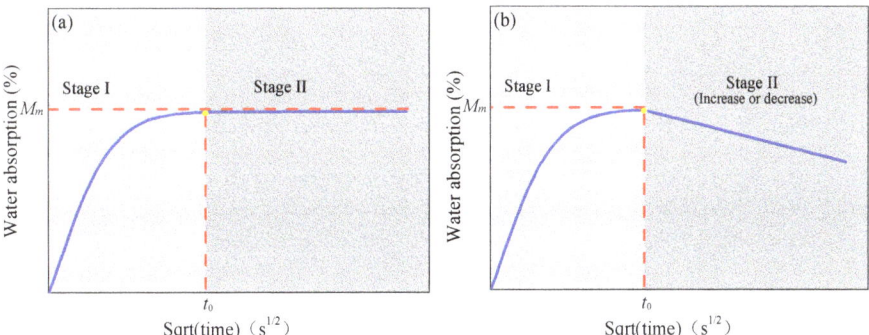

Figure 13. Typical model of water absorption of composites: (**a**) Fick model; (**b**) Two-stage model.

According to the ageing mechanism reported in the literature [49,50], the following assumptions can be used to establish the water absorption model. First, the specimens with different thicknesses have the same diffusion rate for the same immersion solution because the diffusion rate is controlled by the concentration gradient of the immersion solution. Second, although the water absorption at the regular time varies with specimen thickness, the ageing mechanism of the composite remains unchanged. Third, the specimens with different thicknesses fabricated by the same material will eventually reach the same water absorption rate under the same immersion solution.

The Fick's and Two-stage models can be used to describe the relationship between specimen thickness and ageing time in stage I (mass gain due to swelling of FRP composites) and stage II (mass loss due to relaxation of FRP composites), respectively. In the two models, the stage I of water absorption is identical. The Fick's model is shown in Equation (5), but it is difficult to find the relationship among different parameters in this expression, which is needed to be simplified as presented in Equation (6) [36,40].

$$M(t) = M_\infty \left\{ 1 - \frac{8}{\pi^2} \sum_{j=0}^{\infty} \frac{\exp[-(2j+1)^2 \pi^2 \left(\frac{Dt}{h^2}\right)]}{(2j+1)^2} \right\} \quad (5)$$

$$M(t) = M_\infty \left\{ 1 - \exp\left[-7.3\left(\frac{Dt}{h^2}\right)^{0.75}\right]\right\} \tag{6}$$

where M_∞ is the effective moisture equilibrium content, D is the diffusion coefficient.

When two specimens with the thicknesses of h_1 and h_2 soaked in the same solution reach the same water absorption at ageing time t_1 and t_2, respectively. The equilibrium equation, i.e., $M(t_1) = M(t_2)$ can be expressed as Equation (7) according to Fick's model, which can be simplified to Equation (8).

$$M_\infty \left\{ 1 - \exp\left[-7.3\left(\frac{Dt_1}{h_1^2}\right)^{0.75}\right]\right\} = M_\infty \left\{ 1 - \exp\left[-7.3\left(\frac{Dt_2}{h_2^2}\right)^{0.75}\right]\right\} \tag{7}$$

$$\exp\left[-7.3\left(\frac{Dt_1}{h_1^2}\right)^{0.75}\right] = \exp\left[-7.3\left(\frac{Dt_2}{h_2^2}\right)^{0.75}\right] \tag{8}$$

According to the previous assumptions, the diffusion coefficient D is identical when the specimens are soaked in the same solution, yields

$$\frac{t_1}{t_2} = \frac{h_1^2}{h_2^2} \tag{9}$$

Therefore, the AF of water absorption in stage I of Two-stage model can be expressed in

$$AF = \frac{t_1}{t_2} = \left(\frac{h_1}{h_2}\right)^2 \tag{10}$$

After the above derivation of AF in Stage I, it can be found that there is a certain relationship between ageing time t and specimen thickness h indeed. To further explore the relationship between water absorption, ageing time t and specimen thickness h in stage II of Two-stage model, the abscissa of water absorption curves can be changed from $t^{1/2}$ to t/h^2, as shown in Figure 14. According to the literature [48], the changing the abscissa (from $t^{1/2}$ to t/h^2) does not change the trends of water absorption in Stage I and Stage II. The water absorption curve of Two-stage model in Stage II was regarded as declining linearly that is determined by the slope of the descending section, $\tan\alpha$, which is only related to the materials of FRP composite and conditional environment [42,51]. It is known clearly that the $\tan\alpha$ of water absorption curves in Stage II is identical for the specimens B_1 B_2 and B_4 due to the same materials of BFRP laminates and conditional environment. Therefore, in the whole Stage II of the Two-stage model, the relationship between water absorption $M(t)$ at any given time and the maximum water absorption M_m can be expressed as

$$M(t) = M_m - \frac{t - t_0}{h^2} \cdot \tan\alpha \tag{11}$$

When the specimens with two thicknesses h_1 and h_2 reach the same water absorption at t_1 and t_2, respectively, the equilibrium equation can be expressed as Equation (12)

$$M_m - \frac{t_1 - t_{01}}{h_1^2} \cdot \tan\alpha = M_m - \frac{t_2 - t_{02}}{h_2^2} \cdot \tan\alpha \tag{12}$$

Equation (12) can be simplified as

$$\frac{t_1 - t_{01}}{h_1^2} = \frac{t_2 - t_{02}}{h_2^2} \tag{13}$$

It has been known by the derivation of Stage I in the two water absorption models that the Equation (10) is applicable when the value of water absorption increases to M_m, i.e.,

$$\frac{t_{01}}{h_1^2} = \frac{t_{02}}{h_2^2} \qquad (14)$$

Thus, the *AF* of stage II in Two-stage model can be expressed by Equation (10).

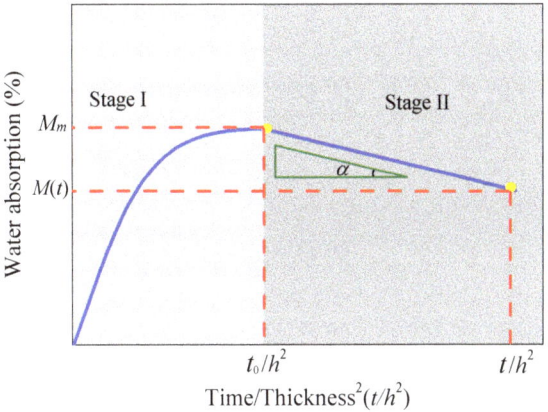

Figure 14. Two-stage model with the abscissa changed to t/h^2.

Therefore, the accelerated factor based on specimen thickness for water absorption is deduced theoretically, which can be applied in Fick's model and Two-stage model.

5.2. Accelerated Factor of Tensile Strength Retention

When the tensile specimens are soaked in the solution, the tensile strength decreases due to the gradual increase of the ageing area. The ageing condition of tensile specimens can be represented by the change of the cross section as shown in Figure 15, which is assumed that the specimen is uniformly aged along the thickness direction on the upper and lower sides. Thus, the actual tensile strength σ can be expressed by Equation (15).

$$\sigma = \frac{\sigma_u \cdot A_u + \sigma_a \cdot A_a}{bh} = \frac{\sigma_u \cdot b \cdot (h-x) + \sigma_a \cdot b \cdot x}{bh} = \sigma_u - (\sigma_u - \sigma_a)\frac{x}{h} \qquad (15)$$

where σ_u and σ_a are the initial (unaged) and residual (aged) tensile strength of the specimen, respectively. A_u and A_a are the unaged area and aged area of the specimen, respectively. b and h are the width and thickness of the specimen, respectively. x is the total ageing depth along the thickness of the laminates.

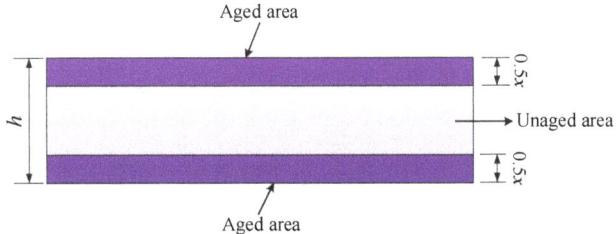

Figure 15. Schematic diagram of ageing along cross section for BFRP laminates.

It has been reported by study [49,52,53] that the total ageing depth x of the laminates is proportional to the square root of the immersion time t, which is as follows regarded as a function of

$$x = \alpha \sqrt{t} \quad (x < h) \tag{16}$$

where α is a constant of the FRP composite that is irrelevant to specimen thickness h.

Then Equation (15) can be transformed as follows

$$\sigma = \sigma_u - (\sigma_u - \sigma_a)\frac{\alpha \sqrt{t}}{h} \tag{17}$$

According to the literature [50,54], for the given resin matrix, fiber, fabrication and immersion condition, the specimens with different thicknesses will decrease to the same strength when the ageing time is long enough. As a result, the equilibrium equation can be expressed as Equation (18) when the tensile strengths of the specimens when two thicknesses, h_1 and h_2, decrease to the same value at ageing time t_1 and t_2, respectively.

$$\sigma_1 = \sigma_u - (\sigma_u - \sigma_a)\frac{\alpha \sqrt{t_1}}{h_1} = \sigma_2 = \sigma_u - (\sigma_u - \sigma_a)\frac{\alpha \sqrt{t_2}}{h_2} \tag{18}$$

Thus, the *AF* of tensile strength retention of BFRP laminates can also be written as Equation (10).

From the previous theoretical derivation, for BFRP laminates with two thicknesses, when the water absorption or tensile strength retention of the BFRP laminates with two different thicknesses reaches the same value, the ratio of the ageing time of the BFRP laminates with the two thicknesses is proportional to the square of the ratio of the corresponding two thicknesses. The square of the ratio of the two thicknesses is considered as accelerated factor (*AF*). As a result, it is feasible to accelerate ageing by reducing the specimen thickness. For example, if the water absorptions or tensile strength retentions of two BFRP laminates with different thicknesses reach a certain same value, the predicted ageing time of thicker specimen (t_1) can be calculated/accelerated by multiplying the real ageing time of thinner specimen (t_2) by the corresponding accelerated factor, $AF = (h_1/h_2)^2$, based on the two thicknesses of the thicker specimen (h_1) and thinner specimen (h_2).

5.3. Model Validation and Discussions

In this study, the ageing accelerated method was taking the BFRP specimen with the thickness $h = 4$ mm as the standard specimen. The *AF*s of BFRP specimens with $h = 1$ mm, $h = 2$ mm, and $h = 4$ mm are calculated by Equation (10) and the results are shown in Table 3. The obtaining of accelerated ageing days is transformed by multiplying the actual ageing days of the thicker specimen by the corresponding *AF* based on the two thicknesses of the thicker specimen and the standard specimen. According to the existing test results, the accelerated time is up to eight years on the water absorption trend and tensile strength retention of BFRP specimen with $h = 4$ mm. It should be noted that the transformed results of the specimens with 1 and 2 mm represent the predicted results of the standard specimen $h = 4$ mm. Figure 16 shows the long-term prediction of water absorption of BFRP specimens with $h = 4$ mm according to the transformed results of 1 and 2 mm specimens in deionized water and alkaline solution. The predicted curves were fitted by using Matlab (2018a Version, MathWorks, Inc., Massachusetts, USA, 2018). It can be seen from Figure 16 that the predicted curves are fitted well with a Two-stage model in whole. It reveals that the predicted law of water absorption in deionized water was better with Two-stage model than that in alkaline solution. Figure 17 shows the long-term prediction of tensile strength retention of BFRP specimens with $h = 4$ mm according to the transformed results of 1 and 2 mm specimens in deionized water and alkaline solution, which is also fitted well with Phani-Bose's model [55] in whole.

Table 3. The accelerated factor *AF* of BFRP laminates with different thicknesses.

	Specimen Thickness		
	1 mm	2 mm	4 mm
Accelerated factor (*AF*)	16	4	1

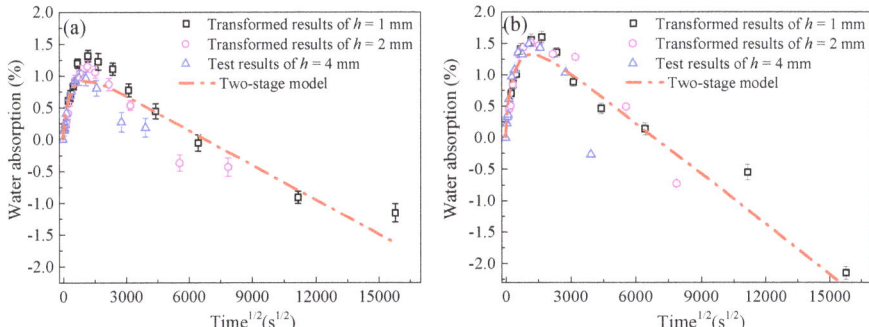

Figure 16. Accelerated ageing results of water absorption based on specimen thickness in (**a**) deionized water and (**b**) alkaline solution.

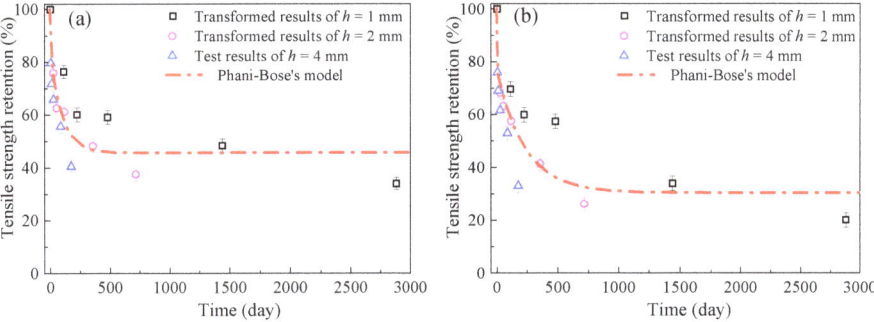

Figure 17. Accelerated ageing results of tensile strength retention based on specimen thickness: (**a**) deionized water and (**b**) alkaline solution.

Compared with the traditional temperature-dependent accelerated ageing method, there are two main advantages of the proposed new specimen thickness-dependent accelerated ageing method in this study. First, the proposed thickness-dependent accelerated ageing method is easy to apply because this method does not need the activation energy, which must be required in the temperature-dependent method. In order to obtain the activation energy in the temperature-dependent method, at least three different temperature environment tests must be conducted [56], which increases the test difficulty. In contrast, the accelerated factor calculating in the proposed method is only dependent on specimen thickness, which would be easy to be conducted. Second, the *AF* of the temperature-dependent accelerated ageing method is limited due to the limitation of T_g of FRP composites, leading to the acceleration times being relatively small. The *AF* of the proposed method is only dependent on specimen thickness and thus eliminates the limitation of T_g. Therefore, greater *AF* can be obtained in the proposed method.

6. Conclusions

In this paper, the BFRP laminates with the thickness of $h = 1$, $h = 2$, and $h = 4$ mm were fabricated by wet-layup method and the influence of thickness on their water absorption and tensile properties were experimentally studied under hygrothermal environment as well as degradation mechanism. A specimen thickness-dependent accelerated ageing method was proposed. The following conclusions can be drawn from the testing results and discussions in this study:

The long-term properties of BFFRP laminates were greatly affected under hygrothermal environment. The water absorption trend of BFRP laminates soaked in both 60 °C deionized water and alkaline solution increased first before reaching their peak water absorption and then decreased with the increase of immersion duration, which was caused by the hydrolysis of the epoxy matrix. The tensile properties of BFRP laminates degraded apparently after ageing, especially in alkaline solution. SEM images show basalt fiber deteriorated due to the solution immersion.

Specimen thickness had a significant influence on the water absorption and tensile strength of BFRP laminates after ageing. When the BFRP laminates with different thicknesses were immersed in the water or alkaline solution for the same ageing time, the water absorption decreased in early stage of immersion and then increased in late stage of immersion as the specimen thickness increased, while the tensile strength retention kept increased during the whole ageing process. The reason is that the ratio of aged area to the total area in the thinner specimens was larger than that of the thicker specimens, leading to the more severe ageing degree of the thinner specimens.

An innovative thickness-dependent accelerated ageing method for water absorption and tensile strength retention of BFRP laminates was proposed, in which the accelerated factors were theoretically deduced based on specimen thickness. The proposed method is in good agreement with test results. Compared with the traditional accelerated ageing method based on temperature, the proposed method is much easier to be conducted and has the potential to obtain a greater accelerated factor.

Author Contributions: Conceptualization, Y.W.; methodology, Y.W. and X.Z.; software, W.Z.; validation, B.W., W.Z.; formal analysis, G.C.; investigation, G.C. and Y.W.; resources, Y.W.; data curation, W.Z.; writing-original draft preparation, W.Z.; writing-review and editing, Y.W., X.Z. and B.W.; visualization, G.C.; supervision, Y.W., X.Z. and G.C.; project administration, Y.W.; funding acquisition, Y.W. All authors have read and agreed to the published version of the manuscript.

Acknowledgments: The authors are grateful for the financial support from the National Key Research and Development Program of China (Project No. 2017YFC0703000), the National Natural Science Foundation of China (Project No. 51778102 and 51978126), the Fundamental Research Funds for the Central Universities of China (Project No. DUT18LK35), and the Natural Science Foundation of Liaoning Province of China (Project No. 20180550763).

Conflicts of Interest: The authors declare no conflict of interest.

Abbreviations

FRP	Fiber-reinforced polymer
BFRP	Basalt FRP
CFRP	Carbon FRP
GFRP	Glass FRP
AFRP	Aramid FRP
RC	reinforced concrete
VARI	vacuum assistant resin infusion
SEM	scanning electron microscopy
AF	accelerated factor

References

1. Wei, Y.; Zhang, Y.; Chai, J.; Wu, G.; Dong, Z. Experimental investigation of rectangular concrete-filled fiber reinforced polymer (FRP)-steel composite tube columns for various corner radii. *Compos. Struct.* **2020**, *244*, 112311. [CrossRef]
2. Bai, Y.L.; Dai, J.G.; Mohammadi, M.; Lin, G.; Mei, S.J. Stiffness-based design-oriented compressive stress-strain model for large-rupture-strain (LRS) FRP-confined concrete. *Compos. Struct.* **2019**, *223*, 110953. [CrossRef]

3. Wang, Y.; Cai, G.; Li, Y.; Waldmann, D.; Si Larbi, A.; Tsavdaridis, K.D. Behavior of circular fiber-reinforced polymer-steel-confined concrete columns subjected to reversed cyclic loads: Experimental studies and finite-element analysis. *J. Struct. Eng.* **2019**, *145*, 04019085. [CrossRef]
4. Zeng, J.; Gao, W.; Duan, Z.; Bai, Y.; Guo, Y.C.; Ouyang, L.J. Axial compressive behavior of polyethylene terephthalate/carbon FRP-confined seawater sea-sand concrete in circular columns. *Constr. Build. Mater.* **2020**, *234*, 117383. [CrossRef]
5. Cao, Q.; Tao, J.; Wu, Z.; Ma, Z.J. Behavior of FRP-Steel confined concrete tubular columns made of expansive self-consolidating concrete under axial compression. *J. Compos. Constr.* **2017**, *21*, 04017037-1-12. [CrossRef]
6. Bai, Y.L.; Yan, Z.W.; Ozbakkaloglu, T.; Han, Q.; Dai, J.G.; Zhu, D.J. Quasi-static and dynamic tensile properties of large-rupture-strain (LRS) polyethylene terephthalate fiber bundle. *Constr. Build. Mater.* **2020**, *232*, 117241. [CrossRef]
7. Wang, Y.; Wang, Y.; Han, B.; Wan, B.; Cai, G.; Chang, R. In situ strain and damage monitoring of GFRP laminates incorporating carbon nanofibers under tension. *Polymers* **2018**, *10*, 777. [CrossRef]
8. Cao, Q.; Li, H.; Lin, Z. Effect of active confinement on GFRP confined expansive concrete under axial cyclic loading. *ACI Struct. J.* **2020**, *117*, 207–216. [CrossRef]
9. Wang, Y.; Chen, G.; Wan, B.; Cai, G.; Zhang, Y. Behavior of circular ice-filled self-luminous FRP tubular stub columns under axial compression. *Constr. Build. Mater.* **2020**, *232*, 117287. [CrossRef]
10. Wei, Y.; Zhang, X.; Wu, G.; Zhou, Y. Behaviour of concrete confined by both steel spirals and fiber-reinforced polymer under axial load. *Compos. Struct.* **2018**, *192*, 577–591. [CrossRef]
11. Zhang, Y.; Wei, Y.; Bai, J.; Zhang, Y. Stress-strain model of an FRP-confined concrete filled steel tube under axial compression. *Thin Walled Struct.* **2019**, *142*, 149–159. [CrossRef]
12. Zhai, K.; Fang, H.; Fu, B.; Wang, F.; Hu, B. Mechanical response of externally bonded CFRP on repair of PCCPs with broken wires under internal water pressure. *Constr. Build. Mater.* **2020**, *239*, 117878. [CrossRef]
13. Wang, Y.; Wang, Y.; Wan, B.; Han, B.; Cai, G.; Li, Z. Properties and mechanisms of self-sensing carbon nanofibers/epoxy composites for structural health monitoring. *Compos. Struct.* **2018**, *200*, 669–678. [CrossRef]
14. Lu, M.; Xiao, H.; Liu, M.; Li, X.; Li, H.; Sun, L. Improved interfacial strength of SiO_2 coated carbon fiber in cement matrix. *Cem. Concr. Compos.* **2018**, *91*, 21–28. [CrossRef]
15. He, J.; Xian, G.; Zhang, Y.X. Effect of moderately elevated temperatures on bond behaviour of CFRP-to-steel bonded joints using different adhesives. *Constr. Build. Mater.* **2020**, *241*, 118057. [CrossRef]
16. Li, J.; Xie, J.; Liu, F.; Lu, Z. A critical review and assessment for FRP-concrete bond systems with epoxy resin exposed to chloride environments. *Compos. Struct.* **2019**, *229*, 111372. [CrossRef]
17. Huang, H.; Jia, B.; Lian, J.; Wang, W. Experimental investigation on the tensile performance of resin-filled steel pipe splices of BFRP bars. *Constr. Build. Mater.* **2020**, *242*, 118018. [CrossRef]
18. Dong, Z.; Wu, G.; Xu, B.; Wang, X.; Taerwe, L. Bond durability of BFRP bars embedded in concrete under seawater conditions and the long-term bond strength prediction. *Mater. Des.* **2016**, *92*, 552–562. [CrossRef]
19. Li, Y.; Wang, Y.; Ou, J. Mechanical behavior of BFRP-steel composite plate under axial tension. *Polymers*. **2014**, *6*, 1862–1876. [CrossRef]
20. Wang, Y.; Wang, Y.; Wan, B.; Han, B.; Cai, G.; Chang, R. Strain and damage self-sensing of basalt fiber reinforced polymer laminates fabricated with carbon nanofibers/epoxy composites under tension. *Compos. Part A Appl. Sci. Manuf.* **2018**, *113*, 40–52. [CrossRef]
21. Dong, Z.; Wu, G.; Zhao, X.L.; Zhu, H.; Wei, Y.; Yan, Z. Mechanical properties of discrete BFRP needles reinforced seawater sea-sand concrete-filled GFRP tubular stub columns. *Constr. Build. Mater.* **2020**, *244*, 118330. [CrossRef]
22. Nguyen, V.D.; Hao, J.; Wang, W. Ultraviolet weathering performance of high-density polyethylene/wood-flour composites with a basalt-fiber-included shell. *Polymers* **2018**, *10*, 831. [CrossRef] [PubMed]
23. Wang, Z.; Zhao, X.L.; Xian, G.; Wu, G.; Singh Raman, R.K.; Al-Saadi, S. Durability study on interlaminar shear behaviour of basalt-, glass- and carbon-fibre reinforced polymer (B/G/CFRP) bars in seawater sea sand concrete environment. *Constr. Build. Mater.* **2017**, *156*, 985–1004. [CrossRef]
24. Quagliarini, E.; Monni, F.; Bondioli, F.; Lenci, S. Basalt fiber ropes and rods: Durability tests for their use in building engineering. *J. Build. Eng.* **2016**, *5*, 142–150. [CrossRef]
25. Li, S.; Hu, J.; Ren, H. The combined effects of environmental conditioning and sustained load on mechanical properties of wet lay-up fiber reinforced polymer. *Polymers* **2017**, *9*, 244. [CrossRef]

26. Sharma, B.; Chhibber, R.; Mehta, R. Seawater ageing of glass fiber reinforced epoxy nanocomposites based on silylated clays. *Polym. Degrad. Stab.* **2018**, *147*, 103–114. [CrossRef]
27. Jiang, X.; Luo, C.; Qiang, X.; Zhang, Q.; Kolstein, H.; Bijlaard, F. Coupled hygro-mechanical finite element method on determination of the interlaminar shear modulus of glass fiber-reinforced polymer laminates in bridge decks under hygrothermal aging effects. *Polymers* **2018**, *10*, 845. [CrossRef]
28. Lu, Z.; Xian, G.; Li, H. Effects of thermal aging on the water uptake behavior of pultruded BFRP plates. *Polym. Degrad. Stab.* **2014**, *110*, 216–224. [CrossRef]
29. Lu, Z.; Xian, G.; Li, H. Effects of exposure to elevated temperatures and subsequent immersion in water or alkaline solution on the mechanical properties of pultruded BFRP plates. *Compos. Part B Eng.* **2015**, *77*, 421–430. [CrossRef]
30. Ma, G.; Yan, L.; Shen, W.; Zhu, D.; Huang, L.; Kasal, B. Effects of water, alkali solution and temperature ageing on water absorption, morphology and mechanical properties of natural FRP composites: Plant-based jute vs. mineral-based basalt. *Compos. Part B Eng.* **2018**, *153*, 398–412. [CrossRef]
31. Xiao, B.; Li, H.; Xian, G. Hygrothermal ageing of basalt fiber reinforced epoxy composites. In *Advances in FRP Composites in Civil Engineering, Proceedings of the 5th International Conference on FRP Composites in Civil Engineering, CICE 2010, Beijing, China, 27–29 September 2010*; Springer: Berlin/Heidelberg, Germany, 2011; pp. 356–359.
32. Wu, G.; Wang, X.; Wu, Z.; Dong, Z.; Zhang, G. Durability of basalt fibers and composites in corrosive environments. *J. Compos. Mater.* **2015**, *49*, 873–887. [CrossRef]
33. Li, X.; Yahya, M.Y.; Nia, A.B.; Wang, Z.; Yang, J.; Lu, G. Dynamic failure of basalt/epoxy laminates under blast—Experimental observation. *Int. J. Impact Eng.* **2017**, *102*, 16–26. [CrossRef]
34. Benmokrane, B.; Manalo, A.; Bouhet, J.C.; Mohamed, K.; Robert, M. Effects of diameter on the durability of glass fiber-reinforced polymer bars conditioned in alkaline solution. *J. Compos. Constr.* **2017**, *21*, 04017040. [CrossRef]
35. Wu, G.; Wang, X.; Wu, Z.; Dong, Z.; Xie, Q. Degradation of basalt FRP bars in alkaline environment. *Sci. Eng. Compos. Mater.* **2015**, *22*, 649–657. [CrossRef]
36. Chen, Y.; Davalos, J.F.; Ray, I. Durability prediction for GFRP reinforcing bars using short-term data of accelerated aging tests. *J. Compos. Constr.* **2006**, *10*, 279–286. [CrossRef]
37. Wang, Z.; Xian, G.; Zhao, X.L. Effects of hydrothermal aging on carbon fibre/epoxy composites with different interfacial bonding strength. *Constr. Build. Mater.* **2018**, *161*, 634–648. [CrossRef]
38. Wang, Y.; Cai, G.; Si Larbi, A.; Waldmann, D.; Tsavdaridis, K.D.; Ran, J. Monotonic axial compressive behaviour and confinement mechanism of square CFRP-steel tube confined concrete. *Eng. Struct.* **2020**, in press.
39. ASTM International. *Standard Test Method for Moisture Absorption Properties and Equilibrium Conditioning of Polymer Matrix Composite Materials (ASTM D5229/D5229-14)*; ASTM International: West Conshohocken, PA, USA, 2014.
40. American Concrete Institute. *Guide Test Methods for Fiber-Reinforced Polymers (FRPs) for Reinforcing or Strengthening Concrete Structures*; A.C.I. Committee 440, Report 440.3R-12; American Concrete Institute: Farmington Hills, MI, USA, 2012.
41. ASTM International. *Standard Test Method for Tensile Properties of Polymer Matrix Composite Materials (ASTM D3039/D3039M-14)*; ASTM International: West Conshohocken, PA, USA, 2014.
42. Hong, B.; Xian, G.; Wang, Z. Durability study of pultruded carbon fiber reinforced polymer plates subjected to water immersion. *Adv. Struct. Eng.* **2018**, *21*, 571–579. [CrossRef]
43. Agwa, M.A.; Taha, I.; Megahed, M. Experimental and analytical investigation of water diffusion process in nano-carbon/alumina/silica filled epoxy nanocomposites. *Int. J. Mech. Mater. Des.* **2017**, *13*, 607–615. [CrossRef]
44. Almeida, J.H.S.; Souza, S.D.B.; Botelho, E.C.; Amico, S.C. Carbon fiber-reinforced epoxy filament-wound composite laminates exposed to hygrothermal conditioning. *J. Mater. Sci.* **2016**, *51*, 4697–4708. [CrossRef]
45. Chin, J.W.; Nguyen, T.; Aouadi, K. Sorption and Diffusion of water, salt water, and concrete. *J. Appl. Polym. Sci.* **1999**, *71*, 483–492. [CrossRef]
46. Garg, M.; Sharma, S.; Mehta, R. Carbon nanotube-reinforced glass fiber epoxy composite laminates exposed to hygrothermal conditioning. *J. Mater. Sci.* **2016**, *51*, 8562–8578. [CrossRef]

47. Wang, Z.; Zhao, X.L.; Xian, G.; Wu, G.; Singh Raman, R.K.; Al-Saadi, S.; Haque, A. Long-term durability of basalt- and glass-fibre reinforced polymer (BFRP/GFRP) bars in seawater and sea sand concrete environment. *Constr. Build. Mater.* **2017**, *139*, 467–489. [CrossRef]
48. Shi, J.; Wang, X.; Wu, Z.; Zhu, Z. Fatigue behavior of basalt fiber-reinforced polymer tendons under a marine environment. *Constr. Build. Mater.* **2017**, *137*, 46–54. [CrossRef]
49. Hojo, H.; Tsuda, K.; Ogasawara, K. Form and rate of corrosion of corrosion-resistant FRP resins. *Adv. Compos. Mater.* **1991**, *1*, 55–67. [CrossRef]
50. Bao, L.R.; Yee, A.F. Effect of temperature on moisture absorption in a bismaleimide resin and its carbon fiber composites. *Polymer* **2002**, *43*, 3987–3997. [CrossRef]
51. Satterfield, M.B.; Benziger, J.B. Non-Fickian water vapor sorption dynamics by nafion membranes. *J. Phys. Chem. B* **2008**, *112*, 3693–3704. [CrossRef]
52. Zhang, X.; Wang, Y.; Wan, B.; Cai, G.; Qu, Y. Effect of specimen thicknesses on water absorption and flexural strength of CFRP laminates subjected to water or alkaline solution immersion. *Constr. Build. Mater.* **2019**, *208*, 314–325. [CrossRef]
53. Wang, Y.; Zhang, X.; Cai, G.; Wan, B.; Waldmann, D.; Qu, Y. A new thickness-based accelerated aging test methodology for resin materials: Theory and preliminary experimental study. *Constr. Build. Mater.* **2018**, *186*, 986–995. [CrossRef]
54. Pi, Z.; Xiao, H.; Du, J.; Liu, M.; Li, H. Interfacial microstructure and bond strength of nano-SiO2-coated steel fibers in cement matrix. *Cem. Concr. Compos.* **2019**, *103*, 1–10. [CrossRef]
55. Phani, K.K.; Bose, N.R. Hydrothermal ageing of CSM-laminate during water immersion—An acousto-ultrasonic study. *J. Mater. Sci.* **1986**, *21*, 3633–3637. [CrossRef]
56. Li, C.; Xian, G.; Li, H. Effect of postcuring immersed in water under hydraulic pressure on fatigue performance of large-diameter pultruded carbon/glass hybrid rod. *Fatigue Fract. Eng. Mater. Struct.* **2019**, *42*, 1148–1160. [CrossRef]

© 2020 by the authors. Licensee MDPI, Basel, Switzerland. This article is an open access article distributed under the terms and conditions of the Creative Commons Attribution (CC BY) license (http://creativecommons.org/licenses/by/4.0/).

Article

Experimental Study on the Performance of GFRP–GFRP Slip-Critical Connections with and without Stainless-Steel Cover Plates

Yang Peng [1,2], Wei Chen [1], Zhe Wu [1], Jun Zhao [1] and Jun Dong [1,*]

[1] College of Civil Engineering, Nanjing Tech University, Nanjing 211816, China; yang.peng@njtech.edu.cn (Y.P.); 201761101650@njtech.edu.cn (W.C.); 662085213177@njtech.edu.cn (Z.W.); 201761101656@njtech.edu.cn (J.Z.)
[2] Nanjing Gongda Construction Technology Co., Ltd., Nanjing 211800, China
* Correspondence: dongjun@njtech.edu.cn; Tel.: +86-136-0140-7837

Received: 3 June 2020; Accepted: 23 June 2020; Published: 26 June 2020

Abstract: Composite structures have become increasingly popular in civil engineering due to many advantages, such as light weight, excellent corrosion resistance and high productivity. However, they still lack the strength, stiffness, and convenience of constructions of fastener connections in steel structures. The most popular fastener connections in steel structures are slip-critical connections, and the major factors that influence their strength are the slip factors between faying surfaces and the clamping force due to the prevailing torque. This paper therefore examined the effect that changing the following parameters had on the slip factor: (1) replacing glass fiber reinforced plastic (GFRP) cover plates with stainless-steel cover plates; (2) adopting different surface treatments for GFRP-connecting plates and stainless-steel cover plates, respectively; and (3) applying different prevailing torques to the high-strength bolts. The impact on the long-term effects of the creep property in composite elements under the pressure of high-strength bolts was also studied with pre-tension force relaxation tests. It is shown that a high-efficiency fastener connection can be obtained by using stainless-steel cover plates with a grit-blasting surface treatment, with the maximum slip factor reaching 0.45, while the effects of the creep property are negligible.

Keywords: GFRP composite structures; slip-critical connection; stainless-steel cover plates; surface treatment; prevailing torque

---

1. Introduction

Recently, composites have become popular as a new type of structural material used in civil engineering because they are light weight, strong, corrosion resistant, and have good designability in section shapes. However, there is still a lack of an effective connecting method in composite structures. In general, similar to steel joints, there are three schemes for composite structure joints [1]: (1) bolted, (2) adhesively bonded, and (3) combined (bolted and bonded) connections. Bolted and adhesively bonded connections each present unique benefits and characteristics. Bolted connections can be installed easily but induce severe stress concentrations. Adhesive bonds achieve good load distribution in composite connections, however, they remain vulnerable to the service temperature, humidity, and other environmental conditions. In combination joints, both adhesives and bolted connections are used, which can provide greater reliability and shear resistance but also increases the difficulty of construction. Unlike slip-critical connections, none of the connections mentioned above can provide with high strength, stiffness, good load distribution and construction convenience simultaneously.

Girão Coelho et al. [2] examined bolted connections and joints in pultruded fiber reinforced polymers and concluded that further research was necessary to improve the design of composite mechanically fastened joints. Unlike metal, the inherently brittle behavior (linear elasticity) of pultruded composite materials compromises its ability to yield and deform to redistribute loads, thus increasing the sensitivity to stress concentrations [3]. As a result, a large portion of current composite structural connections were inadequate because they were created by simply duplicating steel structural connections. Feo et al. [3] introduced four typologies of connections for composite structures and selected a slip-critical connection to conduct shear tests. This connection has a somewhat complicated layout, with two steel connecting plates, double pultruded fiber reinforced plastic (PFRP) cover plates, and double steel cover plates. They found that the slip factor for this type of connection can surpass 0.37. To date, no type of effective slip-critical connection for composite structures has been proposed. Mottram et al. [4] proved that the slip factor for slip-critical connections with all composite components is relatively low, ranging from 0.14 to 0.27, which fails to meet the application standard in civil engineering. Additionally, simple bolted assemblies for PFRP plates would lose more than 40% of the bearing capacity during their service life-time. To address these issues, Hashimoto et al. [5] added stainless-steel cover plates with a grit-blasting treatment to (glass fiber reinforced plastic) GFRP–GFRP slip-critical connections and concluded that the slip factor for this type of connection can reach 0.4, and 86% of its strength would remain after one year under the bolt pre-tension. Stranghöner et al. [6] also examined the influence of different surface treatments in combination with austenitic stainless-steel bolting assemblies in slip-critical connections. The results demonstrated that the influence of surface treatments on the slip-resistant behavior of preloaded bolted connections with all steel elements is significant. However, little research has been dedicated to induced grit-blasting treatments for composite plates within a slip-critical connection or to the influence of cover plates. Feo et al. [3] found that applying the tightening procedure to assemblies including PFRP plates with the torque method recommended by EN1090-2 (which is for steel structural connections) led to a similar accuracy. However, insufficient data are available regarding the influence of various prevailing torques on the slip factor. Due to these disadvantages, slip-critical connections are not permitted by many composite structure design guide books, such as the American society of civil engineers (ASCE) design guide for fiber reinforced polymer (FRP) composite connections [1] and the Composites-Design Manual [7]. Studies on slip-critical connections involving composite materials are currently uncommon, and we are still far from a complete understanding of the mechanism of load transfer by friction forces with GFRP and stainless-steel faying surfaces.

In this paper, we focus on to the effects of the cover plate, surface treatment, and prevailing torque on the slip factor and long-term performance of slip-critical connections in composite structures. Shear tests and pre-tension relaxation tests were completed under quasi-static monotonic loads. Test results were presented in terms of load–displacement curve, slip load and slip factor. We performed our analysis to probe into the optimal configuration of slip-critical composite connections. Based on the adhesion theory of friction, explanations of the working mechanism of the connection are offered.

2. Experimental Details

2.1. Design Procedure

The condition of the faying surface and the bolt pre-tension force are the two main factors that influence the slip-resistant behavior of slip-critical connections in steel structures [8]. Thus, we assumed that these two factors were important parameters requiring study in regard to the connections in composite structures.

The symmetrical butt joint is the most common specimen configuration used in slip-critical connections. There are three configurations of butt joints proposed by different standards. EN 1090-2 specifications [9] suggest a standard test specimen with two cover plates, four slip planes, and four high-strength bolts. The Research Council on Structural Connections (RCSC) specifications [10] suggest a standard test specimen with two cover plates, two slip planes, and one high-strength bolt. The AS 4100 specifications [11] suggest a standard test specimen with two cover plates, four slip planes, and two high-strength bolts. Through a thorough literature review, we determined that the standard specimen proposed by the EN1090-2 specifications is the most popular [12]. However, despite its common use, this standard specimen is unsuitable for the engineering practice of slip factor measurements. By considering the convenience of the assembly and ease of measurement, the butt joint test specimen proposed by the AS 4100 specifications [11] was chosen for the purposes of this study.

2.2. Specimen Configuration

2.2.1. Geometrical Size and Mechanical Properties

The connecting components were composed of GFRP plates, which were 300 mm × 40 mm × 6 mm, the bolt-hole diameter D was 8 mm, and the hole center was 60 mm from the end of the plate. The bolt hole diameter and bolt hole spacing selection were based on American Society of Civil Engineers design guide for FRP composite connections [1]. The GFRP plates were made via a pultrusion process. The fiber volume fraction was 40%. The fiber was oriented at angles of 45°/90°/0°. All the information was provided by the manufacturers. The austenitic stainless-steel 304 plates measured 205 mm × 40 mm × 5 mm, the bolt-hole diameter D was 8 mm, and the center of the hole was 40 mm from the end of the plate. The mechanical properties of the plates are presented in Table 1. The mechanical properties of the stainless-steel plates and those of the GFRP plates were based on the technical data provided by the manufacturers. The washer had an outer diameter D_0 of 14 mm and an inner diameter d_0 of 6 mm. The material of the high-strength bolt was stainless-steel, grade A4-80, with a nominal diameter of 6 mm. The size of the test specimen is shown in Figure 1.

Table 1. Mechanical properties of plates.

Materials	Tensile Strength (MPa)	Elastic Modulus (GPa)	Vickers Hardness (kgf/mm^2)
Glass fiber reinforced plastic (GFRP)	300	15	/
Austenitic stainless-steel 30408	515	193	140

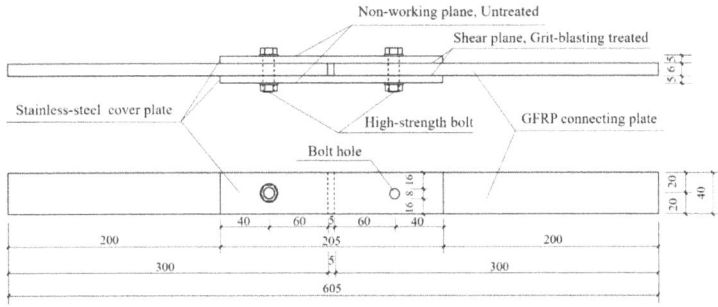

Figure 1. Specimen size (unit:mm).

2.2.2. Surface Preparation

A grit-blasting treatment was introduced in our tests to increase the surface hardness and roughness. Using compressed air, a high-speed jet of air was formed, which sprayed abrasives at high speed onto the surface of the specimen to be treated. The impact and cutting action of the abrasives helped to clean the surface to a certain degree, change the roughness, and enhance the mechanical properties. It was determined that the grit-blasting caused an increase in the Vickers micro-hardness values of all of the materials, which led to an increase in the ultimate tensile strength [13]. Based on the tribological theory, the friction factor can increase with the increment of the surface roughness and strength [14]. The surfaces of the test specimens were treated by grit-blasting using a HZ-1616r-1 sand blaster. The key parameters of the grit-blasting treatment are listed in Table 2.

Table 2. Main parameters of the grit-blasting treatment.

Blasting Treatment	Abrasives Type	Nozzle Pressure (bar)	Blasting Angle (°)	Distance of Spray Gun (mm)	Blasting Time (s)
Manual	Brown corundum	8	90	100	30

The grit-blasting treatment was carried out on the stainless-steel plate using #24 grit and #60 grit abrasive according to Kobayashi et al. [15]. The numbers #24 and #60 represent the size of the abrasive, the higher the number, the smaller the grit size and the smoother the surface will be after blasting [13]. Therefore, the surface roughness of the stainless-steel plates treated with the #24 grit abrasive will be greater than that of the stainless-steel plates treated with #60 grit abrasive. The different surface conditions captured by commercial camera under daylight are presented in Figure 2. When the surfaces of the plates become rougher, the diffuse reflection ratio on the surface will increase, and the mirror reflection ratio will decrease. Therefore, under the same light source, the visual perception of brightness is different, and the rougher the surface, the dimmer the appearance [16]. The blasting time for the GFRP plate should be reasonably controlled to prevent initial damage. The surface of the untreated GFRP plate was left in its original state, while the surface of the GFRP plate treated with the grit abrasive was glossy and somewhat rough.

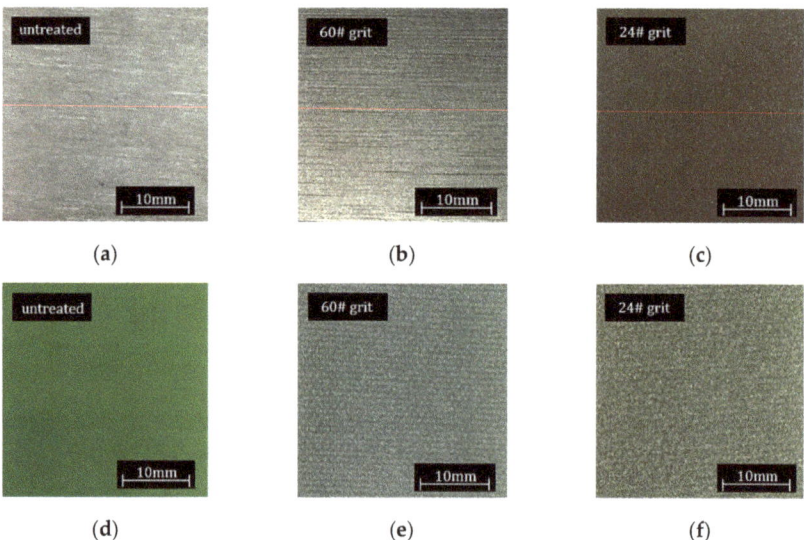

Figure 2. Plate surfaces with and without grit-blasting treatment. (**a**) Untreated stainless-steel surface; (**b**) stainless-steel surface treated with 60# grit-blasting; (**c**) stainless-steel surface treated with 24# grit-blasting; (**d**) untreated GFRP surface; (**e**) GFRP surface treated with 60# grit-blasting; (**f**) GFRP surface treated with 24# grit-blasting.

When implementing the grit-blasting treatment, only the faying planes were treated, and the non-working planes contacting bolt washers remained untreated, as shown in Figure 1. Several combinations of connecting methods were developed in this paper and their abbreviations are listed in Table 3. The optimal matching method of various connections were predicted.

Table 3. Abbreviations for each connection method.

Material of Connecting Plates (Surface Treatment)	Material of Cover Plates (Surface Treatment)	Abbreviation of Connection Methods
GFRP (untreated)	GFRP (untreated)	G-G
GFRP (untreated)	Stainless-steel (untreated)	G-S
GFRP (untreated)	Stainless-steel (#60 grit blasting)	G-S60#
GFRP (untreated)	Stainless-steel (#24 grit blasting)	G-S24#
GFRP (#60 grit blasting)	Stainless-steel (#24 grit blasting)	G60#-S24#

2.3. Slip Load Definition and Slip Factor Calculation

The load–displacement curve was calculated upon conclusion of the tests. The slip load was defined according to this curve. In the specifications given by the RCSC, the load–displacement curves can be classified into three types [10], as shown in Figure 3. Failure of a slip-critical connection is defined at different levels of slip according to the load–displacement curve type acquired by shear tests [17]. Examining the load–displacement curves, (a) is the type of curve where the slip load occurs at the maximum load. There is a drop in the load after the slip load. This drop in load is related to the major slip [2] and the drop marks the failure of the connection. Next, (b) is the type of curve where the slip load becomes an inflection point and changes suddenly, and the inflection point is also the failure point of the connection. Finally, (c) is the type of curve where the slip rate changes gradually. In this case, the slip load and failure point are a same fixed point, at which the slip equals 0.02 in (~0.5 mm). In this type of curve, minor slippage occurs before the slip load [2].

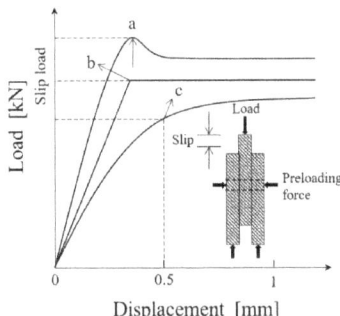

Figure 3. Three types of slip load curves [10].

Referring to the Chinese technical specification for stainless-steel structures (CECS 410) [18], the slip factor, μ, is calculated according to the following formula:

$$\mu = \frac{F}{n_f \sum_{i=1}^{m} P_i} \tag{1}$$

where, F is the slip load; n_f is the number of friction planes; m is the number of high-strength bolts; and P_i is the pre-tension force of the high-strength bolts, calculated using Formula (2).

The Chinese standard JGJ82-2011 [19] provides the details of the torque–tension relationship when tightening high-strength bolt assemblies.

$$T = kdP \tag{2}$$

where, T is the torque applied to the nut, k is the torque coefficient, d is the nominal bolt diameter, and P is the specified bolt preload. If the high-strength bolts adopt the same value of torque coefficient, the required pre-tension can be obtained by applying a certain amount of torque to the bolt.

2.4. Shear Tests

2.4.1. Experimental Setup and Specimen Assembly

According to the Chinese standard, JGJ82-2011 [19], the value of the torque coefficient k of the high-strength bolts ranged from 0.11 to 0.15. The torque coefficient k of the stainless-steel A4-80 high strength bolts was 0.15 according to Chen et al. [20]. The diameter of the high strength bolt was 6 mm. According to ISO3506-1:2009 [21], the torque limit of M6 stainless-steel A4-80 high-strength bolts was 15 N·m. However, due to the low stiffness and strength of the GFRP plates, a prevailing torque higher than 15 N·m may lead to the excessive pre-tensioning of the bolts, subsequently damaging the GFRP plate during the bolt tensioning process [22]. The value recommended by the authors in [23] was used to pre-tension the bolts, and the prevailing torque T was 7 N·m, insuring a level of shear resistance and avoiding initial damage to the components. The values of the bolt preloading force and prevailing torque were listed in Table 4.

Table 4. Test parameters of the high strength bolt.

Type of Bolt	Diameter d (mm)	Preloading Force P (kN)	Torque Coefficient k	Prevailing Torque T (N·m)
A4-80	6	7.78	0.15	7

A UTM5305 microcomputer-controlled electro-hydraulic servo universal testing machine was chosen to perform the shear tests, as the machine could firmly clamp the specimen with hydraulic fixtures and the maximum output load could reach 300 kN. The specimens were assembled first, and then clamped vertically into the fixture of the electronic universal testing machine, and the axis of the specimen was strictly aligned with the fixture center, as shown in Figure 4. The test specimens had tension applied monotonically with a speed of 2 mm/min. The tests stopped once the applied tension sharply increased. The load and displacement were recorded synchronously with a data acquisition system.

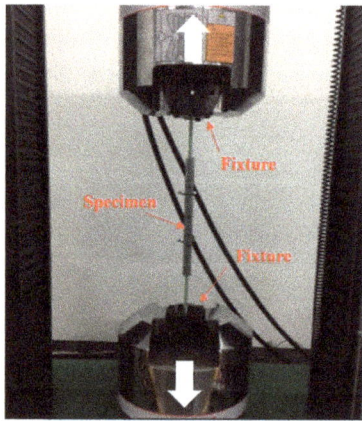

Figure 4. Loading equipment and testing setup.

2.4.2. Test Variables

In order to evaluate the influence of the prevailing torque and surface characteristics of the test plates on the slip factor, the following variables were focused on:

(1) The hardness of the plate surface. The hardness of the stainless-steel plate is much higher than that of the GFRP plate. We hypothesized that the higher hardness could help achieve a higher

slip factor. Two configurations of the plates were used to study the maximum slip factor: (a) a butt joint of two GFRP plates with GFRP cover plates, and (b) a butt joint of two GFRP plates with stainless-steel cover plates.

(2) The roughness of the plate surface. The roughness of the plate surface after grit-blasting was different depending on the grit of the abrasive used. Different slip factors can be achieved depending on the surface roughness. The #24 and #60 grit abrasives were chosen for this study. The surface roughness caused by the #24 grit abrasive was higher than that caused by the #60 grit abrasive. To avoid initial surface damage, the low hardness GFRP plates were only blasted with the #60 grit abrasive.

(3) The prevailing torque. Creep deformation can occur in GFRP plates when they are under a high constant pressure. The creep deformation can reduce the friction capacity of the GFRP plate exposed to this pressure. Based on tribology, the slip factor may be influenced by the pressure on the friction plane. The prevailing torque determines the pre-tension for high-strength bolts. We expected that the pre-tension could influence the contact pressure on the friction planes of butt joints. Therefore, the prevailing torque can be an important influence factor on the slip factor of slip-critical connections. When it appears as a variable, the prevailing torque used was 5 N·m, 6 N·m, 7 N·m, 8 N·m, 9 N·m, and 10 N·m, respectively (in other shear tests, the prevailing torque used was a constant 7 N·m).

To study these variables, we performed shear tests for 6 sets of 22 specimens. The specimens and corresponding parameters are listed in Table 5.

Table 5. Specimens in the shear tests.

Connection Methods	Connecting Plate Materials (Surface Treatment)	Cover Plate Materials (Surface Treatment)	Prevailing Torque T (N·m)	Specimen
G-G	GFRP (untreated)	GFRP (untreated)	7	G-G-1/2
G-S	GFRP (untreated)	Stainless-steel (untreated)	7	G-S-1/2
G-S60#	GFRP (untreated)	Stainless-steel (#60 grit blasting)	7	G-S60#-1/2
G-S24#	GFRP (untreated)	Stainless-steel (#24 grit blasting)	7	G-S24#-1/2
G60#-S24#	GFRP (#60 grit blasting)	Stainless-steel (#24 grit blasting)	7	G60#-S24#-1/2
G-S24#	GFRP (untreated)	Stainless-steel (#24 grit blasting)	5	T-5-1/2
			6	T-6-1/2
			7	T-7-1/2
			8	T-8-1/2
			9	T-9-1/2
			10	T-10-1/2

2.5. Pre-Tension Force Relaxation Tests

GFRP has a creep property [24], that is, when two GFRP plates are subjected to bolt pre-tension forces for a period of time, a stress relaxation occurs in the GFRP. This leads to a continuous reduction in the bolt pre-tension force, resulting in a reduction in the friction capacity and strength of the slip-critical connections. Therefore, it is necessary to examine the effect of the creep deformation of GFRP plates on the strength of the slip-critical connections.

We conducted 2 sets of tests containing 16 specimens. The first test carried out was a standing treatment of the G-G connection and G-S24# connection over different time periods. Then, shear tests were carried out to verify the effect of the creep deformation of the GFRP plates on the strength by examining the residual ratio of the slip load. The assembled test specimens were left to stand for 1 day, 7 days, and 30 days before the subsequent tests. The bolt pre-tension relaxation tests and shear tests were conducted in May and June at Nanjing, China. Before the shear tests, specimens were left to stand in a dry indoor environment and the room temperature was between 19 and 26 °C. The temperature fluctuation in this range would not influence the mechanical properties of GFRP materials according to [25]. The specimens are listed in Table 6.

Table 6. Specimens in the pre-tension force relaxation tests.

Connection Methods	Connecting Plate Materials (Surface Treatment)	Cover Plate Materials (Surface Treatment)	Relaxation Time (Days)	Specimen
G-G	GFRP (untreated)	GFRP (untreated)	0	A–0-1/2
			1	A–1-1/2
			7	A–7-1/2
			30	A–30-1/2
G-S24#	GFRP (untreated)	Stainless-steel (#24 grit blasting)	0	B–0-1/2
			1	B–1-1/2
			7	B–7-1/2
			30	B–30-1/2

3. Results and Discussion

3.1. The Effects of Cover Plates

The slip factors and slip loads are listed in Table 7. Two sets of tests were performed, as shown in the table. Each kind of test was repeated two times using two identical specimens to grant the reliability of the tests. The load–displacement curves of the two tests series are shown in Figure 5. The load–displacement curve of the G-S connection can be described by the (c) type curve from Figure 3. The load–displacement curve of the G-G connection can be described by the (b) type curve from Figure 3. The initial stiffness of the G-S curve was larger than that of the G-G curve. These characteristics can be explained by the adhesion theory of friction [26]. The friction force consists of two components, namely adhesion and deformation [26]. The adhesion force is the rupture resistance of the junctions, which is induced between the two contact surfaces. The deformation force is the resistance that is caused by the plastic deformation of interlocking asperities when the contact surfaces slide with each other. In the G-G connection test, the connecting plates and cover plates are all GFRP, the hardness is the same, and one side cannot be embedded into the other. The asperities on the surface of the GFRP plates were in an elastic contact state. The deformation of asperities in polymers is primarily elastic. The amount of the junction induced by the plastic deformation of the asperities was small. Thus, barely no scratches were formed after shear tests and only minor plastic deformation points could be seen, as shown in Figure 6. The significant proportion of the friction force with polymers came from the deformation force [26]. The surface of the stainless-steel plate was harder than that of the GFRP plate. The asperities on the surface of the stainless-steel could penetrate and plough into the GFRP surface, which produced grooves if the shear strength of the GFRP was exceeded. In Figure 5, the load–displacement curve of the G-S connection was above the curve of the G-G connection. This is due to the larger deformation force of the G-S connection.

Table 7. The slip load and slip factor values of tested connections with different cover plates.

Specimen	Slip Load F (kN)	Slip Factor μ
G-G-1	2.64	0.17
G-G-2	2.78	0.18
Average value	2.71	0.18
G-S-1	3.16	0.20
G-S-2	3.17	0.20
Average value	3.17	0.20

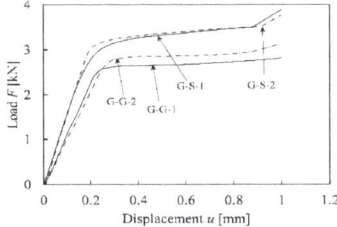

Figure 5. The load–displacement curves of specimens with different cover plates.

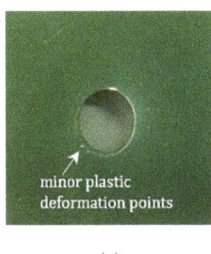
(a)

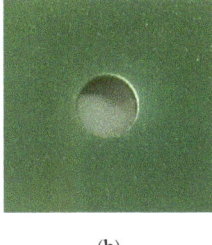
(b)

Figure 6. Plate surfaces of G-G connection after shear test. (a) GFRP connecting plate surface after shear test; (b) GFRP cover plate surface after shear test.

According to Bhushan [26], the friction factor, μ_a, of the adhesion force with polymers can be represented by the following equation:

$$\mu_a = \frac{\tau_0}{p} + \alpha \qquad (3)$$

where, τ_0 is the shear strength of the polymers, p is the contact pressure on the actual contact area, α is a constant, and the value of α for polymers is 0.2. In the polymers, the τ_0/p is much smaller than α [26], the friction factor can be approximately 0.2 when the deformation force is ignored. The experimental average value of the slip factor with a G-G connection was 0.18, which is close to 0.2. This demonstrates that the deformation component can be ignored in G-G connections. The average value of the slip factor with G-S connections was larger than that of the G-G connections. This is related to the higher deformation force, as explained above.

3.2. The Effects of the Grit-Blasting Surface Treatment

Three sets of nine specimens were applied with shear tests. The slip load and slip factor values acquired are shown in Table 8. The load–displacement curves obtained from the shear tests of each connection method are shown in Figure 7. The load–displacement curve of the G60#-S24# connection in Figure 7 can be described by the typical (b) type curve from Figure 3. The specimen showed a gradual change in the load as the displacement was increased. The load–displacement curves of the G-S24# and G-S60# connections in Figure 7 can be described by the (a) type curve in Figure 3. These two types of specimen showed a major slip and reduction in the load. The reduction rate of the load with the G-S24# connection specimens was much larger than that of the G-S60# connection specimens. The slip load increased with the increase in the grit size (from #60 to #24). As previously explained, the main component of the friction force with the GFRP connections is likely the deformation force. The grit-blasting surface treatment increases the number of asperities and roughness on the faying surface of stainless-steel cover plate. The deformation force increased with the increase in the number of asperities and the roughness of the harder stainless-steel cover plate. The asperities on the surface of the stainless-steel could penetrate and plough into the GFRP surface, which produced grooves if the shear strength of the GFRP was exceeded. As is shown in Figure 8, the grooves were formed on the GFRP connecting plate and the GFRP powder ploughed by embedded asperities were left on the stainless-steel cover plate surface. A load drop did not occur in the load–displacement curve of the G60#-S24# connection compared to the load–displacement curves of the G-S24# and G-S60# connections, which may be related to the ratchet mechanism [26]. The ratchet mechanism describes a situation where two rough surfaces contact each other and plastic deformations do not happen. The asperities of one mating surface climb up and down on the other mating surface. The motion mode of the ratchet mechanism resembles a Coulomb friction model. The roughness of the surface of the GFRP plate increased after being treated by grit-blasting. The rough surface of the GFRP plate contacted the rough surface of the stainless-steel plate and the asperities being produced under contact stress elastically deformed. The ratchet mechanism appeared in the G60#-S24# connection, which accounts for the maintenance of the connection bearing capacity. As shown in Figure 9, barely any trace could be found after the shear test due to the ratchet mechanism that is elastic deformation of the asperities recovered after unloading.

Table 8. The slip load and slip factor values of the tested connections with different grit-blasting treatments.

Specimen	Slip Load F (kN)	Slip Factor μ
G-S60#-1	5.72	0.37
G-S60#-2	5.73	0.37
Average value	5.73	0.37
G-S24#-1	6.48	0.42
G-S24#-2	6.42	0.41
Average value	6.45	0.42
G60#-S24#-1	6.75	0.43
G60#-S24#-2	6.72	0.43
Average value	6.74	0.43

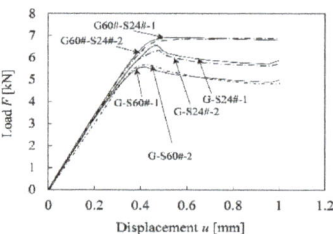

Figure 7. The load–displacement curves of specimens with different grit-blasting surface treatments.

Figure 8. Plate surfaces of G-S24# connection after shear test. (a) GFRP connecting plate surface after shear test; (b) stainless-steel cover plate surface after shear test.

 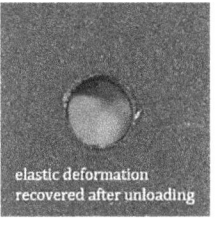

(a) (b)

Figure 9. Plate surfaces of G60#-S24# connection after shear test. (**a**) GFRP connecting plate surface after shear test; (**b**) stainless-steel cover plate surface after shear test.

The slip factor of the G60#-S24# connection was the highest and the slip factor of the G-S60# connection was the lowest among the three connection types. The difference between the slip factors of three connections was minimal. Hashimoto et al. [5] tested the GFRP slip-critical connection with stainless-steel cover plates. In their experiments, the surfaces of stainless-steel cover plates were treated by grit-blasting. They found that the slip factor was 0.449. This value is close to the results from our tests, as listed in Table 8. Feo et al. [3] found that the slip factor value of the GFRP slip-critical connection with structural-steel cover plates was around 0.37, which is again similar to the results of our tests. It is clear that increasing the roughness of the faying surface using a grit-blasting treatment is a valid way to increase the slip load of GFRP slip-critical connections. The slip factor of the G60#-S24# slip-critical connection was slightly higher than that of the G-S24# slip-critical connection, indicating that grit-blasting treatment for GFRP plates was not greatly efficient or economic when stainless-steel plates were already blast treated. The grit-blasting treatment may damage the surface of the GFRP plates, therefore we do not propose any grit-blasting treatment for the GFRP plates.

3.3. The Effects of Prevailing Torque

Based on the phenomenon and analysis above, the G-S24# connection was proven to be the optimal configuration with both efficient shear resistance and economy. As a result, the G-S24# connection only was selected to examine the influence of the prevailing torque on the slip factor of slip-critical connections. The prevailing torques used in the tests were 5 N·m, 6 N·m, 7 N·m, 8 N·m, 9 N·m, and 10 N·m. Each set of tests was performed with two identical specimens. The slip load and slip factor are listed in Table 9 and the load–displacement curves are shown in Figure 10. The load–displacement curves of the G-S24# connection with different prevailing torques in Figure 10 can be described by the (a) type curve from Figure 3. These curves are similar to the load–displacement curve of the G-S24# connection from Figure 7.

Table 9. The slip load and factor values of tested connections with different prevailing torques.

Specimen	Prevailing Torque T (N·m)	Slip Load F (kN)	Average Value	Growth Rate Compared with the Former (%)	Slip Factor μ	Average Value	Growth Rate Compared with the Former (%)
T-5-1	5	4.90	4.92	/	0.44	0.44	/
T-5-2		4.94			0.44		
T-6-1	6	5.93	5.89	19.72	0.45	0.45	2.27
T-6-2		5.85			0.44		
T-7-1	7	6.37	6.40	8.66	0.41	0.41	−8.89
T-7-2		6.43			0.41		
T-8-1	8	6.78	6.78	5.94	0.39	0.39	−4.88
T-8-2		6.77			0.39		
T-9-1	9	7.13	7.09	4.57	0.36	0.36	−7.69
T-9-2		7.05			0.35		
T-10-1	10	7.39	7.41	4.51	0.33	0.33	−8.33
T-10-2		7.43			0.33		

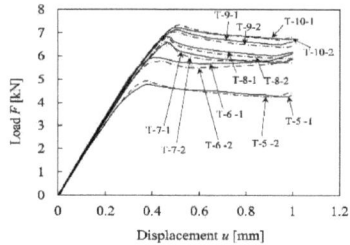

Figure 10. The load–displacement curves of connections with different prevailing torques.

From Table 8, it was clear that as the prevailing torque was increased, the slip load of the connections increased gradually, and the slip factor slightly decreased. Each time the prevailing torque was increased, the slip load also increased, however, the magnitude of the increment was lessened each time. Lacey et al. [12] demonstrated that the slip factor decreased with an increase in contact pressure in steel–steel connections. They explained that the surface asperities were flattened by the applied contact pressure, which decreased the deformation component of the friction force.

3.4. The Effects of Pre-Tension Force Relaxation

The GFRP plates experience a creep deformation under sustained loads [27]. When a GFRP plate is subjected to the preloading force of high-strength bolts during the service period, creep deformation is induced, and the pre-tension force of the high-strength bolts should relax. The slip load decreases as the pre-tension force is relaxed [17]. Table 10 summarizes the variation in the results of the G-G slip-critical connection, and the G-S24# slip-critical connection was due to the creep deformation of the GFRP plates. We found that the slip load decreased as the preloading time increased in both connections. The residual rate of the shear resistance of the G-S24# connection was much higher than that of the G-G connection. After one day of preloading time, the slip load of the G-S24# slip-critical connection decreased by 6%. After 30 days, it decreased by only 9%. The test results showed that compared to the G-G connection, the G-S24# connection still had good slip resistance, even after a long period of preloading time.

Table 10. Slip load variation under the pre-tension of high-strength bolts.

Connection Methods	Specimen	Preloading Time (day)	Slip Load F (kN)	Average Value (kN)	Residual Rate (%)
G-G	A–0-1 A–0-2	0	2.95 2.64	2.78	/
	A–1-1 A–1-2	1	2.40 2.44	2.42	87
	A–7-1 A–7-2	7	2.36 2.28	2.32	83
	A–30-1 A–30-2	30	2.17 2.23	2.20	79
G-S24#	B–0-1 B–0-2	0	6.32 6.26	6.29	/
	B–1-1 B–1-2	1	5.82 5.94	5.88	94
	B–7-1 B–7-2	7	5.79 5.73	5.76	92
	B–30-1 B–30-2	30	5.72 5.68	5.70	91

Kobayashi et al. [15] performed bolt pre-tension relaxation tests for up to half a year to investigate the slip load reduction of GFRP slip-critical connections with steel cover plates. The configuration of the G-S24# connection in our paper is similar to that of Kobayashi's results [15]. The residual rate of slip load from Kobayashi's tests [15] and our own tests are presented in Figure 11. From this figure, the G-G connection would lose more than 20% of shear resistance within a month. The residual rate curve's decreasing rate of slip load of the G-S24# connection was lower than that of Kobayashi's tests within 30 days, and the downward trend seemed to be stable. The slip load in Kobayashi's tests [15] dropped 16% after half a year. Therefore, we speculate that the slip load of the G-S24# connection would be reduced by no more than 16% after standing for half a year, retaining the vast majority of its original bearing capacity.

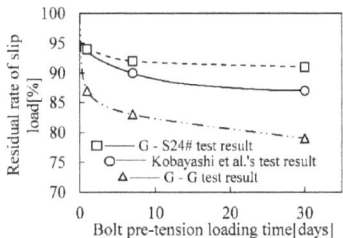

Figure 11. Comparison of the slip load residual rate after the bolt pre-tension relaxation tests.

4. Conclusions

We conducted 22 shear tests with different cover plates, prevailing torques, and grit-blasting surface treatments, and we investigated the slip load and slip factor of the GFRP–GFRP slip-critical connections. We conducted 16 pre-tension force relaxation tests. The long-term performance of the GFRP–GFRP slip-critical connections was investigated. An explanation of the force transformation mechanism of the GFRP–GFRP slip-critical connections, based on the adhesion theory of friction for this type of connection, was discussed. Based on the experimental and theoretical analysis of test results, the following conclusions can be drawn:

(1) The slip factor of the G-S24# connection with grit-blasting treatment stainless-steel cover plates can steadily reach 0.45; this slip factor value could satisfy the requirements of engineering applications. The main component of the friction force with the GFRP–GFRP slip-critical connections was likely the deformation force. A large deformation force could steadily transfer the force between the stainless-steel cover plates and GFRP plates. The grit-blasting surface treatment increased the number of asperities and roughness on the faying surface of stainless-steel cover plates and increased the deformation force. The grit-blasting surface treatment of the GFRP plate could damage the mechanical performance of GFRP plate and the slip factor increase can be small; therefore, the grit-blasting surface treatment of the GFRP plate should be avoided.

(2) The slip factor changed with the variation in the prevailing torque effects of the G-S24# connection. When the prevailing torque equaled 6 N·m, the maximum slip factor of 0.45 occurred. The prevailing torque was not large. A large prevailing torque can flatten the asperities and reduce the deformation force. The reduction in the deformation force caused the slip load reduction.

(3) Compared to the G-G connection, the G-S24# connection demonstrated good long-term slip resistance. The residual rate of the slip load of the G-S24# connection was 91% after 30 days of preloading time. The mechanical performance of the G-S24# connection can be relied upon over the service life by considering the relaxation effect.

Author Contributions: Formal analysis, Y.P.; software, J.Z.; supervision, J.D.; validation, Y.P.; writing—original draft, Y.P., Z.W.; writing—review and editing, Y.P., W.C. All authors have read and agreed to the published version of the manuscript.

Funding: Project of National Natural Science Foundation of China: 51408307 and Project of Nanjing Gongda Construction Technology Co., Ltd.: 2019RD06.

Acknowledgments: This research was funded by Project of National Natural Science Foundation of China grant number [51408307] and Project of Nanjing Gongda Construction Technology Co., Ltd. grant number [2019RD06].

Conflicts of Interest: The authors declare no conflict of interest.

References

1. Mosallam, A.S. *Design Guide for FRP Composite Connections*; American Society of Civil Engineers: Palo Alto, CA, USA, 2011; ISBN 978-0-7844-0612-0.
2. Coelho, A.M.G.; Mottram, J.T. A review of the behaviour and analysis of bolted connections and joints in pultruded fibre reinforced polymers. *Mater. Des.* **2015**, *74*, 86–107. [CrossRef]
3. Feo, L.; Latour, M.; Penna, R.; Rizzano, G. Pilot study on the experimental behavior of GFRP-steel slip-critical connections. *Compos. Part B Eng.* **2017**, *115*, 209–222. [CrossRef]
4. Mottram, J.T.; Lutz, C.; Dunscombe, G.C. Aspects on the behavior of bolted joints for pultruded fiber reinforced polymer profiles. In *Advanced Polymer Composites for Structural Applications in Construction*; Hollaway, L.C., Chryssanthopoulos, M.K., Moy, S.S.J., Eds.; Woodhead publishing limited: Cambridge, UK, 2004; pp. 384–391. ISBN 1-85573-736-7.
5. Hashimoto, K.; Sugiura, K. Mechanical consideration on frictional behavior and maximum strength of GFRP members connected by high strength frictional bolted joint. *J. Struct. Eng.* **2012**, *A 58A(0)*, 935–945. (In Japanese)
6. Stranghöner, N.; Afzali, N.; De Vries, P.; Schedin, E.; Pilhagen, J. Slip factors for slip-resistant connections made of stainless steel. *J. Constr. Steel Res.* **2019**, *152*, 235–245. [CrossRef]
7. Quinn, J.A. *Composites-Design Manual*, 3rd ed.; James Quinn Associates Ltd.: Liverpool, UK, 2002; ISBN 0-9534654-1-1.
8. Mottram, J. Friction and load transfer in bolted joints of pultruded fibre reinforced polymer section. In *FRP Composites in Civil Engineering - CICE 2004*; CRC Press: Boca Raton, FL, USA, 2004; pp. 845–850.
9. European Committee for Standardization. *BS EN 1090-2:2008+A1, Execution of Steel Structures and Aluminium Structures-Part 2: Technical Requirements for Steel Structures, vol. 1*; European Committee for Standardization: Brussels, Belgium, 2011.
10. RCSC. *Specification for Structural Joints Using High-strength Bolts*; Research Council on Structural Connections: Chicago, IL, USA, 2009.
11. SAI Global Limited. *AS 4100-1998 (R2016) Steel Structures*; Standards Australia, SAI Global Limited: Sydney, Australia, 2016.
12. Lacey, A.; Chen, W.; Hao, H.; Bi, K. Experimental and numerical study of the slip factor for G350-steel bolted connections. *J. Constr. Steel Res.* **2019**, *158*, 576–590. [CrossRef]
13. Akinlabi, E.T.; Akinlabi, S.A.; Ogunmuyiwa, E. Characterizing the effects of sand blasting on formed steel samples. *Int. Sch. Sci. Res. Innov.* **2013**, *7*, 2216–2219.
14. Bouledroua, O.; Meliani, M.H.; Azari, Z.; Sorour, A.A.; Merah, N.; Pluvinage, G. Effect of Sandblasting on Tensile Properties, Hardness and Fracture Resistance of a Line Pipe Steel Used in Algeria for Oil Transport. *J. Fail. Anal. Prev.* **2017**, *17*, 890–904. [CrossRef]
15. Kobayashi, K.; Hino, S.; Yamaguchi, K.; Ohmoto, T. Experimental study on strength of GFRP and steel plates connection using adhesively-bonded and bolted joint. *J. Struct. Eng.* **2009**, *55A*, 1140–1149. (In Japanese)
16. Van De Hulst, H.C.; Twersky, V. Light Scattering by Small Particles. *Phys. Today* **1957**, *10*, 28. [CrossRef]
17. Heistermann, C.; Veljkovic, M.; Simões, R.; Rebelo, C.; Da Silva, L.S. Design of slip resistant lap joints with long open slotted holes. *J. Constr. Steel Res.* **2013**, *82*, 223–233. [CrossRef]
18. Association for Engineering Construction Standardization CECS 410: 2015. *Technical Specification for Stainless-Steel Structures*; China Association for Engineering Construction Standardization: Beijing, China, 2015. (In Chinese)

19. *JGJ 82-2011. Technical Specification for High Strength Bolt Connection of Steel Structures*; Ministry of Housing and Urban Rural Development of the People'S Republic of China: Beijing, China, 2011. (In Chinese)
20. Chen, Z.; Peng, Y.; Su, W.; Qian, F.; Dong, J. Experimental investigation for anti-slipping performance of stainless steel slip-resistant connections with particles embedded in connected plates. *Constr. Build. Mater.* **2017**, *152*, 1059–1067. [CrossRef]
21. International Organization for Standardization. *ISO 3506-1:2009(E), Mechanical Properties of Corrosion-Resistant Stainless-Steel Fasteners —Part 1: Bolts, Screws and Studs*; ISO: London, UK, 2009.
22. Xie, M.J. *Connection of Composite Materials*; Shanghai Jiaotong University Press: Shanghai, China, 2011; ISBN 9787313163998. (In Chinese)
23. Wang, Y.Q.; Guan, J.; Zhang, Y.; Yang, L. Experimental research on slip factor in bolted connection with stainless-steel. *J. Shenyang Jianzhu Univ.* **2013**, *29*, 769–774. (In Chinese)
24. Jonathan, H.E. Creep and mechanical properties of carbon fibre reinforced PEEK composite material. Master's Thesis, University of Manitoba, Winnipeg Manitoba, MB, Canada, 10 October 1992.
25. Wu, C.; Bai, Y.; Mottram, J.T. Effect of Elevated Temperatures on the Mechanical Performance of Pultruded FRP Joints with a Single Ordinary or Blind Bolt. *J. Compos. Constr.* **2016**, *20*, 04015045. [CrossRef]
26. Bhushan, B. *Introduction to Tribology*; Wiley: Columbus, OH, USA, 2013.
27. Li, S.; Hu, J.; Ren, H. The Combined Effects of Environmental Conditioning and Sustained Load on Mechanical Properties of Wet Lay-Up Fiber Reinforced Polymer. *Polymer* **2017**, *9*, 244. [CrossRef] [PubMed]

© 2020 by the authors. Licensee MDPI, Basel, Switzerland. This article is an open access article distributed under the terms and conditions of the Creative Commons Attribution (CC BY) license (http://creativecommons.org/licenses/by/4.0/).

Article

Safe and Sustainable Design of Composite Smart Poles for Wireless Technologies

Donato Di Vito [1], Mikko Kanerva [2,*], Jan Järveläinen [3], Alpo Laitinen [4], Tuomas Pärnänen [2], Kari Saari [4], Kirsi Kukko [4], Heikki Hämmäinen [5] and Ville Vuorinen [4]

1. Faculty of Information Technology and Communication Sciences, Tampere University, FI-33014 Tampere, Finland; Donato.divito@tuni.fi
2. Faculty of Engineering and Natural Sciences, Tampere University, FI-33014 Tampere, Finland; Tuomas.parnanen@tuni.fi
3. Premix Oy, FI-05200 Rajamäki, Finland; Jan.jarvelainen@premixgroup.com
4. Department of Mechanical Engineering, Aalto University, FI-00076 Espoo, Finland; Alpo.laitinen@aalto.fi (A.L.); Kari.saari@aalto.fi (K.S.); Kirsi.kukko@aalto.fi (K.K.); Ville.vuorinen@aalto.fi (V.V.)
5. Department of Communications and Networking, Aalto University, FI-00076 Espoo, Finland; Heikki.hammainen@aalto.fi
* Correspondence: Mikko.kanerva@tuni.fi; Tel.: +358-40-718-8819

Received: 20 September 2020; Accepted: 21 October 2020; Published: 28 October 2020

Abstract: The multiplicity of targets of the 5G and further future technologies, set by the modern societies and industry, lacks the establishment of design methods for the highly multidisciplinary application of wireless platforms for small cells. Constraints are set by the overall energy concept, structural safety and sustainability. Various Smart poles and Light poles exist but it is challenging to define the design drivers especially for a composite load-carrying structure. In this study, the design drivers of a composite 5G smart pole are determined and the connecting design between finite element modelling (FEM), signal penetration and computational fluid dynamics (CFD) for thermal analysis are reported as an interdisciplinary process. The results emphasize the significant effects of thermal loading on the material selection. The physical architecture, including various cutouts, is manipulated by the needs of the mmW radios, structural safety and the societal preferences of sustainable city planning, i.e., heat management and aesthetic reasons. Finally, the paint thickness and paint type must be optimized due to radome-integrated radios. In the future, sustainability regulations and realized business models will define the cost-structure and the response by customers.

Keywords: tubular composites; finite element analysis; computational fluid dynamics; wireless communication; signal attenuation

1. Introduction

1.1. Wireless Outdoor Platforms

The application spectrum enabled by the fast 5G development is about to cover a multiplicity of wireless technologies and services. The selection of the frequencies for '5G' has been globally discussed and, in Europe, the focus is on the 3.6 GHz and 26 GHz bands. The Electronic Communications Committee 'ECC' conducted a survey already 2017 that suggested the bands of 24.25–27.5 GHz, 40.5–43.5 GHz and 66–76 GHz as the prioritized bands. The higher end of the radio frequency (RF) band range directly affects the radio configuration and energy usage. The needs for higher data rates and the available RF bands have led to the concepts of small cell networks, the necessary low latency, new business environment with the end to end networks [1–3] and the emphasis on cost distribution. Besides, new terms, such as Smart pole or Light pole, known as concepts for 5G-enabling poles (5GPs) and heavier 5G gantries have been exhibited. Several demo designs or even demo sites with certain pole designs are built and running—yet many of them lack 5G operation or having a partial 5G operation. In few years, the number of 5GPs proposals in public has increased from a few to number of designs. Yet, the overall 5GP concept along with data management is still to be explored.

The strategic importance of 5GPs as a platform for 5G outdoor small cells finally stems from its costs and regulations applied. Because of the large number of required 5G base station sites, the cost of deployment and operation is high even for national mobile network operators (MNOs). A simple wholesale site contract on 5GPs with the city council is lucrative compared to hand-picking and tailoring contracts for non-uniform sites on private buildings. This encourages MNOs toward 5GP sharing for the high frequency deployments.

On the other hand, the increasing pressure to unify and beautify the city antenna 'jungle' also supports the sharing of well-designed 5GPs. Therefore, the national regulators may allow extending the monopolies of light pole and grid networks with another monopoly, a neutral host company operating the 5GP pole system [4]. According to initial studies, the cost of the structural part (the pole shaft) is significant—between 15–25% of the total deployment cost. This cost share varies depending on the amount of electronics of various services per pole and on the density of fully configured 5G smart light poles [5]. Interestingly, the faster price erosion of electronics seems to gradually increase the relative value of the poles shafts. The 5G data pricing models tend to be even more complex since it is currently not clear who sells and what kind of products.

Clearly, the modularity of devices plays a role in the prospective design of services per 5GP site. The modularity is also a tool to handle the cost structure per type of site and at individual 5GPs. The integration of a 5GP requires connections to data, power and possibly cooling network of the city or suburb. The development of design tools combining the wireless network and city planning is essential and is a significant design phase in the future. A sophisticated smart pole, a 5GP, can be considered as aesthetically fitted integrated structure, which embodies various devices including the radios within the main structure of the pole.

1.2. Design Drivers for a Smart Pole Structure

The general, main design drivers, are illustrated in Figure 1. The physical frame, referred to as pole or shaft, is needed primarily to carry the (RF) radios and other electrical devices in a functional and maintenance-friendly way. Since the transmitting and receiving electronics require protection against weather—irrespective if they are fully integrated inside the pole or not—the selected structural materials must possess known, specified interference with RF signals to allow for a dense pole population operating over the specified frequency range. Especially, when customers, such as cities, require for an operational lifetime of 20...50 years for each pole, durability is essential as well.

For many countries, the safety in terms of vehicle crash must be accounted for in the design of the 5G pole's shaft. The crash design affects primarily the pole sites with a high traffic density. Thus, an electrical vehicle (EV) charging stations or similar services mounted low are not allowed for these sites. The safety and overall sustainable operation of 5G is essential in general [6], since the citizens and their considerations justify the realized 5GPs at urban areas. The sustainability of the manufacture and minimum usage of material in 5GPs depends clearly on the realized, future operation time. Whenever material can be recycled, the selection will affect the sustainability in the big picture of future operation [7]. In an even wider perspective, the sustainability of an individual pole and manufacture covers only part of the truth. The thermal management and energy efficiency of the devices attached to the pole have a cardinal role. For certain pole sites, centralized cooling might form an important effect on the over-lifetime carbon and energy foot print of the entire wireless platform.

For modern, dense-packed electronics [8], thermal properties and heat management are an essential part of the system design to prevent overheating during the anticipated operation. Fluid dynamics and heat transfer are the key fields of science regarding thermal management.

The state-of-the-art numerical approach for simulating heat transfer with solid-liquid or solid-air interfaces is computational fluid dynamics (CFD). Using CFD simulations, thermal assessment on 5GPs can also be carried out. In a previous work, air cooling of high-power electronics was investigated inside a tubular pole-like structure [9]. However, the flow control can be challenging in complex environments, such as inside a tight-packed smart pole shaft.

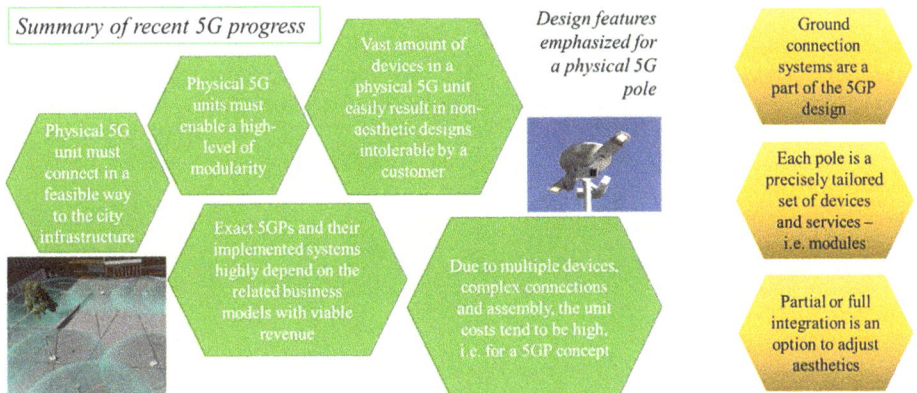

Figure 1. Summary of the design drivers for a physical platform for various wireless devices.

1.3. The Advantages and Sustainability of Composite Materials

The advantages of composite materials in terms of mechanical properties and weight are well-known. Similarly, in a smart pole, the mechanical properties, i.e., high stiffness to prevent large sway due to the wind and sun as well as the low weight for sustainable transportation and overhaul are advantages. The overall advantage of composite materials is the ability to be precisely tailored [10]. Composites have been successfully used for various shaft structures with fully composite or hybrid material lay-up [11,12]. The raw material costs and manufacture can be affected by the selection of proper fibre and matrix. Polyester resins as well as polypropylene have been applied in composites as the matrix component [13,14] in order to have a suitable balance between costs and performance. The smart pole shafts could also be made of natural fibre reinforced composites [15,16]. However, the use of natural materials requires clearly more information about sustainable fillers for fire retardancy [17] and also about the susceptibility of natural fibres to moisture [18] in long term outdoor operation.

In general, glass fiber reinforced polymers have been applied in challenging applications, such in wind turbine blades [19] where the composite reaches the extreme requirements of fatigue life and durability. Although polyesters represent the 'low' performance of composites, their benefit is the well-known behaviour in various environments [20,21] and lower material costs. From the point of view of enclosure functionality, fibrous composites can be hybridized with metal sheets to control the electromagnetic response [22]. With fiber-metal hybrids, enclosures can be made to totally protect sensitive electronics against external, harmful or unwanted radiation and signals [23]. As an alternative, particle inclusions [24] can be used to control the signal penetration.

Whenever signal penetration is necessary through the enclosure or shaft wall, the material selection becomes challenging. Polymers, typically used as matrix in composites, incur low electromagnetic attenuation in terms of dielectric loss—especially for frequencies below the gigahertz-regime. For the higher frequencies, already the type and grade of the polymer blend must be well optimized [25,26]. Reinforcing fibers are generally not especially advantageous in terms of signal penetration—carbon fibers and all conductive fibers lead to very high attenuation. Even when using glass or polymeric (e.g., aramid) fibers, the more or less sporadic orientation at a micro-scale and their multi-interface configuration within the matrix result in reflections and attenuation. Moreover, any accumulated moisture has significant effects on attenuation. Anyhow, a finite amount of attenuation can be accepted while zero penetration for metal alloys is not an option. The tailoring of composites refers to, along with the mechanical and electromagnetic properties, thermal properties. Requirements for the matrix and fiber selection can also include limits for the thermal conductance and expansion. It should be emphasized that the thermal expansion in a composite material can be tailored by the selection of fibre and matrix as well as by the decision of lay-up.

In this study, we focus on a physical application for wireless technologies, i.e., an integral smart pole. The research targets to offer the physical platform and process for the latest 5G implementations with a multiplicity of functional requirements along with the structural safety. Whenever the (5G) wireless devices and services require highly precise and stable location in a pole, meaning minimum sway in various environments, the mechanical design must be united with the functional and thermal design. The main phases of the process described in this study are shown in Figure 2.

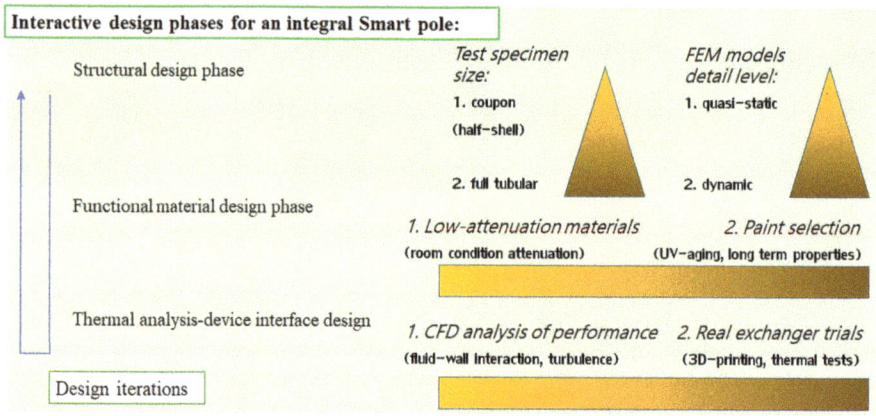

Figure 2. A graph of the main analysis phases for an integral smart pole allowing current 5G services.

2. Materials and Methods

2.1. Pole Structure

The pole structure for experimental tests in this study was a hollow pull-winded (specialized pultrusion) composite profile. The reinforcing fiber was a commercial E-glass fiber yarn (4800 tex, filament diameter ≈ 23 µm, Europe) for the structural layers and ECR-glass for the surfacing mat. Glass fibre reinforced plastic (GFRP) was formed by using polyester (Norsodyne P 46074, Polynt, Italy) as the matrix constituent. The shaft was pultruded by Exel Composites Plc. (Finland) and a lay-up of [0°, 85°, 0°] was finally applied with nominal layer thicknesses of 2.8, 0.4, 2.8 mm, respectively. The shaft had a constant outer diameter of $D_{outer} = 168$ mm and as-received nominal wall thickness of $t_{wall} = 6$ mm. The surfacing mat was used due to aesthetic reasons and had its thickness comparable to the standard manufacture deviation in the pole thickness.

2.2. Impact Dynamics of the Shaft GFRP

Quasi-static (QS) response by indentation was measured from half-circular (180°) panels (see Figure 3a) of the GFRP profile ($D_{outer} = 168$ mm, see Section 2.1). The panel specimen (length of 250 mm) was supported by two half-circular steel sections, which were set 200 mm apart (defining the measurement area). The upper side of the specimen was supported in the areas of steel sections resulting in a semi-rigid boundary condition. The specimen was loaded in a universal testing machine (30 kN load cell, model 5967, Instron, High Wycombe, UK) by a hemispherical head (radius 10 mm). The loading was subjected in the middle of the specimen by using a test rate of 1.0 mm/min. 3D Digital image correlation (DIC) was used to record and analyze (Davis 8.4 software, LaVision, Göttingen, Germany) the displacement on the lower surface of the GFRP panel. In addition, the contact force was measured by the load cell located over the loading head.

The testing was continued on the panel specimens loaded by a falling/drop-weight impactor (FWI Type 5, Rosand, Leominster, Herefordshire, UK) (see Figure 3b). Instead of the semi-rigid support, an open boundary was used for the half-circular specimen, i.e., the upper side of the specimen was able to deform freely during the dynamic loading. The specimen was loaded by a hemispherical impactor head (radius 10 mm) weighting 7.67 kg with an impact energy of 100 J. The contact force of the impactor head was measured by a piezo-electric load cell located above the impactor (Type 9031A, Kistler, Winterthur, Switzerland). The displacement of the impactor head was calculated analytically from the contact force-time response.

In the last test step, a tubular shaft specimen was tested using the drop-weight impactor (Figure 3c). A shaft length of 780 mm was used in a cantilever support mode and loaded with a half-circular shaped (2D) impactor head weighting 8.49 kg. The impact energy in the testing was 100 J. Similar to the testing of the half-circular specimen, the contact force of the impactor head was measured by the piezo-electric load cell and the displacement of the impactor head was calculated analytically from the contact force-time response.

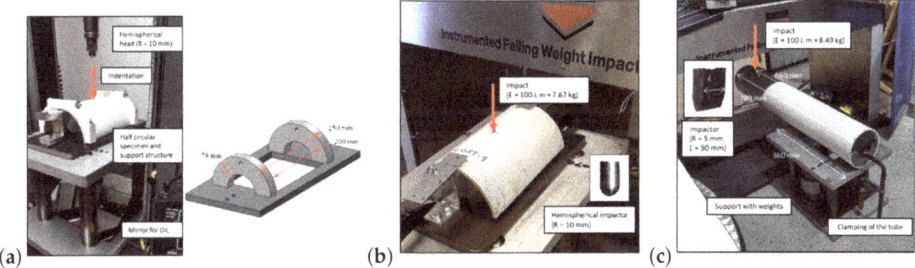

Figure 3. Testing of GFRP mechanics: (**a**) the QS indentation of a half-circular specimen with a semi-rigid boundary; (**b**) the drop-weight impact testing of a half-circular specimen with an 'open' boundary, and; (**c**) the drop-weight impact testing of a tubular pole specimen.

2.3. Finite Element Modelling and Mechanical Analysis

In this study, two different shaft cross-sections, i.e., shaft designs, were analyzed: (1) traditional circular profile and (2) a near-rectangular diamond-shaped profile. The near-rectangular cross-section, as illustrated in Figure 4, can accommodate radios, other devices and cabling while optimizing the coverage of the transmitting devices around the surroundings of the pole. Moreover, such design can be further divided into internal slots to manage the cabling and piping inside the pole and to separate different equipment by standard modules.

The composite shaft structure—the two cross-sections—were modelled by using the ABAQUS standard/explicit (2017) software code. In order to widely characterize the behaviour and loading conditions, different 3D models of the shaft were analyzed with finite element analysis (FEA). The full-scale model of the pole was simulated to characterize deformation under a mechanical load and thermal expansion due to temperature increase by solar radiation. Furthermore, material was removed in specific pole locations of the model to simulate the machining processes needed to accommodate and mount electronic devices (i.e., via so-called maintenance cutouts and doors) in the root and top section of the pole. In this case, curvature radii as well as dimensions of the holes were analyzed, together with their effect on the pole's structural integrity for the static loads mentioned above. The results obtained were then used to properly design the cutouts for electronic components and maintenance doors also at the base of the 5GP. In order to keep the consistency of the results between the different analyses, continuum shell elements (SC8R) were used in all of the finite element models of the pole shaft. The mechanical properties of the laminate and the local impact performance have been reported elsewhere [27,28]; the elastic constants used in the modelling here are shown in Table 1 and the strength values for the laminate are 500 MPa, 200 GPa, 50 MPa, 100 MPa and 40 MPa for the longitudinal tensile strength, longitudinal compressive strength, tensile transverse strength, compressive transverse strength and shear strength, respectively.

Table 1. Elastic constants of the GFRP (unidirectionally reinforced layer) used in this work for the pole shaft FEA [27,28].

Constant	Parameter	Value (Units)
Axial Young's modulus	E_{11}	35 GPa
Transverse Young's modulus	E_{22}	7.0 GPa
Out-of-plane Young's modulus	E_{33}	4.5 GPa
Poisson's ratios	ν_{12}, ν_{23}	0.3
Poisson's ratio	ν_{13}	0.05
Shear modulus	G_{12}	3.6 GPa
Shear modulus	G_{13}	3.6 GPa
Shear modulus	G_{23}	2.0 GPa

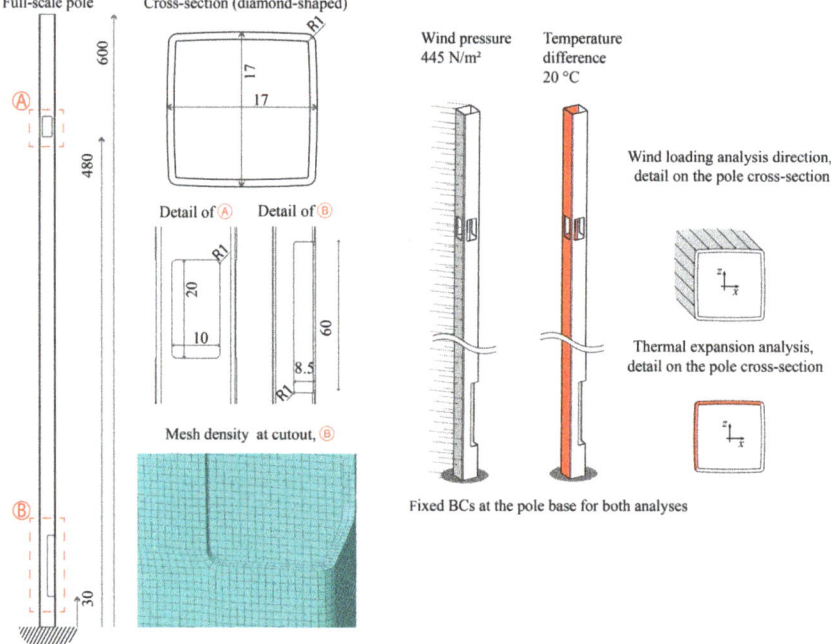

Figure 4. The dimensions (units: cm), loading, boundary conditions and mesh density applied to FEA.

For most of the composite materials, the thermal expansions are lower than those for any steel alloy. Thermal expansion coefficients of fibrous composites depends on the applied constituent materials and the lay-up of reinforcement, along with many factors of measurements [29,30]. In this study, the linear thermal expansion coefficients (CTEs) of the GFRP composite were determined using an experimental arrangement and an iterative FEA of the CTE values based on the residual strain, as described in Section 2.4 and Appendix A. The boundary conditions and loads (wind load distribution), shown in Figure 4, reflect a standard specification [31]. The wind loads represent the primary mechanical load against which the pole structure must be verified. The wind reference speed is given by the standard, e.g., here it is 21 m/s for the terrain category II of Scandinavia. The services for positioning systems are typically well balanced for pole sway yet device-specific performance can set a limit for the maximum allowed shaft bending. In this study, car-crash performance was not evaluated but impact-type loads subjected to the pole shaft were deemed an important local failure type.

As depicted in the figure, for both thermal expansion and wind load analyses the poles were assumed to be completely fixed to the ground, i.e., the degrees of freedom were prevented at the base of the pole (0 in every direction), while the loadings were applied over the half surface of the pole outer surface (volume). In detail, the distributed loads were applied on the opposite side of the cutout (in order to test the pole integrity under the highest moment). Respectively, the applied wind load distribution and the temperature difference (thermal load) were equal to 445 N/m² and 20 °C. The nominal element size (edge length) considered in the analysis was 5 mm, and each of the static analyses was roughly constituted by 250,000 elements.

2.4. Experimental-Numerical CTE Determination

The experimental setup for determining the CTEs of GFRP consisted of a coupon (projection 252 mm× 60 mm) cut off from the tubular composite pole (D_{outer} = 168 mm, see Section 2.1). The coupon was clamped (clamping length 64 mm) to a robust holder from the other end and the free end was subjected to mechanical loading. The mechanical load was subjected by a wire and a free-hanging mass (1.230 kg) attached at a distance of 56 mm from the free end. A strain gauge (KFGS-5-120-C1, Kyowa, Japan) was glued according to the manufacturer's instructions at a 138 mm distance from the clamped end and middle in the transverse direction. The arrangements are illustrated in Figure 5. Finally, the entire setup was placed in a digitally controlled oven. The test included two steps: (1) loading the coupon mechanically; (2) heating the oven in steps (24 °C…60 °C). Each heating step was launched after the strain reading from the gauge had essentially settled.

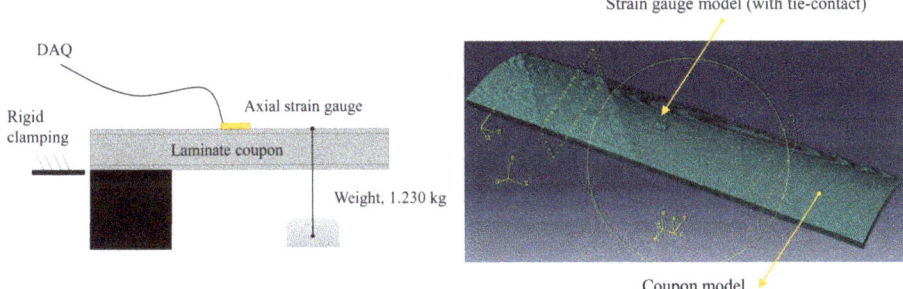

Figure 5. Experimental arrangements and the finite element model of the test coupon to determine thermal expansion coefficients.

The finite element model assembly consisted of a coupon model and a gauge model. The composite was modelled using the material constants in Table 1. Because a solid element type (C3D8R) was used, a cylindrical material coordinate system was set so that the radial direction corresponded to the ABAQUS axis nomination '1', the perimeter direction to the nomination '2' and the axial direction to the nomination '3'. The gauge was modelled as a strip of isotropic polyimide (Kapton®) with a Young's modulus value of 4.0 GPa and Poisson's ratio of 0.3 [32]. A CTE value of 1.17×10^{-5} 1/°C was used for the gauge based on the adoptable thermal expansion given by the manufacturer (i.e., the gauge follows corresponding expansion in terms of strain reading (zero)). The gauge was meshed by using parabolic (C3D20R) elements. The gauge model was attached to the coupon model by so-called tie-constraints. The model was run by a point load and thermal field ($\Delta T = 36$ °C) in two steps.

The computed strain to match with the experimental strain gauge reading was calculated based on the residual axial stress in the gauge after the thermal load. By presuming that the composite is stiffer than the gauge, i.e., the gauge follows the expansion of the substrate (coupon), we have:

$$\varepsilon_{g,residual} = \sigma_{g,residual} / E_g, \tag{1}$$

where $\varepsilon_{g,residual}$ is the residual strain of the gauge in the axial direction, $\sigma_{g,residual}$ is the computed (FEA) residual stress in the axial direction, and E_g is the Young's modulus of the gauge. In Equation (1), it is presumed that the length-change of the gauge is simply due to the 'external' force by the expansion (contraction) of the composite coupon. In reality, the gauge has a finite stiffness and the force balance-given length-change is partly due to the thermal expansion of the gauge (which does not induce stresses). The gauge stresses were recorded from three elements and the average value was calculated.

2.5. Signal Attenuation

The sections of 5GPs that provide the needed weather protection to the (5G) radios, are typically referred to as 'radomes'. Low-attenuation and low-permittivity materials are an especial family of polymers that could be employed for the radomes. Radomes mainly protect against moisture and ultraviolet (UV) radiation (e.g., the reference standard UL 746C), as illustrated in Figure 6. The selected material samples' attenuation was measured by using a split-post dielectric resonator (SPDR) at 2.45 GHz (QWED, Warsaw, Poland) with a Microwave Frequency Q-Meter (QWED, Warsaw, Poland). The sample size was 60 mm × 60 mm (thickness 2.5–3.0 mm). The measurements were made at a constant signal frequency of 2.45 GHz and sample-specific thickness was measured to determine attenuation. An especial polymer blend (PREPERM, Premix, Finland) was selected as a candidate material for these details of the pole.

To account for environmental ageing during the anticipated pole operation, a series of samples were conditioned in a UV-chamber. The chamber had UVA-340 fluorescence lamps (Q-Lab, Farnworth, Bolton, UK) with a peak intensity at 340 nm. Each sample set involved five test samples and a set of samples was removed from the chamber and measured at pre-set time intervals. Accelerated aging cycles of 0, 432 h, 864 h and 1728 h (0, 18, 36, 72 days) were analyzed. Due to the durability requirements of sustainable long-term application, black and white paints on the outer surface of the radome materials were surveyed in addition to non-painted samples. The temperature at the chamber varied between 22 °C and 36 °C.

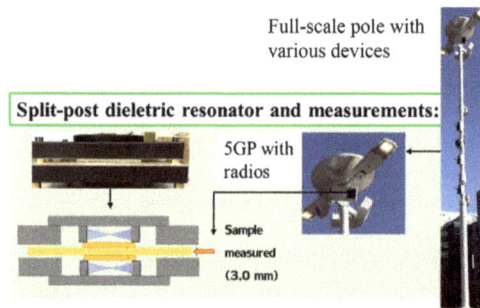

Figure 6. Attenuation measurements of radome/pole structure's material samples representing signal windows; radome and measurement setup by using the SPDR method.

2.6. Heat Exchangers

The thermal performance of a smart pole is governed by the heat sources, i.e., the devices related to the wireless communication and other services of the product. Therefore, the heat exchanger for the transfer of thermal energy is a crucial component of the system. Here, aluminum heat exchangers were designed and analyzed for a liquid cooling concept. The motivation of heat exchangers in 5G and other smart pole applications is well-known. The 3D printing was the chosen manufacturing technique because it allows more freedom of design and more complicated shapes. The 3D printing powder was AlSi10Mg and the parts were manufactured by FIT Prototyping GmbH (Lupburg, Germany) by using a 3D printing device SLM 500 (SLM Solutions, Lübeck, Germany), which uses a Powder Bed Fusion technique. The exact geometry of the exchanger is given in Appendix B. The cooling channels were directly 3D printed inside the part to form single integral component.

2.7. Computational Fluid Dynamics

The design power range of 100–500 W, as the operating power, required by the 5G radios alone, can be categorized as high-power electronics (HPE). For uniform spatial coverage, multiple directional radios are required leading to ≈1 kW power consumption. The power range is of the same order as the graphics processing units (GPUs) for which liquid cooling systems are commonly employed to maintain moderate temperature levels at components. Hence, one of the ultimate needs of the pole thermal management is the creation of an efficient cooling concept for its HPE. As a common concept for current 5GPs, the initial design was based on air cooling.

For the CFD investigations, the Reynolds Averaged Navier Stokes (RANS) and Large-Eddy Simulations (LES) were utilized [9,33]. In the air cooling simulations, the target was to solve the Navier-Stokes equations along with a transport equation for temperature using the standard, incompressible pimpleFoam solver of the open source CFD code OpenFOAM. The liquid cooling heat transfer simulations were performed with CHT analysis using the standard chtMultiRegionFoam solver in OpenFOAM were the conservation of mass, momentum and energy are calculated simultaneously in both, the liquid and the solid domains. For the heat management, the cooling of four radio units was analyzed. Appropriate cooling capacity requires proper heat exchange and design with optimized internal channeling either inside pole shaft (air cooling) or liquid cooling. Also, a strategy for connecting the cooling medium flow through the four heat exchangers is needed. The 3D printed aluminium heat exchanger was tested for a version family 'V1'. Due to weight, a version family 'V2' and water-cooled heat exchangers were analyzed with the following objectives: (1) minimum material costs, (2) low surface temperatures, (3) more compact size than V1 versions, and (4) the system should be functional even if lukewarm water is available. The simulation parameters are given in Table 2. The design concept with flow and heat values are given in Figure 7. The model consisted of 11 million cells in the fluid domain and three million cells in the solid domain—hexagonal cells except at the edges/interfaces cubic cells were used.

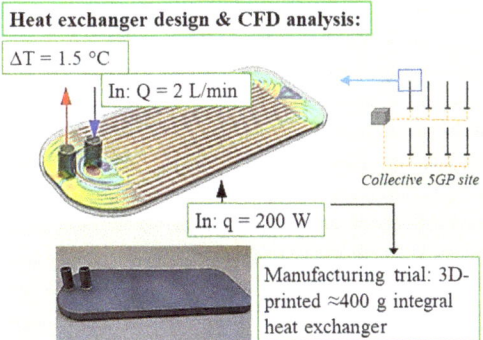

Figure 7. The cooling design for an integral device configuration in the smart pole(s): the estimated input data for thermal analysis of a heat exchanger by using CFD.

Table 2. Input selections used for the CFD analysis of heat management with 3D printed heat exchangers.

Fluid Domain	Solid Domain
Turbulence model used: k-ω SST	
Inlet:	**Heated surface:**
Velocity 0.4 m/s	Fixed temperature gradient
Turbulent intensity 4.00%	Zero pressure gradient
$k = 0.00038$ m^2/s^2, $\omega = 51.57$ 1/s	**Fluid-solid walls:**
Temperature 300 K	Temperature calculated
Pressure with zero gradient	Zero pressure gradient
Walls:	**Outer surfaces:**
Velocity—no slip	Zero temperature gradient (adiabatic)
Turbulent variables—wall functions	Zero pressure gradient
Zero pressure gradient	
Temperature calculated	
Outlet:	
Zero velocity gradient	
Zero gradient of turbulent variables	
Fixed pressure (atm pressure)	
Zero temperature gradient	

3. Results

3.1. Impact Dynamics of the Selected GFRP Shaft

During the design process, prior to manufacture of full-scale products, experimental validation and qualification was started in phases. For a composite 5GP, even for no-traffic sites, the critical damage is impact-type loads at the shaft root. In this study, an experimental campaign was realized by a step-by-step approach starting from QS indentations on GFRP panel sections and, further, to full-scale impact tests on the tubular shaft (more details in Section 2.2). The QS testing serves as a limit case (reference and control) for impact-concerned design since dynamic effects are omitted. In general, the deformations of the curved specimens clearly localized close to the contact areas. The localization for the GFRP panel can be seen in Figure 8, where the deformations of the lower surface in the indentation case (at the maximum loading moment) are evident. The localization challenges the sizing process of the pole since the ultimate (fracture) behaviour starts playing a big role in the GFRP's deformation response.

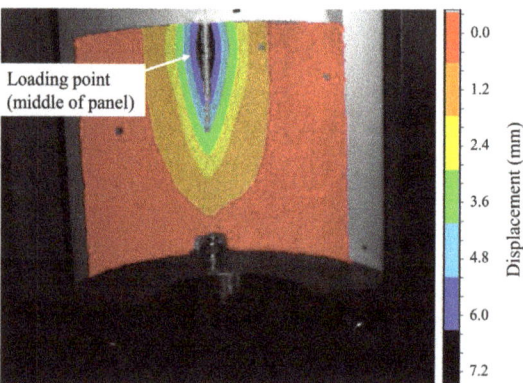

Figure 8. The mechanical testing of a panel specimen: the DIC-resolved displacement field during the indentation loading of the half-circular specimen at the moment of maximum load.

The contact force versus displacement of the loading head was defined for different loading configurations during the testing campaign (see Figure 9). The results show clearly the effects of specimen size and support on the load response. For the similar size half-circular specimen, the trend of the loading in the QS indentation and drop-weigh impact is essentially similar regardless of the difference in the support (semi-rigid or open boundary). However, when comparing the impact response of the tubular specimen to the half-circular panel, the indentation and impact cases showed clear differences in terms of maximum load and ultimate deflection. Figure 9 also includes local FEA of the GFRP panel (failure criteria applied) [27]. It was confirmed, as is typical for composites, that the load-carrying capability after the damage onset remains. The first failure mode due to the impact was internal delamination and often visible crack in the axial direction of the shaft.

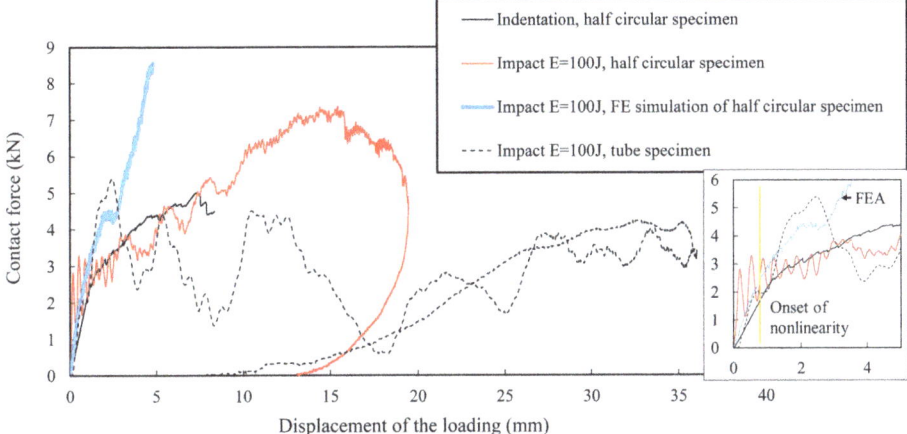

Figure 9. The performance of the GFRP laminate and tube: the contact force-displacement response based on experiments and FEA (linear material [27]). The inset graph on the right shows the detailed data at the beginning of the tests/analyses and predicted displacement level for damage onset.

3.2. Finite Element Analysis

FEA was used to predict the effects of cutouts (maintenance doors) when the pole is subjected to the wind load and thermal load (e.g., due to radiation from the sun). For most of the design cases, the critical details are the cutouts designed for connections (cabling in and out from the pole) or maintenance, typically located at the shaft root with a high level of bending moment. Figure 10 shows the failure analysis for the diamond-shaped cross-section at the root cutout, where the maximum stress criterion predicts the onset of damage. The exact type and form of the failure criterion required analysis for the curved shape and different strain-rates [34]. Finally, the confirmation of the selection was made by comparing with the experiments (see Section 3.1). A 5 mm-wall (for the load-carrying layers) thickness represented a safe and low-weight solution.

For the composite shaft, the thermal strains (per material's coefficients of thermal expansion) are governing the absolute deformation of the composite pole for the anticipated, average operation environment, i.e., non-storm weather. For this reason, additional thermal analyses were carried out together with the wind load analysis (Figure 11). The pole design with the circular cross-section experienced 15–64% higher deformations (51 mm compared to 84 mm with wind load and 64 mm compared to 73 mm with thermal load for the diamond-shaped and circular cross-section, respectively). Due to the anticipated services of the project (see Section 4.2), the level of deformations was not seen problematic and the lower costs of the circular cross-section overran the mechanical advantages of the diamond-shaped cross-section.

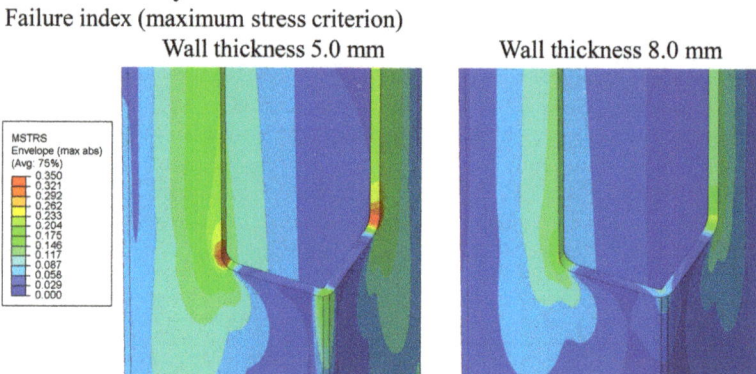

Figure 10. FEA for the failure index when comparing two different pole shaft laminates and a diamond-shaped cross-section (5 mm and 8 mm-wall thickness); corners (hotspot) at a root cutout are shown.

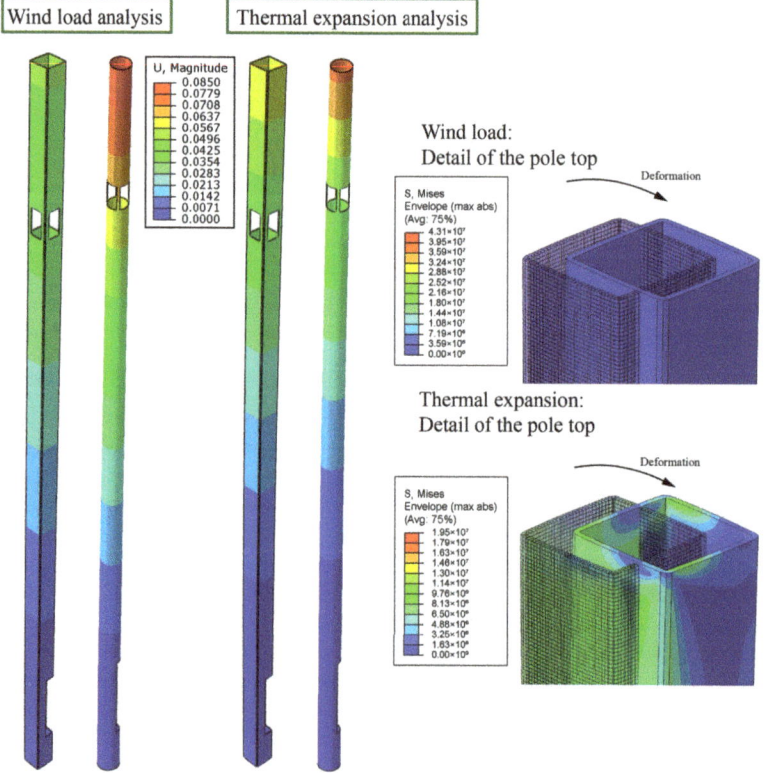

Figure 11. FEA for the deformation when the pole shaft is subjected to the standard wind load (**left side image**) and a thermal load (**right side image**). Details of the highest deformation (pole top) are shown on the right.

3.3. Signal Attenuation at the GHz-Regime

Any enclosing of the (5G) radios inside a housing or shaft requires analyzing the RF signal attenuation due to the surrounding enclosure or radome wall. The design of the details, i.e., radomes or 'signal windows' in cutouts, must satisfy experimental verification regarding the attenuation. In this study, the radomes were designed not be load-carrying parts so that reinforcements (fibers) were not needed for the radomes. PREPERM polymer (see Section 2.5) was analyzed here, and attenuation defined in terms of dielectric loss (DL). In particular, the long-term properties of the radome materials in outdoor environments were not well-known. Therefore, the effects of UV radiation (from the sun) were considered in this study. In reality, the signal windows might need to be painted that makes the measurements with painted samples necessary. According to the measurement results, shown in Figure 12, the minimum paint design is crucial for a high signal penetration. The effect of a paint layer had a significant impact on the signal attenuation (i.e., dielectric loss (DL)) and the treatment increased the DL levels 90–175%. Slightly lower increase was measured as the UV degradation was increased (longer UV exposure time); the color of the paint did not observably affect the attenuation.

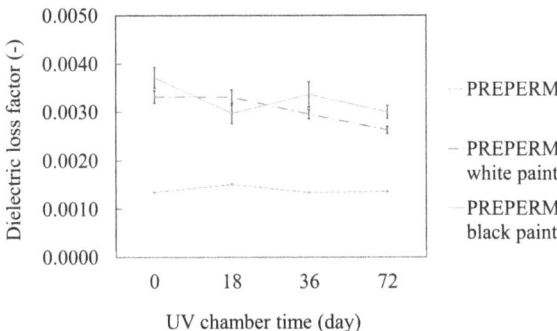

Figure 12. Measured attenuation in terms of DL; results over the 72-day accelerated UV-ageing period.

3.4. Thermal Management and Multiple Radio Analysis

The CFD analysis process was divided into: (1) CFD investigations, (2) feasibility study of 3D printed heat exchangers, and (3) experiments on a 3D printed heat exchanger operating at 200 W. The air-cooling systems by standard plane and pin-fin designs led to heat exchangers with a weight of 3.5–3.8 kg per piece and their volume was overly large compared to the space available inside the demo 5GPs. With room temperature air, the system's maximum surface temperatures remained below 51 °C that is substantially below the 65 °C critical allowable tempareture. Concerning the CFD simulations, the constant surface temperature-boundary condition was noted to be misleading as: (1) the surface temperature is not known, and (2) the inserted power cannot be fixed. In the experimental validation, the simulated and experimental velocity fields (using Laser Doppler Anemometry) agreed well. Conjugate heat transfer (CHT) studies are proposed as the next design steps of the design so that temperature transport in both fluid and the solid can be accounted for.

One of the key drivers for a cooling concept is the compact size and centralized thermal control at an entire 5GP site. A liquid cooling was a tempting option to be analyzed since the density ratio of water to air is 1000:1 while the specific heat ratio is 4:1. Furthermore, water is much better heat conductor than air with the heat conductivity ratios of approximately 60:1. Motivated by the experiences from the air cooling investigations, we utilized the CHT solver in OpenFOAM called chtMultiRegionFoam. With this procedure, the incompressible Navier-Stokes equations were solved for the fluid phase while the convection-diffusion equation was solved for temperature both in the solid and fluid.

Based on the results, the walls could be made thinner (e.g., 1 mm) to result in a less than 200 g mass of the exchanger while the tested exchanger version (wall thickness 2 mm) had a ≈393 g mass. It was confirmed by simulations that the system can maintain surface temperatures below 65 °C for a range of mass flow rates for cooling water's inflow temperature below 50 °C (see Figure 13). Figure 13b shows the inlet temperature of the cooling water as a function of the mass flow rate that maintains the heat exchangers' heated surface temperatures at 65 °C. The serial and parallel configurations indicate whether the heat exchangers (inside) of the 5GP would be connected in series or in parallel to cool all the four radios. The highest allowed inlet water temperature for the serial and parallel configurations of liquid cooling heat exchanger is based on the assumption that the temperature difference between the inlet water and the heated surface is independent of the absolute temperature—based on the mass flow rate sensitivity analysis that has confirmed the independency of transfer performance on the inlet Reynolds number [33]. Furthermore, the functionality of the heat exchanger was tested experimentally, and the measured surface temperatures were in a good agreement with the CFD results.

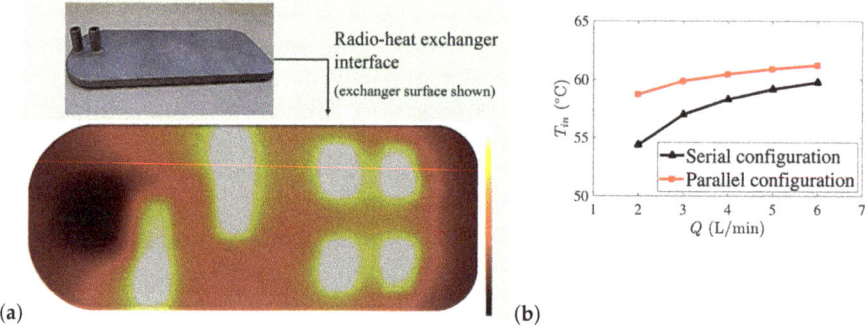

Figure 13. CFD analysis of a 3D printed heat exchanger and the cooling strategy: (**a**) simulated heat exchanger (outside) surface temperature (color range represents $\Delta T = 10$ °C) when water flows into and out from a 3D printed heat exchanger via embedded rows of channels; (**b**) the highest allowed inlet temperature as a function of mass flow rate to maintain below 65 °C surface temperatures per cooling strategy.

4. Discussion

4.1. Design Process and Interconnections of Results

The FEA combined with the experimental campaign resulted in the GFRP wall thickness of 5 mm (load-carrying layers). For a six-meter pole and GFRP's density of 1670 kg/m³, the mass of the entire pole would be 28.8–29.6 kg (from circular to diamond-shaped cross-section, respectively). For a similar steel pole, the mass would increase 380% (steel pole mass 153 kg). Because 5GPs are more deformation (sway) critical than strength critical, a steel pole could be made six times thinner in thickness in theory. If a practical minimum wall thickness would be two millimeters to prevent instability, the steel pole continue to be 60% heavier (i.e., GFRP leads to a minimum of 37% weight saving). The significantly lower mass of the GFRP pole directly makes its handling easier and lowers transportation emissions for a global 5GP usage.

Because of the bending moment and the following stress concentrations at the root of the 5GP shaft, the wall thickness could be increased at the pole root. Alternatively, the pole root in turn would be an ideal location for an additional, fully or partly load-carrying structure, i.e., a wide shaft tube. A wide housing at the root would lower the mechanical stresses and embody a neat space for standard power and data connection modules. A root housing could also accommodate devices, e.g., an EV charge station for some urban 5GP sites. Due to the linearity and strength of composite materials, there are clearly more options, e.g., large maintenance doors, as was given by the FEA at the cutout corners in this study. At the top of the pole, cutouts are also necessary for radio radomes. Large cutouts for signal windows (covered by a non-reinforced polymer) make the GFRP pole even lighter. Any other than composite/polymer design could be an obstacle for receiving (indoor/outood) devices [26] and to fit the 5GP for individual customer needs. The attenuation measurements indicated a 90–175% increase of signal attenuation due to a surface treatment on radome materials—this means that as large cutouts and as thin as possible radomes are needed in 5GPs, even when using GFRP for the pole shaft.

The CFD analysis of the heat management presented that liquid cooling is an efficient technology for pole-integrated radios. Liquid cooling requires piping between individual radios as well as in and out from the pole. The added pipe lines would require further space inside the pole; the power and data cabling with various connector appliances define the necessary pole diameter in general. Any larger diameter will lead to a heavier pole—again emphasizing the benefits of GFRP.

The results of this study showed that the interaction between functionality (i.e., large cutouts allowing large radomes and maintenance doors), heat management for integrated devices and mechanical design with a minimum material usage and safe structure prefer a composite design. Due to the higher manufacturing costs of composite structures compared to traditional steels, the markets of 5GPs will define the amount of device integration in the future.

4.2. Assembly and Future Applications

Currently, smart poles as commercial products are complex systems with various stakeholders involved. The division between a product owner, seller, data handling, etc., has not yet settled. As a European solution, the consortiums of Luxturrim5G, Neutral Host and Luxturrim5G+ [35] ventures have defined a 5GP overall concept that must deal and handle all the issues of future wireless platforms: legislation, radios, big data handling for 'Smart cities', viable business concepts, the open data platform and safe physical integration at urban areas. The full-scale 5GPs were mounted at the Karaportti site (Espoo, Finland) in the autumn 2019. The 5GPs of the site finally included the following services: 60 GHz WiGig radios, video and audio surveillance, weather monitoring, and EV charging. Part of the devices and all the power and data cabling were pole shaft integrated. The connection with the city infrastructure was analyzed using a 3D planning tool (AURA, Sitowise Oy, Espoo, Finland) prior to the excavation work. In 2020, activities continue by mounting the future 5GPs with 60 GHz WiGig and 26 GHz radios, as well as new services for traffic monitoring, autonomous driving and public safety.

5. Conclusions

Several potential designs of 5GPs have been proposed for the physical device frame and service platforms within the current industry of wireless communication technologies. This study focuses on the physical structure of a 5G smart light pole and its multidisciplinary design process. The work includes an interacting research of a GFRP composite pole structure with finite element (FE) analysis and experimental verification, signal attenuation measurements of the latest low-attenuation materials and metal 3D printing combined with high-fidelity CFD computation to understand the heat management inside the densely device-integrated 5G pole. Based on the results, the work revealed the following specific novelties related to the next-era wireless application platform:

- A full-composite glass fibre reinforced 5G pole was FE-modelled and analysed against standard wind and thermal load. The findings showed that a mechanically safe and functional (stiff) GFRP shaft results in significant weight savings (37–80%) compared to traditional steel shafts;
- RF signal attenuation at a GHz-regime (2.45 GHz) was found to increase significantly (90–175%) due to any paint layer while long-term UV degradation in the polymer structure led only to a nominal decrease of attenuation in terms of dielectric loss;
- Entirely integral one-piece heat exchangers were designed with CFD analysis of the fluid-solid interaction for heat transfer, and printed. It was found that a parallel liquid cooling of four

radio units is rather insensitive to the flow rate (range 2...6 L/min) and as high as ≈60 °C inlet temperature can be allowed to keep the device surfaces at or below a critical 65 °C.

Author Contributions: Conceptualization, M.K. and V.V.; methodology, T.P. and K.S. and H.H.; software, D.D.V. and A.L. and M.K.; validation, D.D.V., A.L. and T.P.; investigation, J.J. and K.K.; writing—original draft preparation, M.K. and D.D.V. and T.P.; writing—review and editing, V.V. and H.H.; visualization, M.K.; project administration, M.K. and H.H. and V.V. All authors have read and agreed to the published version of the manuscript.

Funding: This work was done as part of the LuxTurrim5G ecosystem funded by the participating companies and Business Finland.

Acknowledgments: Researchers J. Jokinen and O. Orell are acknowledged for their help and support. Nokia Bell Labs (J. Salmelin and P. Wainio) are acknowledged for the collaboration and design process development. Exel Composites (M. Lassila and K. Sjödahl) is acknowledged for the material and manufacture support.

Conflicts of Interest: The authors declare no conflict of interest. The funding agency had no role in the interpretation of data.

Appendix A. Thermal Expansion of the GFRP Composite

The results of the CTE determination for GFRP are shown in Figure A1. The error between the experiment and FEA only for the mechanical load was 13% (FE model stiffer, $\varepsilon_g = 4.18 \times 10^{-5}$). For the combination of mechanical load and thermal load, the target deviation (error) was kept the same. The experimental strain developed slowly per temperature step, and an extrapolated value of $\varepsilon_{g,residual} = -1.6 \times 10^{-4}$ was selected as the target. Finally, an iterated solution was met by using the values of $CTE_{axial} = 1.15 \times 10^{-5}$ 1/°C and $CTE_{transverse} = 2.5 \times 10^{-5}$ 1/°C ('axial' corresponding to the '3' direction and 'transverse' corresponding to the '2' and '1' directions in the FEA). This iteration was deemed acceptable due to the fact that there are no non-linearities accounted for in the FEA that in reality might appear (e.g., at gauge-coupon interface or in the gauge). Additionally, Equation (1) does not take into account the deformation of the substrate (coupon) by the expansion of the gauge. These effects would lower the calculated strain with the selected CTE values [36]. It should be noted that the transverse CTE of GFRP affects the gauge strain (Equation (1)) due to the Poisson's effect (in FEA). The iterated CTE values are in a good agreement with the literature values for GFRP [37,38].

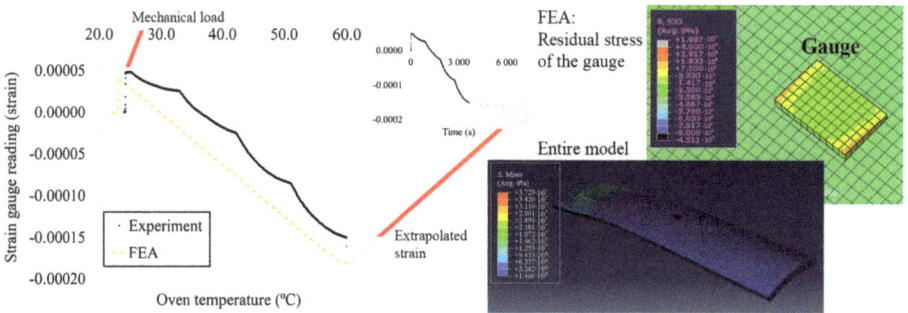

Figure A1. The experimental and the finite element analysis results used for the determining of GFRP's thermal expansion coefficients.

Appendix B. Heat Exchanger Design

The dimensions of the 3D printed integral heat exchanger are given in Figure A2.

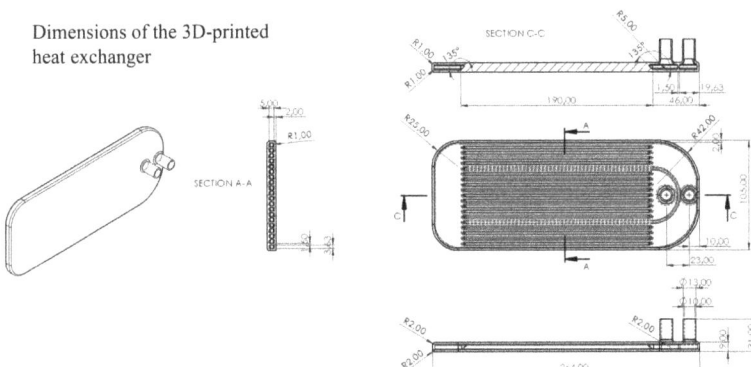

Figure A2. Technical drawing of the heat exchanger. Dimensions are in millimeters.

References

1. Henry, S.; Alsohaily, A.; Sousa, E. 5G is Real: Evaluating the Compliance of the 3GPP 5G New Radio System with the ITU IMT-2020 Requirements. *IEEE Access* **2020**, *8*, 42828–42840. [CrossRef]
2. Munoz, R.; Mangues-Bafalluy, J.; Vilalta, R.; Verikoukis, C.; Alonso-Zarate, J.; Bartzoudis, N.; Georgiadis, A.; Payaro, M.; Perez-Neira, A.; Casellas, R.; et al. The CTTC 5G end-to-end experimental platform: Integrating heterogeneous wireless/optical networks, distributed cloud, and IoT devices. *IEEE Veh. Technol. Mag.* **2016**, *11*, 50–63. [CrossRef]
3. Bangerter, B.; Talwar, S.; Arefi, R.; Stewart, K. Networks and devices for the 5G era. *IEEE Commun.* **2014**, *52*, 90–96. [CrossRef]
4. Landertshamer, O.; Benseny, J.; Hämmäinen, H.; Wainio, P. Cost model for a 5G smart light pole network. In Proceedings of the CTTE-FITCE, Ghent, Belgium, 25–27 September 2019; Volume 1.
5. Benseny, J.; Walia, J.; Hämmäinen, H.; Salmelin, J. City strategies for a 5G small cell network on light poles. In Proceedings of the CTTE-FITCE, Ghent, Belgium, 25–27 September 2019; Volume 1.
6. International Commission on Non-Ionizing Radiation Protection. Guidelines for limiting exposure to electromagnetic fields (100 kHz to 300 GHz). *Health Phys.* **2020**, *118*, 483–524. [CrossRef] [PubMed]
7. Seligar, G. *Sustainability in Manufacturing: Recovery of Resources in Product and Material Cycles*; Springer: Berlin, Germany, 2007.
8. Francesco, F.; Smierzchalski, M.; Ettorre, M.; Aurinsalo, J.; Kautio, K.; Lahti, M.; Lamminen, A.; Säily, J.; Sauleau, R. An LTCC beam-switching antenna with high beam overlap for 60-GHz mobile access points. In Proceedings of the 2017 IEEE International Symposium on Antennas and Propagation and USNC-URSI Radio Science Meeting, San Diego, CA, USA, 9–14 July 2017.
9. Peltonen, P.; Saari, K.; Kukko, K.; Vuorinen, V.; Partanen, J. Large-Eddy Simulation of local heat transfer in plate and pin fin heat exchangers confined in a pipe flow. *Int. J. Heat Mass Transf.* **2019**, *134*, 641–655. [CrossRef]
10. Kanerva, M.; Lassila, M.; Gustafsson, R.; O'shea, G.; Aarikka-Stenroos, L.; Hemilä, J. Emerging 5G Technologies Affecting Markets of Composite Materials, 2018. White Paper, Exel Composites Plc. Available online: https://www.luxturrim5g.com/publications (accessed on 15 August 2020).
11. Gubran, H. Dynamics of hybrid shafts. *Mech. Res. Commun.* **2005**, *32*, 368–374. [CrossRef]
12. James Prasad Rao, B.; Srikanth, D.; Suresh Kumar, T.; Sreenivasa Rao, L. Design and analysis of automotive composite propeller shaft using FEA. *Mater. Today Proc.* **2016**, *3*, 3673–3679. [CrossRef]
13. Reyes, V.G.; Cantwell, W. The mechanical properties of fibre-metal laminates based on glass fibre reinforced polypropylene. *Compos. Sci. Technol.* **2000**, *60*, 1085–1094. [CrossRef]
14. Kanerva, M.; Sarlin, E.; Hoikkanen, M.; Rämö, K.; Saarela, O.; Vuorinen, J. Interface modification of glass fibre-polyester composite-composite joints using peel plies. *Int. J. Adhes. Adhes.* **2015**, *59*, 40–52. [CrossRef]
15. Manuele, G.; Bringiotti, M.; Lagana, G.; Nicastro, D. *Fiber Glass and 'Green' Special Composite Materials as Structural Reinforcement and Systems; Use and Applications from Milan Metro, Brenner Tunnel up to High Speed Train Milan–Genoa*; CRC Press: Boca Raton, FL, USA, 2019; pp. 2625–2634.

16. Pandey, J.; Ahn, S.; Lee, C.; Mohanty, A.; Misra, M. Recent advances in the application of natural fiber based composites. *Macromol. Mater. Eng.* **2010**, *295*, 975–989. [CrossRef]
17. Khalili, P.; Liu, X.; Zhao, Z.; Blinzler, B. Fully biodegradable composites: Thermal, flammability, moisture absorption and mechanical properties of natural fibre-reinforced composites with nano-hydroxyapatite. *Materials* **2019**, *12*, 1145. [CrossRef] [PubMed]
18. Crawford, B.; Pakpour, S.; Kazemian, N.; Klironomos, J.; Stoeffler, K.; Rho, D.; Denault, J.; Milani, A. Effect of fungal deterioration on physical and mechanical properties of hemp and flax natural fiber composites. *Materials* **2017**, *10*, 1252. [CrossRef]
19. Brøndsted, P.; Lilholt, H.; Lystrup, A. Composite materials for wind power turbine blades. *Annu. Rev. Mater. Res.* **2005**, *35*, 505–538. [CrossRef]
20. Bélan, F.; Bellenger, V.; Mortaigne, B. Hydrolytic stability of unsaturated polyester networks with controlled chain ends. *Polym. Degrad. Stab.* **1997**, *56*, 93–102. [CrossRef]
21. Partini, M.; Pantani, R. FTIR analysis of hydrolysis in aliphatic polyesters. *Polym. Degrad. Stab.* **2007**, *92*, 1491–1497. [CrossRef]
22. Atxaga, G.; Marcos, J.; Jurado, M.; Carapelle, A.; Orava, R. Radiation shielding of composite space enclosures. In Proceedings of the International Astronautical Congress, Naples, Italy, 1–5 October 2012.
23. Kanerva, M.; Koerselman, J.; Revitzer, H.; Johansson, L.; Sarlin, E.; Rautiainen, A.; Brander, T.; Saarela, O. Structural assessment of tungsten-epoxy bonding in spacecraft composite enclosures with enhanced radiation protection. In Proceedings of the European Conference on Spacecraft Structures, Materials and Mechanical Testing, Braunschweig, Germany, 1–4 April 2014; ESA SP-727.
24. Gaier, J.; Hardebeck, W.; Bunch, J.; Davidson, M.; Beery, D. Effect of intercalation in graphite epoxy composites on the shielding of high energy radiation. *J. Mater. Res.* **1998**, *13*, 2297–2301. [CrossRef]
25. Jarvelainen, J.; Karttunen, A.; Aurinsalo, J.; Huhtinen, I.; Hujanen, A. Characterization of mmWave radomes for base stations and automotive radars. In Proceedings of the 13th European Conference on Antennas and Propagation, Krakow, Poland, 31 March–5 April 2019; EuCAP: Krakow, Poland, 2019; Volume 1.
26. Karttunen, A.; Le Hong Nguyen, S.; Koivumäki, P.; Haneda, K.; Hentilä, T.; Asp, A.; Hujanen, A.; Huhtinen, I.; Somersalo, M.; Horsmanheimo, S.; et al. Window and wall penetration loss on-site measurements with three methods. In Proceedings of the 12th European Conference on Antennas and Propagation, London, UK, 9–13 April 2018; Institution of Engineering and Technology IET: London, UK, 2018; Volume CP741.
27. Rodera Garcia, O. Damage Onset Modelling of Curved Composite Laminates. Master's Thesis, Tampere University, Tampere, Finland, 2018. Available online: http://urn.fi/URN:NBN:fi:tty-201810032382 (accessed on 4 September 2020).
28. Pournoori, N.; Guilherme, C.; Orell, O.; Palola, S.; Hokka, M.; Kanerva, M. Adiabatic heating and damage onset in a pultruded glass fiber reinforced composite under compressive loading at different strain rates. *Int. J. Impact Eng.* **2021**, *147*, 103728. [CrossRef]
29. Lanza di Scalea, F. Measurement of thermal expansion coefficients of composites using strain gages. *Exp. Mech.* **1998**, *38*, 233–241. [CrossRef]
30. Devendra, K.; Rangaswamy, T. Thermal conductivity and thermal expansion coefficient of GFRP composite laminates with fillers. *Mech. Confab* **2013**, *2*, 39–44.
31. SFS. *Standard SFS-EN 40-7 CEN/TC 50 Lighting Columns and Spigots*; Finnish Transport Infrastructure Agency (FTIA): Helsinki, Finland, 2003; refers to EN 12767:2019.
32. He, W.; Goudeau, P.; Le Bourhis, E.; Renault, P.O.; Dupre, J.; Doumalin, P.; Wang, S. Study on Young's modulus of thin films on Kapton by microtensile testing combined with dual DIC system. *Surf. Coat. Technol.* **2016**, *308*, 273–279. [CrossRef]
33. Laitinen, A.; Saari, K.; Kukko, K.; Peltonen, P.; Laurila, E.; Partanen, J.; Vuorinen, V. A computational fluid dynamics study by conjugate heat transfer in OpenFOAM: A liquid cooling concept for high power electronics. *Int. J. Heat Fluid Flow* **2020**, *85*, 108654. [CrossRef]
34. Di Vito, D.; Pärnänen, T.; Jokinen, J.; Orell, O.; Kanerva, M. Lateral indentation and impact analyses on curved composite shells. In Proceedings of the 7th International Conference on Fracture Fatigue and Wear, Ghent, Belgium, 9–10 July 2018; Lecture Notes in Mechanical Engineering; Wahab, M., Ed.; Springer: Singapore, 2018; pp. 171–183, [CrossRef]
35. Heino, M.; Salmelin, J.; Wainio, P. LuxTurrim5G—Building the Digital Backbone for a Smart City. White Paper, 2020. Available online: https://www.luxturrim5g.com/publications (accessed on 15 September 2020).

36. Kanerva, M.; Antunes, P.; Sarlin, E.; Orell, O.; Jokinen, J.; Wallin, M.; Brander, T.; Vuorinen, J. Direct measurement of residual strains in CFRP-tungsten hybrids using embedded strain gauges. *Mater. Des.* **2017**, *127*, 352–363. [CrossRef]
37. Sousa, J.; Correia, J.; Firmo, J.; Cabral-Fonseca, S.; Gonilha, J. Effects of thermal cycles on adhesively bonded joints between pultruded GFRP adherends. *Compos. Struct.* **2018**, *202*, 518–529. [CrossRef]
38. Ran, Z.; Yan, Y.; Li, J.; Qi, Z.; Yang, L. Determination of thermal expansion coeffcients for unidirectional fiber-reinforced composites. *Chin. J. Aeronaut.* **2014**, *27*, 1180–1187. [CrossRef]

Publisher's Note: MDPI stays neutral with regard to jurisdictional claims in published maps and institutional affiliations.

© 2020 by the authors. Licensee MDPI, Basel, Switzerland. This article is an open access article distributed under the terms and conditions of the Creative Commons Attribution (CC BY) license (http://creativecommons.org/licenses/by/4.0/).

Article

A Numerical Assessment on the Influences of Material Toughness on the Crashworthiness of a Composite Fuselage Barrel

A. Riccio [1,*], S. Saputo [1], A. Sellitto [1] and F. Di Caprio [2]

1. Department of Engineering, Università degli studi della Campania "Luigi Vanvitelli", via Roma 29, 81031 Aversa (CE), Italy; salvatore.saputo@unicampania.it (S.S.); andrea.sellitto@unicampania.it (A.S.)
2. C.I.R.A. Italian Aerospace Reserach Center, via Maiorise, 81043 Capua (CE), Italy; f.dicaprio@cira.it
* Correspondence: aniello.riccio@unicampania.it; Tel.: +39-0815010407

Received: 13 February 2020; Accepted: 13 March 2020; Published: 16 March 2020

Abstract: In the present work, a numerical study on the dynamic response of a composite fuselage barrel, in relation to crashworthiness, has been investigated. The aim of this work is to investigate the influence of the material fracture toughness on the capability of a composite fuselage barrel to tolerate an impact on a rigid surface. Three different material configurations with different intra-laminar fracture energy values were considered to take into account variations in material toughness. Indeed, the dynamic behaviour of the analysed fuselage barrel has been numerically simulated by means of the FE (Finite Element) code Abaqus/Explicit. The effects of intralaminar fracture energy variations on the impact deformation of the barrel has been evaluated comparing the numerical results in terms of displacements and damage evolution for the three analysed material configurations.

Keywords: crashworthiness; finite element analysis (FEA); composites; progressive failure analysis (PFA)

1. Introduction

In recent years, several steps forward in increasing civil aircraft safety levels have been taken in relation to crashworthiness during an emergency landing. In this context, crashworthiness has become of fundamental relevance for the design and certification of an aircraft. The main aim in designing crashworthy aircrafts is to ensure the ability to absorb as much energy, deformations and breakage as possible during an impact event, without compromising the occupants' safety. Hence, the structure, under deformations and/or breakage conditions, must preserve the living space of passengers and crew, allowing escape routes from the aircraft [1]. Indeed, the entire structure should be able to guarantee the transfer to passengers of acceleration which can be tolerated by the human body [2–5]. In the past, the assessment of the crashworthiness performance of aircrafts was verified exclusively by several costly full scale experimental testing campaigns, as in the case of the Airbus A320 [6,7], Boing 707 [8–10], Boing 737 [11,12] and NAMC YS [13,14]. According to the literature, the experimental drop tests have allowed to identify the structural components which are generally involved in the absorption of energy during an impact event: circumferential frames close to the vertical supports and fasteners where large structural deformations occurred [15,16].

Nowadays, the use of advanced numerical tools can contribute to reduce the number of expensive experimental testing campaigns. Numerical developments focused on crash simulations of civil aircraft fuselages. Due to the crash phenomenon and fuselage structure's inherent complexity, numerical models with increasing levels of accuracy were progressively introduced by several authors. Basic models, using concentrated masses positioned at different locations and beams were used to start exploring the mechanical response of a fuselage during a crash phenomenon. Subsequently, models with non-linear springs and concentrated masses were introduced [17,18].

Today, the use of finite element simulations with explicit methodologies allows to mimic, with good accuracy, the evolution of the crash phenomenon of a fuselage during a fall, allowing to identify both the damages and their development [19–23].

The high complexity of the phenomenon under examination is amplified by the complexity of the modern aircrafts. In particular, the huge size of the fuselage structure, the number of different sub-components characterised by several sources of non-linearity (geometry, materials, contacts, etc.) make the validation of numerical models very complicated.

In addition to all these conditions, new aircraft generations are characterised by a further source of complexity represented by the use of fibre-reinforced composite materials. Composite fibre-reinforced material, by their nature, are subjected to intra-laminar and inter-laminar damages, which are not easily numerically analysed [24].

The use of fibre-reinforced composite materials, even with the intrinsic advantages in terms of weight-performing ratio, is source of further complexities in terms of components crash behaviour. Indeed, the absence of a material plastic phase [25] and the presence of different interacting mechanisms, such as fibre and/or matrix breakage, delamination and fibre matrix de-bonding, for absorbing impact energy [26–28] makes the crash behaviour even more complex with respect to conventional metallic materials. Furthermore, a crash events response is a highly geometrical and materially nonlinear event [29]. Hence the crash behaviour of a complex composite structure is strongly influenced by the failure characteristics and strength of its material system. Consequently, the use of effective theories for the prediction for the onset and evolution of the damage becomes mandatory to understand and describe the complex mechanism of breakages in composite materials during a crash event.

In the literature, several theories are proposed to accurately describe the failure mechanisms for laminated aeronautical structures [30–36]. The theory of continuum damage mechanics (CDM) is mainly adopted for the assessment of intralaminar damages. According to this theory, the stiffness reduction can be physically compared to the presence of distributed micro-fractures or defects. The presence of micro damage, phenomenologically, can be represented by a variation of internal state variables of the material. Alternative methodologies for the simulation of degradation of the material properties (MPDM) based on a thermo-dynamic model can be found in [37–40]. These methodologies demonstrated to be very effective in conjunction with finite element modelling.

In general, the intra-laminar damages evolution in composite materials and then in complex composite structures is strictly influenced by fracture toughness for the four fracture modes (matrix and tensile and compressive fibre). Moreover, the experimental determination of these fracture toughness-related parameters is rather complex [41–43].

The purpose of this paper is to numerically evaluate the influence of these fracture toughness-related parameters on the mechanical behaviour of a composite fuselage section during a crash event. The Abaqus/Explicit platform has been adopted to perform the numerical simulations of the mechanical behaviour of the fuselage barrel subjected to a crash event on a rigid ground. The structural subcomponents of the fuselage were detailed discretised by adopting finite elements with a three-dimensional formulation. Three different configurations with material systems characterised by different in-plane fracture toughness values were analysed and compared to each other to assess the effects on the energy absorption capabilities of the structures and on the consequent deformations and stresses distributions.

In Section 2, the theoretical background is introduced, while in Section 3, the description of the adopted finite element model is provided. In Section 4, the numerical results obtained from the different analysed configurations are introduced and discussed.

2. Theoretical Background: Material Damage Model

Among several criteria for the prediction of the initiation and evolution of intra-laminar damage for structures made of unidirectional fibre-reinforced composite materials, the Hashin criteria are probably the most adopted ones. These criteria, based on the separation of the failure modes by introducing separated equations, allow to clearly identify the damage of the matrix and of the fibre under tension or compression stress conditions. Indeed, according to Hashin failure criteria [44], in Equations (1)–(4), the limit values for the onset of the damage for the fibre traction (F_{ft}), fibre compression (F_{fc}), matrix traction (F_{mt}) and matrix compression (F_{mc}) are, respectively, introduced.

$$F_{ft} = \left(\frac{\sigma_{11}}{X_T}\right)^2 + \left(\frac{\sigma_{12}}{S_L}\right)^2 = 1 \tag{1}$$

$$F_{fc} = \left(\frac{\sigma_{11}}{X_C}\right)^2 = 1 \tag{2}$$

$$F_{mt} = \left(\frac{\sigma_{22}}{Y_T}\right)^2 + \left(\frac{\sigma_{12}}{S_L}\right)^2 = 1 \tag{3}$$

$$F_{mc} = \left(\frac{\sigma_{22}}{2S_T}\right)^2 + \left[\left(\frac{Y_C}{2S_T}\right)^2 - 1\right] \cdot \frac{\sigma_{22}}{Y_C} + \left(\frac{\sigma_{12}}{S_L}\right)^2 = 1 \tag{4}$$

In Equations (1)–(4) σ_{11}, σ_{22}, σ_{12} are the components of the effective stress tensor, respectively, along fibre direction, matrix direction and shear while X_T, X_C, Y_T, Y_C, S_L and S_T are, respectively, the fibre tensile, fibre compressive, matrix tensile, matrix compressive and shear strengths in longitudinal and transversal directions. The evolution of the damage for each failure mode is modelled according to the bilinear law schematically represented in Figure 1.

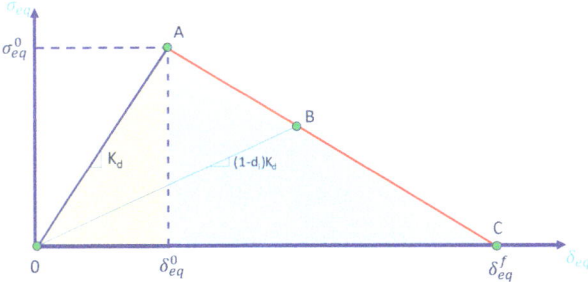

Figure 1. Constitutive relation (damage onset and evolution).

According to Figure 1, it is possible to identify the onset and evolution damage phases for each damage mode. In particular, point A of Figure 1 identifies the condition of Hashin's criterion satisfaction.

At this condition, the element has absorbed an amount of energy represented by the yellow triangle area. Beyond this condition, the evolution of the damage and the corresponding loss of stiffness of the material is simulated by introducing a parameter allowing to reduce linearly the characteristics of the undamaged material. The phase of the damage evolution is represented by the segment in Figure 1 connecting point A to point C representing the completely damaged element condition. Actually, the element is considered completely damaged when it has absorbed an energy represented by the sum of the yellow and blue triangles areas. Indeed, this global area is representative of the fracture toughness of simulated material system.

In order to evaluate the degradation status of the element, a material degradation coefficient d_i is introduced which is evaluated independently for matrix and fibre under traction and compression conditions. The degradation coefficient d_i definition is introduced in Equation (5).

$$d_i = \frac{\delta_{i,eq}^t \left(\delta_{i,eq} - \delta_{i,eq}^0\right)}{\delta_{i,eq}\left(\delta_{i,eq}^t - \delta_{i,eq}^0\right)}; \quad \delta_{i,eq}^0 \leq \delta_{i,eq} \leq \delta_{i,eq}^t; \quad i \in (f_c, f_t, m_c, m_t) \tag{5}$$

where $\delta_{i,eq}$ is equivalent displacement; $\delta_{i,eq}^0$ is equivalent displacement on set damage; $\delta_{i,eq}^t$ is equivalent displacement on full damage (Point C in Figure 1). Moreover, f_c, f_t, m_c, m_t are fibre failure compression, fibre failure tension, matrix failure compression and matrix failure tension.

In Equation (6), the definition of the maximum equivalent displacement, reached at point C of Figure 1, is introduced.

$$\delta_{i,eq}^t = \frac{2 G_c}{\sigma_{i,eq}^0} \tag{6}$$

In Equation (6), $\sigma_{i,eq}^0$ and $\delta_{i,eq}^0$ are, respectively, the equivalent stress and displacement at the Hashin's limit condition (point A), and G_c is the material fracture toughness, i.e., the area of the global triangle in Figure 1. In Table 1, the relations adopted to evaluate the equivalent stress and displacement are introduced.

Table 1. Equivalent stress and displacement definitions.

Load Condition	Equivalent Stress	Equivalent Displacement
Fibre Tension	$\frac{L_c(\langle \sigma_{11}\rangle\langle\varepsilon_{11}\rangle + \sigma_{12}\cdot\varepsilon_{12})}{\delta_{ft,eq}}$	$L_c\sqrt{\langle\varepsilon_{11}\rangle^2 + \varepsilon_{12}^2}$
Fibre Compression	$\frac{L_c\langle-\sigma_{11}\rangle\langle-\varepsilon_{11}\rangle}{\delta_{fc,eq}}$	$L_c\langle-\varepsilon_{11}\rangle$
Matrix Tension	$\frac{L_c(\langle \sigma_{22}\rangle\langle\varepsilon_{22}\rangle + \sigma_{12}\cdot\varepsilon_{12})}{\delta_{mt,eq}}$	$L_c\sqrt{\langle\varepsilon_{22}\rangle^2 + \varepsilon_{12}^2}$
Matrix Compression	$\frac{L_c(\langle-\sigma_{22}\rangle\langle-\varepsilon_{22}\rangle + \sigma_{12}\cdot\varepsilon_{12})}{\delta_{mc,eq}}$	$L_c\sqrt{\langle-\varepsilon_{22}\rangle^2 + \varepsilon_{12}^2}$

In Table 1, L_C and $\langle\rangle$ are, respectively, the element characteristic length and the Macauley bracket operator [45].

3. Fuselage Barrel Test Case: Geometrical Description and Finite Element Model

As already mentioned, in this paper, a composite fuselage barrel has been adopted as a numerical test case to assess the influence of the fracture toughness characteristics on the mechanical behaviour of a composite structure undergoing a crash event. In this paragraph, the numerical test case is introduced by providing details on the geometry, materials, boundary conditions and on the finite element discretisation.

3.1. Geometrical Description

The geometry of the fuselage section adopted as a test case for the impact numerical analyses is shown in Figure 2. The considered fuselage barrel does not take into account geometric variations related to the tail plane, nose and wing attachments. This assumption has allowed to neglect the stress concentration effects at a global level while stress concentration effects were considered at sub-components level.

The fuselage is composed of different structural sub-components (shown with different colours in Figure 2). The overall dimensions of the fuselage section are reported in Figure 2, together with an indication on the impact angle between the fuselage and the impacted rigid plane, which has been set to 3 degrees.

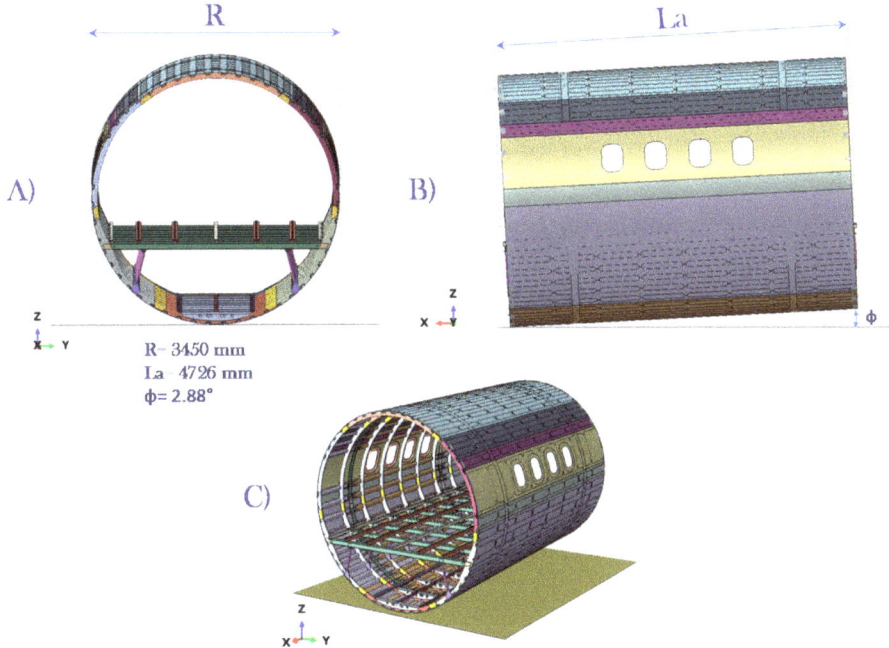

Figure 2. Fuselage section components. (**A**): Frontal view. (**B**): Lateral view. (**C**): Isometric view.

3.2. Material System

Figure 3 allows to highlight the material systems adopted for the different subcomponents of the fuselage. Three different material systems were adopted: namely a fibre-reinforced composite material system with long unidirectional fibres (UNIDIRECTIONAL UD), a woven fabric material system and AL2025 aluminium. The related material mechanical properties are introduced in the following Tables 2–4.

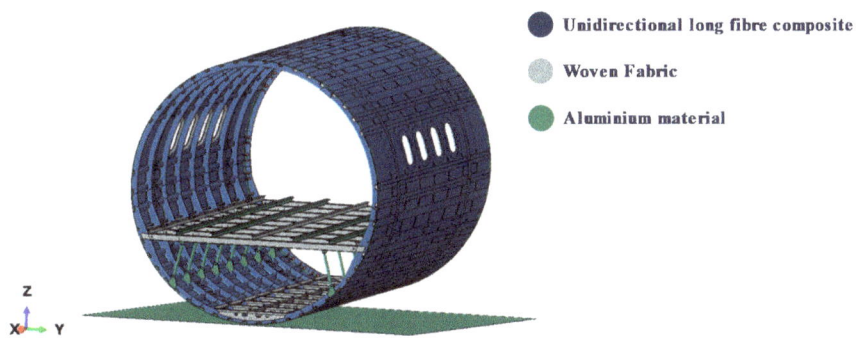

Figure 3. Material components.

Table 2. UNIDIRECTIONAL (UD) material system properties.

Fibre Composite Material	
Young's Modulus, E_{11} [MPa]	137,500
Young's Modulus, E_{22} [MPa]	8200
Shear Modulus, G_{12} [MPa]	3950
Shear Modulus, G_{13} [MPa]	3950
Shear Modulus, G_{23} [MPa]	3950
Poisson's Ratio, $\nu_{12} = \nu_{13} = \nu_{23}$	0.35
Fibre Tensile Strength F_{1t} [MPa]	1890
Fibre Compressive Strength F_{1c} [MPa]	1008
Matrix Tensile Strength F_{2t} [MPa]	86.5
Matrix Compressive Strength F_{2c} [MPa]	112
In-Plane Shear Strength, S_{12} [MPa]	95
Out-Plane Shear Strength, S_{23} [MPa]	100
Density [ton/mm^3]	$1.9\,e^{-9}$
Ply Thickness [mm]	0.129

Table 3. Woven fabric material system properties.

Woven Fabric Material	
Young's Modulus, E_{11} [MPa]	55,000
Young's Modulus, E_{22} [MPa]	55,000
Shear Modulus, G_{12} [MPa]	3363
Shear Modulus, G_{13} [MPa]	3363
Shear Modulus, G_{23} [MPa]	3363
Poisson's Ratio, $\nu_{12} = \nu_{13} = \nu_{23}$	0.30
Fibre Tensile Strength F_{1t} [MPa]	650
Fibre Compressive Strength F_{1c} [MPa]	650
Matrix Tensile Strength F_{2t} [MPa]	650
Matrix Compressive Strength F_{2c} [MPa]	650
In-Plane Shear Strength, S_{12} [MPa]	150
Out-Plane Shear Strength, S_{23} [MPa]	150
Density [ton/mm^3]	$1.97\,e^{-9}$
Ply Thickness [mm]	0.25

Table 4. Aluminium Al 2024 mechanical properties

Aluminium Al2025	
Young's Modulus, E [MPa]	70,000
Poisson's ratio, $\nu_{12} = \nu_{13} = \nu_{23}$	0.33
Yield stress [MPa]	369
Ultimate Tensile Stress [MPa]	469
Density [ton/mm^3]	$2.7\,e^{-9}$

For the aluminium plates, a thickness of 8 mm was chosen.

Table 5 shows the stacking sequences for the composites' made sub-components. A discrete coordinate system has been introduced in the numerical model for each composite barrel sub-component. Fibres are oriented according to the fuselage longitudinal direction, while the normal axis has been chosen according to the element normal direction.

As already mentioned, three different material systems configurations, characterised by different fracture toughness energies, were considered for the simulations in order to assess the influence of fracture characteristics on the evolution of the damage and hence on the dynamic response of the whole fuselage.

Table 5. Sub-components stacking sequences.

	Stacking Sequence
Skin	$[90/45/0/45]_s$
Stringer	$[45/45/0/0/90/0/0]_s$
Frames	$[90/45/0/45/-45/90/45/45/0/-45/45/-45]_s$
Components in Woven Fabric Material	$[0/45/0/-45]_s$

Each configuration, identified with a number (I,II,III), is associated to different values of the fracture toughness energies for the matrix and fibre under tensile and compression conditions for both the UNIDIRECTIONAL UD and the woven composite material systems. Moreover, tensile and compression fracture energy values for the woven fabric were assumed identical for all the considered configurations.

As it can be observed from Table 6, Configuration II can be considered the most toughened configuration while configurations I is characterised by the lowest toughness. Finally, configuration III is characterised by an intermediate/low toughness. Moreover, for all the material configurations, the degradation of the matrix toughness in both tension and compression was not taken into account. In fact, the small matrix toughness variations during propagation have a significant effect only on inter-laminar damages which have not been considered in the frame of the presented application.

Table 6. Fracture toughness energies.

		G_{1C}^T [kJ/m^2]	G_{1C}^C [kJ/m^2]	G_{2C}^T [kJ/m^2]	G_{2C}^C [kJ/m^2]
I	UNIDIRECTIONAL UD	24	10	3	6
I	Woven	8	8	8	8
II	UNIDIRECTIONAL UD	44	16	3	6
II	Woven	12	12	12	12
III	UNIDIRECTIONAL UD	26	12	3	6
III	Woven	6	6	6	6

3.3. Finite Element Description

The finite element model of the fuselage section is introduced in Figure 4A,B. The entire model consists of 1,976,157 nodes and 995,858 elements. The elements used for the aluminium components are discretised with three-dimensional elements with eight nodes and reduced integration scheme available in the Abaqus library (SC8R); the elements have a dimension of about 10 mm × 10 mm. Planar shell elements with rigid body constraint were adopted to model the impact plate between representing the impacted rigid ground. With reference to the computational grid of the composite components of the fuselage section, eight nodes continuum shell elements with a reduced integration scheme were adopted. These elements are general-purpose shells allowing finite membrane deformation and large rotations, and thus, they are suitable for nonlinear geometric analysis. These elements include the effects of transverse shear deformations and the effects of thickness change. Connections between the sub-components were simulated with tie-constraints allowing to couple a master surface with a slave one by taking into account with all the degrees of freedom. The connections between the vertical stanchions and the passenger floor/frames were simulated by fastener elements allowing to introduce breakage criteria based on maximum force in each reference direction.

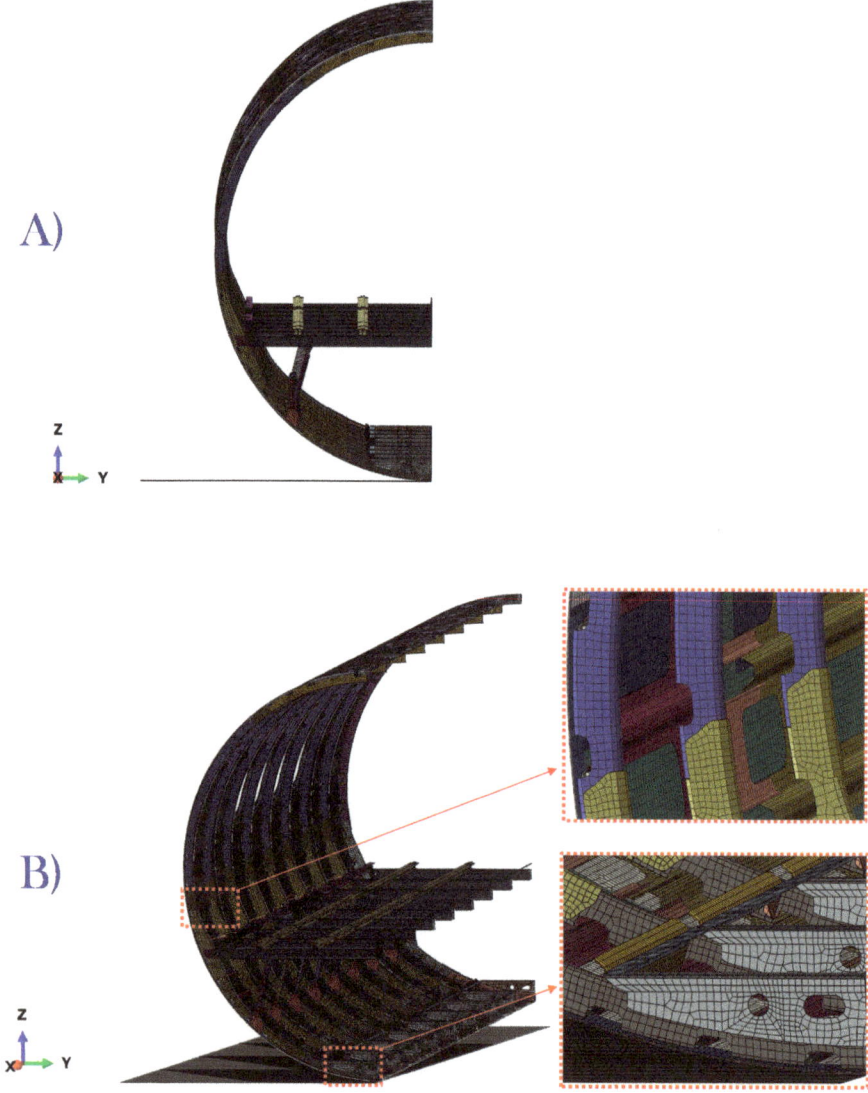

Figure 4. Fuselage mesh model subsection: (**A**) Frontal view. (**B**) Isometric view, Stringer view, Stringer section.

3.4. Boundary Conditions

The fuselage section, considered in the frame of the numerical simulations, has a mass of approximately 940 kg with several mass points added to simulate external non-structural components: row of seats and occupants and balancing mass. The balancing mass has been opportunely chosen in order to create an unbalancing effect between the right and the left side of the fuselage barrel during the simulated drop test. This unbalancing effect has been introduced in order to assess the effects of the variations in material toughness on lateral damage distribution during the impact simulation. A general contact type "all-with-self" available in Abaqus/Expicit was adopted with a friction coefficient of 0.3.

In addition, the weight of the three seats and the two dummies was applied in the centre of gravity of the seats and rigidly connected in the points of attachment of the seats to the passenger floor. Figure 5 shows the points of connection of the various masses.

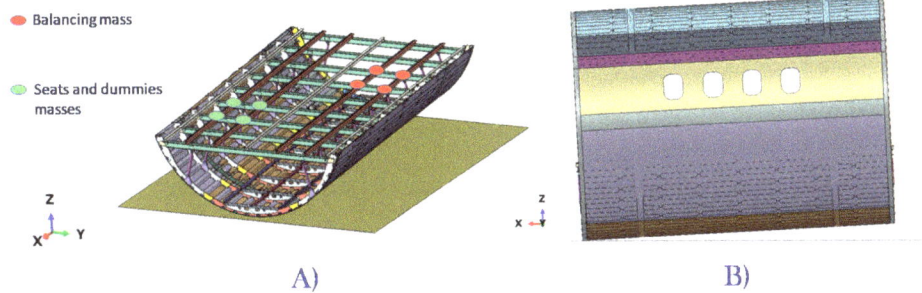

Figure 5. Added masses and relative positions. (**A**) Passenger floor isometric view. (**B**) Lateral view.

The impact simulation of the fuselage section with the ground has been carried out by considering a rigid plane bounded in space and by applying an initial velocity to the entire fuselage section equal to 9900 mm/s. The initial velocity has been evaluated by considering the fuselage section dropping from a height of 5000 mm with respect to the rigid ground surface.

4. Numerical Results

In this section, the numerical results obtained for the three analysed configurations according to Table 6 are introduced. The different configurations were compared in terms of maximum displacement along the drop direction and in terms of qualitative deformation of the whole fuselage section.

For the three different configurations, the history of the vertical displacement of the control points A and B, as identified in Figure 6, was evaluated. The control points are positioned on the crossbeams of the passenger floor in the middle of the beams where the seat rails are fixed, as shown in Figure 6.

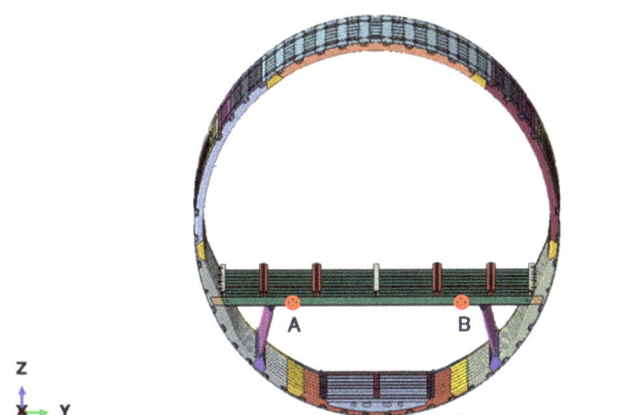

Figure 6. Point displacement positions.

4.1. Configuration I

Figure 7A shows the vertical displacements of the control points A and B obtained for the configuration I which is the configuration with the lowest toughness. As shown in Figure 7A, for this configuration, the maximum value of the vertical displacement at control point was reached (about 250 mm). Moreover, a discrepancy in the displacements between the two points can be noticed. This is generated by the aforementioned unbalancing effect of the balancing mass between the left and the right side of the fuselage. However, the configuration with the lowest toughness is able to ensure a quasi-symmetric distribution of the damage, and deformations between the left and the right side of the fuselage are small despite of the unbalancing effect of the introduced mass.

Figure 7B–D shows the deformed structure at the maximum point of deflection. In particular, Figure 7B shows the frontal view of the deformed fuselage section. Figure 7C shows the isometric view, while Figure 7D shows a zoomed view of the deformed and broken structure in the cargo area. Indeed, the cargo area can be considered the most stressed one during a crash event. This consideration leads to pay special attention to this area in a crashworthy design in order to maximise the dissipation of the kinetic energy.

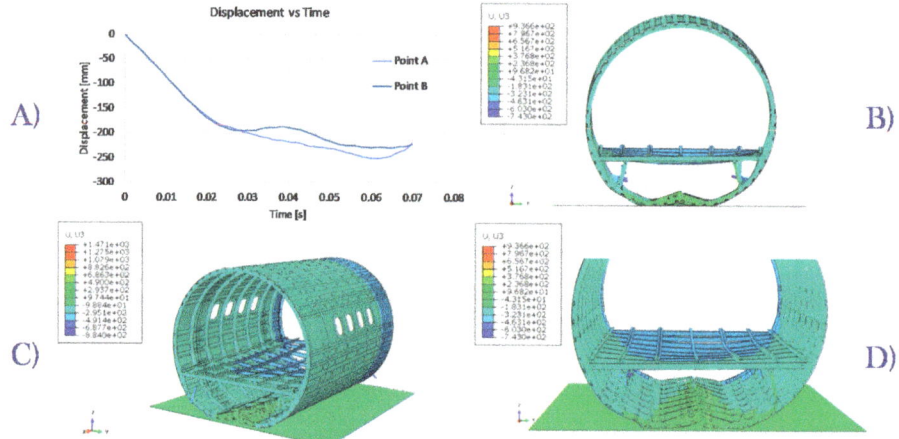

Figure 7. (**A**) Displacement vs. time. (**B**) Frontal view. (**C**) Isometric view. (**D**) Cargo area details.

4.2. Configuration II

Figure 8A shows the vertical displacements of the control points A and B obtained for the configuration II which is the configuration with the highest toughness. As shown in Figure 8A, for this configuration, the minimum value of the vertical displacement at the control point is reached (about 220 mm). Moreover, the maximum discrepancy in the displacements between the two points (A and B) can be noticed. Indeed, the configuration with the highest toughness amplifies the asymmetry in damage distribution and deformations between the left and the right side of the fuselage due to the unbalancing effect of the introduced mass.

Figure 8B–D shows the deformed structure at the maximum point of deflection. In particular, Figure 8B shows the frontal view of the deformed fuselage section. Figure 8C shows the isometric view, while Figure 8D shows a zoomed view of the deformed and broken structure in the cargo area. Indeed, due to the toughening effect, the damage seems to be less distributed in the whole fuselage and more concentrated in the cargo area if compared to configuration I (Figure 7D).

Appl. Sci. **2020**, *10*, 2019

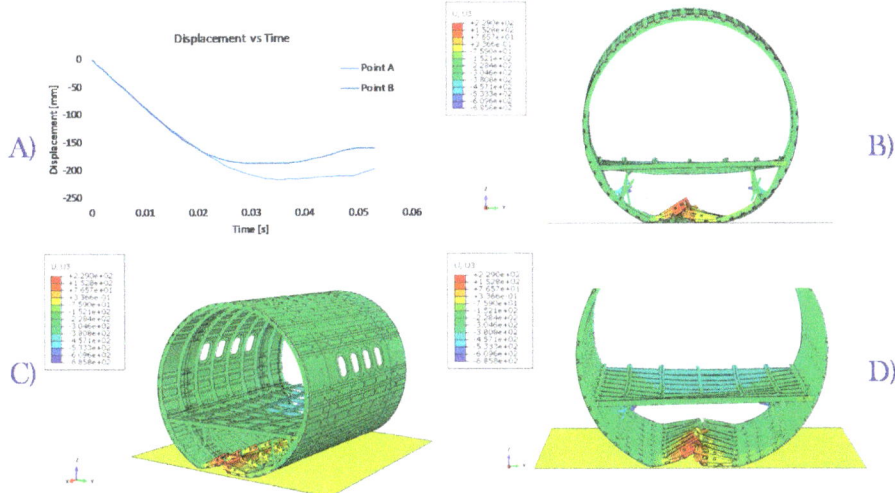

Figure 8. (**A**) Displacement vs. time. (**B**) Frontal view. (**C**) Isometric view. (**D**) Cargo area details.

4.3. Configuration III

Figure 9A shows the vertical displacements of the control points A and B obtained for the configuration III, which is the configuration with the intermediate low toughness. As shown in Figure 9A, for this configuration, as expected, the minimum value of the vertical displacement at control point is almost identical to the one observed for configuration I (about 250 mm). The same considerations can be repeated for the magnitude of the discrepancy in the displacements between the two points (A and B) due to the unbalancing effect of the introduced mass.

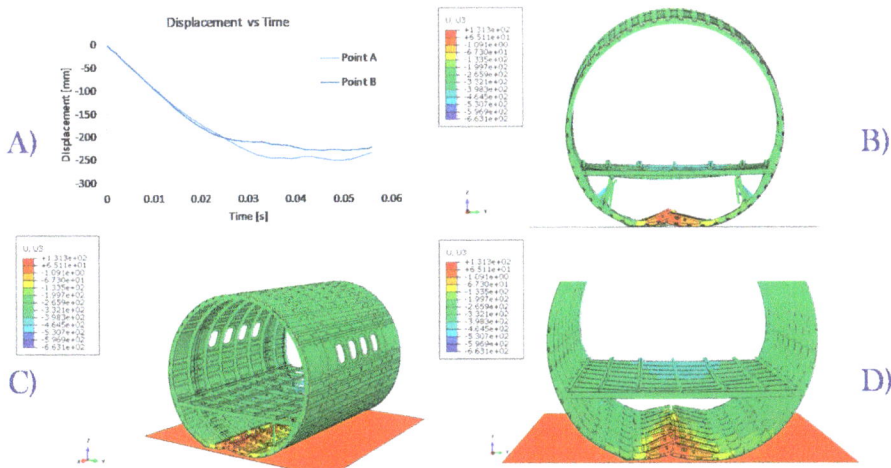

Figure 9. (**A**) Displacement vs. time. (**B**) Frontal view. (**C**) Isometric view. (**D**) Cargo area details.

Figure 9B–D highlights the damaged structure with particular focus on the maximum deflection point. Moreover, in Figure 9B, a frontal view of the deformed fuselage section is reported. In Figure 9C, an isometric view of the considered fuselage section is exhibited, while a zoomed view of the damaged cargo area of the structure is shown in Figure 9D. Indeed, for configuration III, the cargo area seems to have a reduced damaged concentration if compared with configuration II.

4.4. Configuration Comparison

In order to appreciate the effect of the fracture toughness on the mechanical behaviour of the fuselage undergoing the crash event, the displacement patterns were superimposed in Figure 10. Indeed, in Figures 10A and 10B, respectively, the vertical displacements of control point A and point B obtained for the three analysed configurations were superimposed. As shown in Figure 10A,B, for all the configurations, the slopes at the beginning of the crash event are identical straight lines. In fact, at the beginning of the crash event, although the structure deforms and the damage has started, the effects of the damage evolution were found almost negligible. Indeed, the effects of the in-plane fracture toughness were found more significant later during the crash event almost at the maximum deflection.

Configurations I and III, characterised by a similar value of fracture energies for both materials, both show a maximum deflection of about 250 mm for control point A and about 230 mm for control point B. For both these configurations, a significant variation of stiffness has been found at 0.022 s for control point A (Figure 10A) and 0.027 s for control point B (Figure 10B). On the other hand, configuration II, characterised by the highest toughness values, shows a maximum deformation of about 220 and 180 mm, respectively, for control points A and B. These maximum values, as expected, are lower if compared to configurations I and III.

Finally, as already remarked, the configuration with the highest toughness, differently from configurations I and III, amplifies the asymmetrical distribution of the damage between the left side (control point A) and the right side (control point B) of the fuselage induced by the unbalancing effect of the introduced mass.

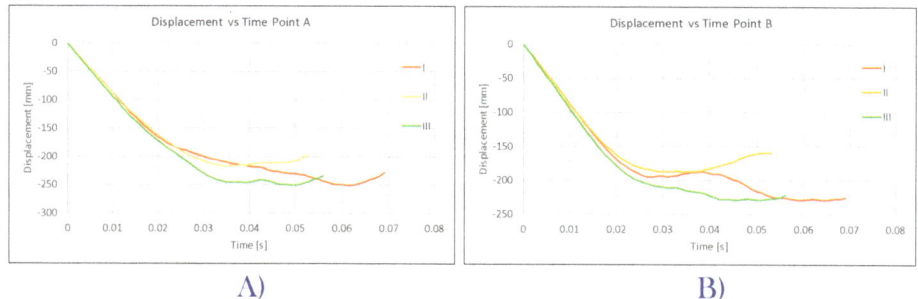

Figure 10. Displacement vs. time for the three considered configurations: (**A**) Point A location. (**B**) Point B location.

Moreover, in Figure 11, a comparison of damage energy dissipation is reported for the three fracture toughness energies. Configuration II dissipates a lower amount of energy if compared to the other configurations. Configurations I and III dissipate a relevant amount of the total energy as damage energy; consequently, a higher value of the maximum displacement was found for this configurations, compared to configuration II.

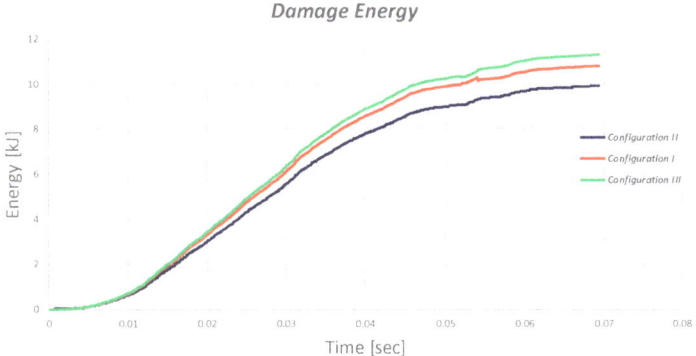

Figure 11. Damage energy dissipation.

5. Conclusions

In the presented work, a numerical study on the influence of the in-plane toughness on the dynamic behaviour of a complex composite fuselage barrel, undergoing a crash event, was attempted. In order to evaluate the effects of intralaminar fracture toughness, three different material systems characterised by different toughness (low toughness, high toughness and intermediate/low toughness) were considered, both for the unidirectional fibre-reinforced composite and for the woven fabric one.

As a result of the performed numerical study, a relevant influence of the in-plane toughness on the global dynamic response of the fuselage barrel was found. Actually, the configuration characterised by the highest toughness showed the lowest vertical deflection and the most significant damage in the cargo area. Moreover, this configuration seems to be more sensitive to the unbalancing of the mass in the lateral direction producing the most significant asymmetry in damage distributions between the left and the right side of the fuselage. On the other hand, the configurations characterised by the lowest in-plane toughness showed the maximum vertical deflection and a more distributed damage evolution in the whole structure, leading to less significant damage in the cargo area. Finally, a low sensitivity to the unbalancing of the mass in the later direction was found.

Author Contributions: All authors equally contribute to this work. All authors have read and agreed to the published version of the manuscript.

Funding: This research received no external funding.

Conflicts of Interest: The authors declare no conflict of interest.

References

1. Heimbs, S.; Vogt, D.; Hartnack, R.; Schlattmann, J.; Maier, M. Numerical simulation of aircraft interior components under crash loads. *Int. J. Crashworthiness* **2008**, *13*, 511–521. [CrossRef]
2. Obergefell, L.A.; Gardner, T.R.; Kaleps, I.; Fleck, J.T. Articulated total body model Enhancements. In *AAMRL-TR-88-043 User's Guide*; Armstrong Aerospace Medical Research Laboratory: Dayton, HO, USA, 1988; p. 2.
3. Ruan, J.; Zhou, C.; Khalil, T.; King, A. *Techniques and Applications of Finite Element Analysis of the Biomechanical Response of the Human Head to Impact*; CRC Press: Boca Raton, FL, USA, 2000.
4. De Jager, M.K.J.; Sauren, A.A.H.J.; Thunnissen, J.G.M.; Wismans, J.S.H.M. A three-dimensional head-neck model: Validation for frontal and lateral impacts. *SAE Trans.* **1994**, *103*, 1660–1676.
5. Mertz, H.J.; Prasad, P.; Nusholtz, G. Head Injury Risk Assessment for Forehead Impacts. *SAE Tech. Pap. Ser.* **1996**, *105*, 26–46.

6. Hashemi, S.M.R.; Walton, A.C. A systematic approach to aircraft crashworthiness and impact surface material models. *Proc. Inst. Mech. Eng. Part G J. Aerosp. Eng.* **2000**, *214*, 265–280. [CrossRef]
7. Kindervater, C.M. The crashworthiness of composite aerospace structures. In Proceedings of the Workshop the Crashworthiness of Composite Transportation Structures, Crowthorne, UK, 3 October 2002.
8. Carden, H.D.; Boitnott, R.L.; Fasanella, E.L. Behavior of composite/metal aircraft structual elements and components under crash type loads. In Proceedings of the 17th Congress of the International Council of Aeronautical Sciences, Stockholm, Sweden, 9–14 September 1990.
9. Jackson, K.E.; Boitnott, R.L.; Fasanella, E.L.; Jones, L.E.; Lyle, K.H. A history of full-scale aircraft and rotorcraft crash testing and simulation at NASA Langley research center. In Proceedings of the 4th Triennial International Aircraft Fire and Cabin Safety Research Conference, Lisbon, Portugal, 15–18 November 2004.
10. Logue, T.V.; McGuire, R.J.; Reinhardt, J.W.; Vu, T.V. *Vertical Drop Test of a Narrow-Body Fuselage Section with Overhead Stowage Bins and Auxiliary Fuel Tank System on Board*; Report DOT/FAA/CT-94/116; U.S. Department of Transportation; Federal Aviation Administration: Atlantic City, NJ, USA, 1995.
11. Abramowitz, A.; Smith, T.G.; Vu, T. *Vertical Drop Test of a Narrow-Body Transport Fuselage Section with a Conformable Auxiliary Fuel Tank Onboard*; Report DOT/FAA/AR-00/56; U.S. Department of Transportation; Federal Aviation Administration: Atlantic City, NJ, USA, 2000.
12. Byar, A.D.; Ko, J.; Rassaian, M. 737 fuselage section drop test simulation using LS-Dyna finite element method, ICRASH 2008. In Proceedings of the International Crashworthiness Conference, Kyoto, Japan, 22–25 June 2008.
13. Kumakura, I.; Minegishi, M.; Iwasaki, K.; Shoji, H.; Yoshimoto, N.; Terada, H.; Sashikuma, H.; Isoe, A.; Yamaoka, T.; Katayama, N.; et al. Vertical Drop Test of a Transport Fuselage Section. In Proceedings of the SAE World Aviation Congress, Phoenix, AZ, USA, 6–8 May 2002.
14. Kumakura, I.; Minegishi, M.; Iwasaki, K.; Shoji, H.; Miyaki, H.; Yoshimoto, N.; Sashikuma, H.; Katayama, N.; Isoe, A.; Hayashi, T.; et al. Summary of Vertical Drop Tests of YS-11 Transport Fuselage Sections. In Proceedings of the SAEWorld Aviation Congress, Montreal, QC, Canada, 9–11 September 2003.
15. Deletombe, E.; Delsart, D.; Fabis, J.; Langrand, B.; Ortiz, R. Recent developments in computer modelling, materials characterisation and experimental validation with respect to crash dynamics. In Proceedings of the 4th Triennial International Aircraft Fire and Cabin Safety Research Conference, Lisbon, Portugal, 15–18 November 2004.
16. Waimer, M.; Kohlgr¨uber, D.; Hachenberg, D.; Voggenreiter, H. The kinematics model—A numerical method for the development of a crashworthy composite fuselage design of transport aircraft. In Proceedings of the 6th Triennial International Aircraft Fire and Cabin Safety Research Conference, Atlantic City, NJ, USA, 25–28 October 2010.
17. Fen, Z.; Hao, P.; Zou, T. Research development of crashworthiness simulation evaluation on civil aircraft. *Procedia Eng.* **2011**, *17*, 286–291.
18. Kindervater, C.M.; Kohlgr¨uber, D.; Johnson, A. Composite vehicle structural crashworthiness—A status of design methodology and numerical simulation techniques. *Int. J. Crashworthiness* **1999**, *4*, 213–230. [CrossRef]
19. Adams, A.; Lankarani, H.M. A modern aerospace modeling approach for evaluation of aircraft fuselage crashworthiness. *Int. J. Crashworthiness* **2003**, *8*, 401–413. [CrossRef]
20. Byar, A.; Awerbuch, J.; Lau, A.; Tan, T. Finite element simulation of a vertical drop test of a Boeing 737 fuselage section. In Proceedings of the 3rd Triennial International Aircraft Fire and Cabin Safety Research Conference, Atlantic City, NJ, USA, 1–4 August 2001.
21. Jackson, K.E.; Fasanella, E.L. Crash simulation of a vertical drop test of a B737 fuselage section with overhead bins and luggage. In Proceedings of the 3rd Triennial International Aircraft Fire and Cabin Safety Research Conference, Atlantic City, NJ, USA, 1–4 August 2001.
22. Rassaian, M.; Byar, A.; Ko, J. Numerical simulation of 737 fuselage section drop test. In Proceedings of the NAFEMS World Congress, Crete, Greece, 16–19 June 2009.
23. Wiggenraad, J.; Michielsen, A.; Santoro, D.; Lepage, F.; Kindervater, C.; Beltrán, F.; Al-Khalil, M. Finite element methodologies development to simulate the behaviour of composite fuselage structure and correlation with drop test. *Air Space Eur.* **2001**, *3*, 228–233. [CrossRef]
24. Middendorf, P.; Heimbs, S. Crash simulation of composite structures. *Compos. Mater* **2007**, *2*, 18–22.
25. Califano, A. *Modelling the Fatigue Behavior of Composites under Spectrum Loading*; AIP Publishing: University Park, MD, USA, 2018.

26. Riccio, A.; Damiano, M.; Raimondo, A.; Di Felice, G.; Sellitto, A. A fast numerical procedure for the simulation of inter-laminar damage growth in stiffened composite panels. *Compos. Struct.* **2016**, *145*, 203–216. [CrossRef]
27. Riccio, A.; Di Costanzo, C.; Di Gennaro, P.; Sellitto, A.; Raimondo, A. Intra-laminar progressive failure analysis of composite laminates with a large notch damage. *Eng. Fail. Anal.* **2017**, *73*, 97–112. [CrossRef]
28. Riccio, A.; Sellitto, A.; Saputo, S.; Russo, A.; Zarrelli, M.; LoPresto, V. Modelling the damage evolution in notched omega stiffened composite panels under compression. *Compos. Part B Eng.* **2017**, *126*, 60–71. [CrossRef]
29. Liu, X.; Guo, J.; Bai, C.; Sun, X.; Mou, R. Drop test and crash simulation of a civil airplane fuselage section. *Chin. J. Aeronaut.* **2015**, *28*, 447–456. [CrossRef]
30. Kuraishi, A.; Tsai, S.W.; Liu, K.K. A progressive quadratic failure criterion, part B. *Compos. Sci. Technol.* **2002**, *62*, 1683–1695. [CrossRef]
31. Spahn, J.; Andrä, H.; Kabel, M.; Muller, R. A multiscale approach for modeling progressive damage of composite materials using fast Fourier transforms. *Comput. Methods Appl. Mech. Eng.* **2014**, *268*, 871–883. [CrossRef]
32. Pineda, E.J.; Waas, A.M.; Bednarcyk, B.A.; Collier, C.S.; Yarrington, P.W. Progressive damage and failure modeling in notched laminated fiber reinforced composites. *Int. J. Fract.* **2009**, *158*, 125–143. [CrossRef]
33. Basu, S.; Waas, A.; Ambur, D.R. Prediction of progressive failure in multidirectional composite laminated panels. *Int. J. Solids Struct.* **2007**, *44*, 2648–2676. [CrossRef]
34. Apalak, Z.G.; Apalak, M.K.; Genç, M.S. Progressive Damage Modeling of an Adhesively Bonded Unidirectional Composite Single-lap Joint in Tension at the Mesoscale Level. *J. Thermoplast. Compos. Mater.* **2006**, *19*, 671–702. [CrossRef]
35. Ridha, M.; Wang, C.H.; Chen, B.; Tay, T. Modelling complex progressive failure in notched composite laminates with varying sizes and stacking sequences. *Compos. Part A: Appl. Sci. Manuf.* **2014**, *58*, 16–23. [CrossRef]
36. Pineda, E.J.; Waas, A. Modelling progressive failure of fibre reinforced laminated composites: Mesh objective calculations. *Aeronaut. J.* **2012**, *116*, 1221–1246. [CrossRef]
37. Pineda, E.J.; Waas, A.M. Numerical implementation of a multiple-ISV thermodynamically-based work potential theory for modeling progressive damage and failure in fiber-reinforced laminates. *Int. J. Fract.* **2013**, *182*, 93–122. [CrossRef]
38. Laurin, F.; Carrere, N.; Maire, J.-F. A multiscale progressive failure approach for composite laminates based on thermodynamical viscoelastic and damage models. *Compos. Part A Appl. Sci. Manuf.* **2007**, *38*, 198–209. [CrossRef]
39. Lapczyk, I.; Hurtado, J.A. Progressive damage modeling in fiber-reinforced materials. *Compos. Part A Appl. Sci. Manuf.* **2007**, *38*, 2333–2341. [CrossRef]
40. Murakami, S.; Kamiya, K. Constitutive and damage evolution equations of elastic-brittle materials based on irreversible thermodynamics. *Int. J. Mech. Sci.* **1997**, *39*, 473–486. [CrossRef]
41. Donadon, M.V.; Falzon, B.G.; Iannucci, L.; Hodgkinson, J.M. Intralaminar toughness characterisation of unbalanced hybrid plain weave laminates. *Compos. Part A Appl. Sci. Manuf.* **2007**, *38*, 1597–1611. [CrossRef]
42. Pappas, G.; Botsis, J. Intralaminar fracture of unidirectional carbon/epoxy composite: Experimental results and numerical analysis. *Int. J. Solids Struct.* **2016**, *85*, 114–124. [CrossRef]
43. Iwamoto, M.; Ni, Q.-Q.; Fujiwara, T.; Kurashiki, K. Intralaminar fracture mechanism in unidirectional CFRP composites. *Eng. Fract. Mech.* **1999**, *64*, 721–745. [CrossRef]
44. Hashin, Z. Failure Criteria for Unidirectional Fiber Composites. *J. Appl. Mech.* **1980**, *47*, 329–334. [CrossRef]
45. Abaqus Theory Manual Version 6.16. Dassault System France 2016. Available online: http://www.abaqus.com (accessed on 5 April 2019).

© 2020 by the authors. Licensee MDPI, Basel, Switzerland. This article is an open access article distributed under the terms and conditions of the Creative Commons Attribution (CC BY) license (http://creativecommons.org/licenses/by/4.0/).

Article

Numerical-Experimental Investigation into the Tensile Behavior of a Hybrid Metallic–CFRP Stiffened Aeronautical Panel

Andrea Sellitto [1],*, Salvatore Saputo [1], Angela Russo [1], Vincenzo Innaro [2], Aniello Riccio [1], Francesco Acerra [2] and Salvatore Russo [2]

[1] Department of Engineering, University of Campania "Luigi Vanvitelli", via Roma 29, 81031 Aversa, Italy; salvatore.saputo@unicampania.it (S.S.); angela.russo@unicampania.it (A.R.); aniello.riccio@unicampania.it (A.R.)
[2] Leonardo Company SpA, Via dell'Aeronautica, 80038 Pomigliano d'Arco, Italy; vincenzo.innaro@finmeccanica.com (V.I.); francesco.acerra@leonardocompany.com (F.A.); salvatore.russo@leonardocompany.com (S.R.)
* Correspondence: andrea.sellitto@unicampania.it; Tel.: +39-081-5010-407

Received: 20 February 2020; Accepted: 5 March 2020; Published: 10 March 2020

Abstract: In this work, the tensile behavior of a hybrid metallic–composite stiffened panel is investigated. The analyzed structure consists of an omega-reinforced composite fiber-reinforced plastic (CFRP) panel joined with a Z-reinforced aluminum plate by fasteners. The introduced numerical model, able to simulate geometrical and material non-linearities, has been preliminary validated by means of comparisons with experimental test results, in terms of strain distributions in both composite and metallic sub-components. Subsequently, the inter-laminar damage behavior of the investigated hybrid structure has been studied numerically by assessing the influence of key structural subcomponents on the damage evolution of an artificial initial debonding between the composite skin and stringers.

Keywords: hybrid structures; metallic/composite joints; plasticity; damage propagation; FEM

1. Introduction

In recent years, the use of composite fiber-reinforced plastic (CFRP) for the manufacturing of aerospace primary structures has unquestionably increased. However, despite the undeniable advantages in terms of weight related to composites, metal parts are still widely used because of the criticalities related to the damage management of the new outstanding composites [1–6]. Therefore, for several structural components, metal/composite hybrid solutions are adopted, due to the effective reduction in weight and costs without strong compromises in terms of safety [7,8].

In the literature, the damage behavior of composite materials has been widely investigated, pointing out the effects in terms of induced damage of impacts with foreign objects [9–14], of the manufacturing or the assembly processes [15–18] and, finally, the effects of service loading conditions.

Indeed, in order to exploit the advantages related to the adoption of composite–metal hybrid solutions, it is mandatory to understand the issues related to the manufacturing and the joining of these different materials [19].

In [20], a comprehensive review on the methods, commonly adopted, to join CFRP and aluminum alloys parts was presented. Among the several methods described, adhesive joints, bolts, and self-piercing rivets (SPR) are, of course, the most affordable. In particular, bonded joints have been found to be the method most adopted to join CFRPs and aluminum alloys parts. Indeed, bonded joints avoid undesirable stress concentrations and does not require an intrusive application, being at the same time weight saving with respect to other joining techniques. However, since the bonding is an irreversible process, the separation of joined components often results in critical material damage, as demonstrated by several investigations on bonded joints found in the literature [21–25]. A CFRP/aluminum double lap-bonded joint was presented in [26] where a non-linear cohesive material model was used to investigate the bonded interface behavior. In [27], the strain distribution in CFRP/aluminum double lap joints was experimentally and numerically investigated. The surface strain within the adhesive layer was measured by using high-magnification Moiré interferometry. Hence, the spatial variation of longitudinal, peel, and shear strains was evaluated in detail and compared to numerical predictions.

Moreover, due to the different thermal expansion coefficients of metal and composite sub-components, the effects of the thermal residual stress in terms of curvature of the composite/metal assembly should be addressed. This issue was investigated in [28], where a thermo-chemical-mechanical constitutive model was used to simulate the curing process of the composites and to evaluate the thermal residual stresses of a metal/CFRP plate. In [29], the mode I and II interfacial fracture behavior of a titanium/CFRP bonded joint was studied by considering the bending-extension coupling induced by the presence of the aluminum beams and the manufacturing-induced residual thermal stresses.

Bolts and rivets are a common alternative to bonded joints [30–35]. The bolt is inserted in a uniaxial common hole in components to be joined. However, the hole drilling may generate damage and defects in the CFRP components, which can evolve due to the service load, affecting the life of the components themselves. In [36], an experimental study on the manufacturing defects induced by drilling aluminum/CFRP stacks was performed, aimed to determine the critical geometric parameters of a carbide drill in order to reduce the fragmentation of the metallic chips while avoiding damages on the composite components. This issue was also addressed in [37], where nano-coated carbide drills were used to reduce the damages resulting in drilling multimaterials made of aluminum alloy and CFRP.

Additionally, a numerical and experimental investigation on the mechanical behavior of CFRP/aluminum bolted joints was introduced in [38], focusing on the effect of the environmental conditions (temperature and moisture) on the mechanical performances of the bolted joint. Hence, different tests were presented by considering extremely hot, hot and humid, and extremely cold environmental conditions.

In [39], the thermal effects on a single lap hybrid metal/composite bolted joint were investigated. Indeed, metal and CFRP are characterized by different thermal expansion coefficients, resulting in uneven load-deformation characteristics and three-dimensional stress field around the bolt hole.

The self-piercing rivet technique is used to join sheet materials [40,41]. Unlike bolted joints, this method does not require a pre-drilled hole, allowing faster joining performed in just one operation. However, this operation has a critical impact on the fatigue life and on the static strength of hybrid CFRP/aluminum components [41–43]. In particular, in [44] the possible damage induced by self-pierce riveting of CFRP with aluminum, which include delamination and fiber/matrix failures in the CFRP region, were investigated, and the influence of the damage on the joint strength was assessed. In [45], the tensile and fatigue properties of CFRP/aluminum self-pierce riveting joints were evaluated. Different joint configurations were investigated, and the influence of the rivets' geometrical parameters on the joint failure was studied.

In general, the choice of the particular joining technique strongly depends on the specific application. In order to take advantage of the benefits of the different joining techniques, hybrid solutions can be found whose performance is enhanced by combining adhesives with rivets or bolts [46–49]. In [50], an experimental investigation on single-lap joined CFRP laminate and AA2024-T6 aluminum sheet was introduced. Bonded, self-pierce riveting, and hybrid SPR/bonded joints were considered. In particular, an appreciable synergism of the bonded and self-pierce riveting joints could be observed in the hybrid solution, which exploited high resistance levels due to the bonding and the high failure energy due to the self-pierce riveting. In [51], a semi-analytical method was developed for the analysis of composite bolted/bonded single lap joints, under coupled in-plane and bending loads. The method was based on Mindlin and Timoshenko beam theories to evaluate the laminate and the bolt displacements, respectively. Several test cases were introduced, including analyses of bolted joints, bonded joints, and hybrid joints. Additional analyses were carried out for the hybrid configuration, considering a debonding in the bonded interface between the laminates.

A few works can be found in the literature on the mechanical behavior, including damage, of hybrid composite–metal structures, composed of differently joined subcomponents. An investigation on a complex bolted structure can be found in [52], where the mechanical behavior of a hybrid composite/aluminum wing-box was assessed. The structure was composed of a number of single-lap bolted joints with a titanium fastener, which were numerically studied by introducing a three-dimensional finite element model (FEM), able to account for material non-linearity such as progressive damage, and plasticity models, respectively, in the composite and metal regions. Two-node connector elements were used to simulate the fasteners behavior. Both twisting and bending loading conditions were considered, together with an applied increased temperature, in order to locally investigate the behavior of the bolted joints. Experimental results were used to validate the numerical model.

However, more investigations are mandatory to fully understand the effect of the joints on the mechanical behavior of a complex structure composed of different subcomponents, as well as the effects of key structural components on the damage behavior of hybrid complex structures.

Hence, in this work the mechanical behavior, including damage onset and evolution, of a tensile loaded hybrid composite/metal panel, is investigated. The investigated structure consists of an omega-reinforced carbon fiber-reinforced plastic (CFRP) panel joined with a Z-reinforced metal plate by fasteners. Numerical models have been introduced, able to take into account the geometrical and material non-linearity related to large displacements and deformations, such as the plasticity on the metal region and the intra-laminar damage on the composite components [53–55]. The introduced numerical models, based, respectively, on tie and fastener formulation to connect the sub-components, have been validated by means of comparisons with data, in terms of strain distribution on both composite and metallic regions, from an experimental tensile test. Finally, the stringer termination debonding has been investigated, by assessing the influence of key structural components on the inter-laminar damage behavior [56,57] of the hybrid structure. In Section 2, the analyzed test case is defined. In Section 3, the experimental setup is briefly described, while in Section 4 the numerical models are introduced. The numerical and experimental results are presented and critically analyzed in Section 5, while the stringer termination debonding is introduced and discussed in Section 6.

2. Test Case Description

The investigated structure consisted of a hybrid composite-metal reinforced panel. According to Figure 1, different regions can be identified. The first metal region consisted of an aluminum plate reinforced with Z-stringers, while the composite region consisted of a plate reinforced with omega stringers. A second metal region was placed between the plate. All the components were connected by means of fasteners. The hybrid structure was loaded in tension, up to 300 kN, by clamping one edge while applying an imposed tensile displacement on the opposite side. In Figure 1, the geometrical description of the structure, including the different materials regions and the boundary conditions, is reported, while in Figure 2 some views of the geometrical model are shown.

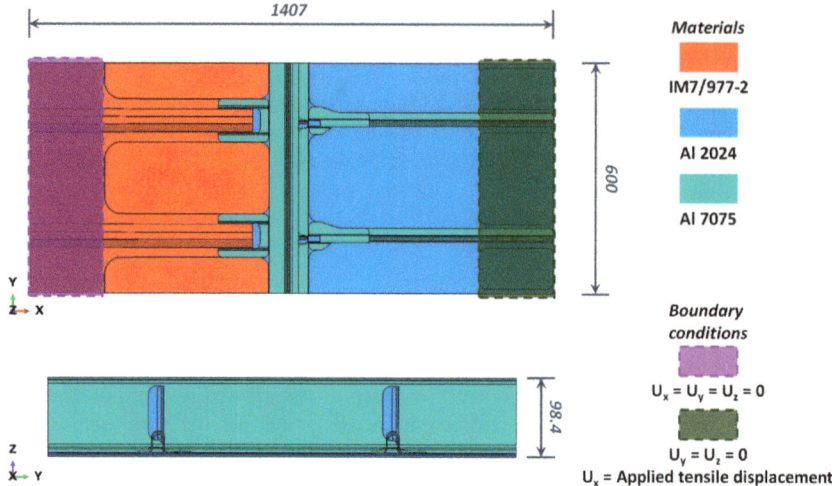

Figure 1. Geometrical description, materials, and boundary conditions of the investigated structure (dimensions in mm).

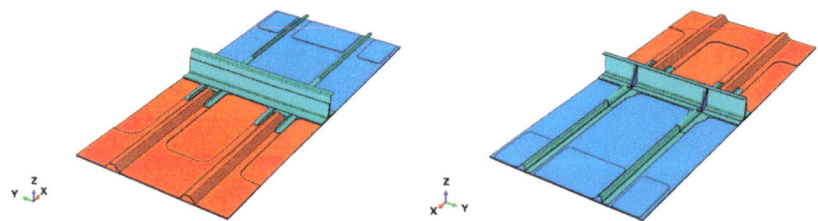

Figure 2. Geometrical model: isometric views.

Three different material systems were adopted to manufacture the panel: AA 2024 and AA 7075 (metal regions), and IMS/977-2 (CFRP region). The mechanical properties of the involved material systems are presented in Tables 1 and 2. In particular, Table 2 reports the mechanical properties of the IMS/977-2 material system including elastic properties, inter-laminar critical values (evaluated by means of the double cantilever beam and end notched flexure tests), and inter-laminar critical values.

Table 1. AA 2024 and AA 7075 mechanical properties.

AA 2024–T42		AA 7075–T62	
E [MPa]	ν [-]	E [MPa]	ν [-]
72,400	0.33	71,700	0.33

Table 2. IMS/977-2 mechanical properties [58].

IMS/977-2								
E_1 [MPa]	$E_2 = E_3$ [MPa]	$G_{12} = G_{13}$ [MPa]	G_{23} [MPa]		ν_{12} [-]	G_{IC} [kJ/m^2]	$G_{IIC} = G_{IIIC}$ [kJ/m^2]	
152,310	8730	3940	2840		0.34	0.18	0.5	
X_T [MPa]	X_C [MPa]	Y_T [MPa]	Y_C [MPa]	S_T [MPa]	S_L [MPa]	G^T_{1C} [kJ/m^2]	G^T_{2C} [kJ/m^2]	G^C_{1C} [kJ/m^2]	G^C_{2C} [kJ/m^2]
2700	1300	55	190	40	102	45	0.298	0.334	3.349

The thickness of each IMS 977-2 ply was 0.188 mm. The number of plies of the composite skin and of the stringers ranged gradually from 8 to 16. The number of plies and the corresponding stacking sequences are reported in detail in Figures 3 and 4 for the skin and the stringers, respectively.

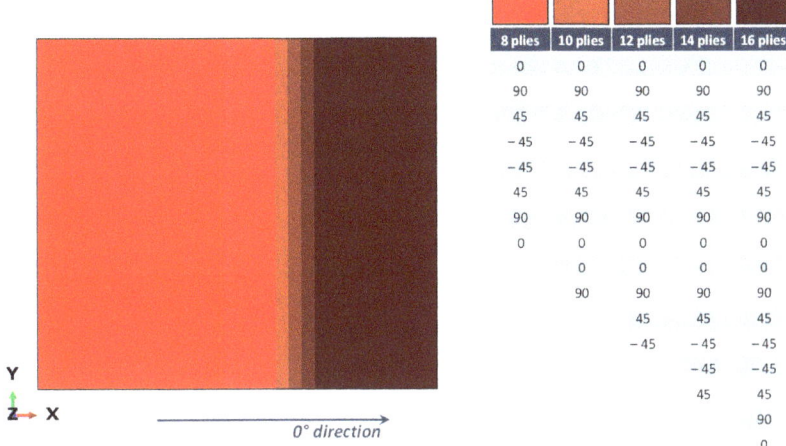

Figure 3. Stacking sequences in the skin.

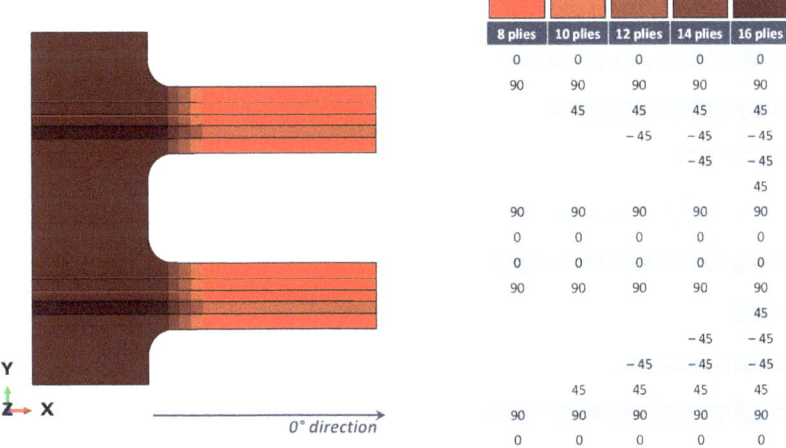

Figure 4. Stacking sequences in the stringers.

3. The Experimental Setup

Fourteen strain gauges were placed along the loading direction to monitor the deformations on the hybrid panel. The locations of the strain gauges are shown in Figure 5.

Indeed, four strain gauges were placed on the composite stringers (SG1, SG2, SG5, and SG6), five strain gauges were placed on the composite panel (SG3, SG4, SG7, SG8, and SG9), and the last five strain gauges were placed on the metal panel (SG10, SG11, SG12, SG13, and SG14).

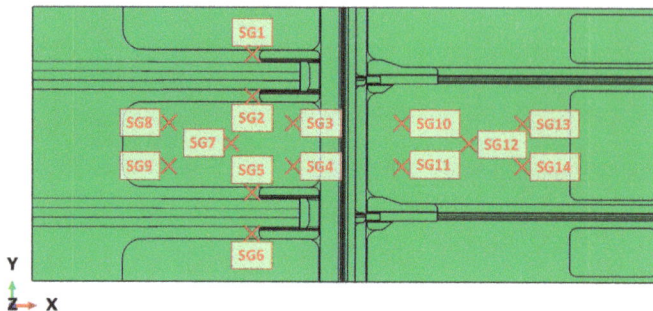

Figure 5. Strain gauge locations.

Figure 6 shows a picture of the hybrid panel including the fixtures used for clamping the panel to the test rig to perform the tensile test. The undergoing experimental test is shown in Figure 7.

Figure 6. Hybrid panel (back side) including clamps.

Figure 7. Test rig.

4. The Numerical Model

Numerical simulations were carried out within the ABAQUS (Finite Element) FE environment. The composite region was discretized by means of 28,348 continuum shell elements with a reduced integration scheme (SC8R), while 27,468 solid elements with a reduced integration scheme (C3D8R) were used to model the metallic components. The mesh size was chosen based on a preliminary mesh sensitivity analysis, which is not reported here for the sake of brevity. Figure 8 shows the finite element model, highlighting the connections among the different sub-components.

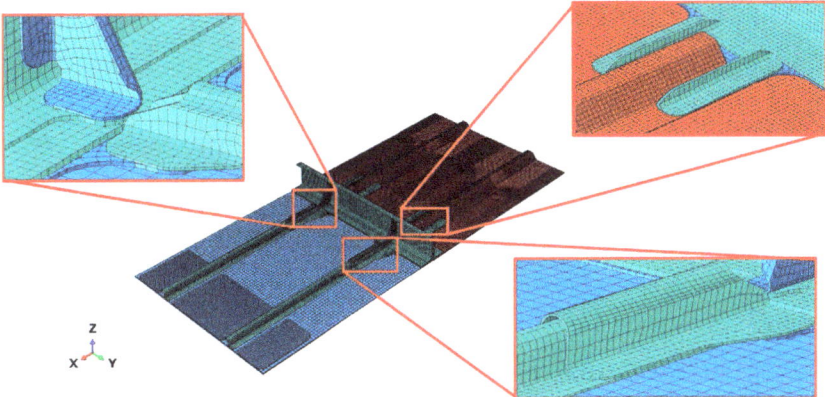

Figure 8. Finite element model (FEM).

The damage onset and evolution were considered in the frame of the preformed tensile analyses. In particular, the Hashin's limit failure criteria was used to simulate the damage in the composite sub-components, since they consider the interaction between the stresses acting on each lamina, distinguishing the failure mechanisms associated to both fibre and matrix failures due to tensile and compressive loads [59]. Moreover, a bilinear plasticity model was considered for the metallic sub-components.

According to Hashin's limit failure approach, separate criteria were introduced by Equations (1)–(4) to determine the damage status within the material:

$$F_{ft} = \left(\frac{\sigma_{11}}{X_T}\right)^2 + \alpha\left(\frac{\sigma_{12}}{S_L}\right)^2 = 1 \quad \text{if} \quad \sigma_{11} > 0 \tag{1}$$

$$F_{fc} = \left(\frac{\sigma_{11}}{X_C}\right)^2 = 1 \quad \text{if} \quad \sigma_{11} < 0 \tag{2}$$

$$F_{mt} = \left(\frac{\sigma_{22}}{Y_T}\right)^2 + \left(\frac{\sigma_{12}}{S_L}\right)^2 = 1 \quad \text{if} \quad \sigma_{22} > 0 \tag{3}$$

$$F_{mc} = \left(\frac{\sigma_{22}}{2S_T}\right)^2 + \left[\left(\frac{Y_C}{2S_T}\right)^2 - 1\right]\left(\frac{\sigma_{22}}{Y_C}\right) + \left(\frac{\sigma_{12}}{S_L}\right)^2 = 1 \quad \text{if} \quad \sigma_{22} < 0 \tag{4}$$

In more detail, Equations (1)–(4), associated respectively to fibre failure in tension and compression and matrix failure in tension and compression, are expressed as a function of the fibre tensile (X_T) and compressive (X_C) strength, of the matrix tensile (Y_T) and compressive (Y_C) strengths, and of the transversal (S_T) and longitudinal (S_L) shear strengths. Once the Hashin's limit values have been reached, the material can be considered damaged, and the stiffness is gradually reduced by a coefficient d, defined for each failure mode, ranging from 0 (undamaged condition) to 1 (completely damaged condition).

For metallic sub-components, according to the introduced bilinear plastic model schematically represented in Figure 9, when stresses overcome the yield threshold, the resulting deformation ε_{tot} can be expressed as the sum of an elastic contribution ε_{el} (evaluated as the ratio between the applied stress σ and the elastic modulus E; $\varepsilon_{el} = \sigma/E$) and a plastic contribution ε_p.

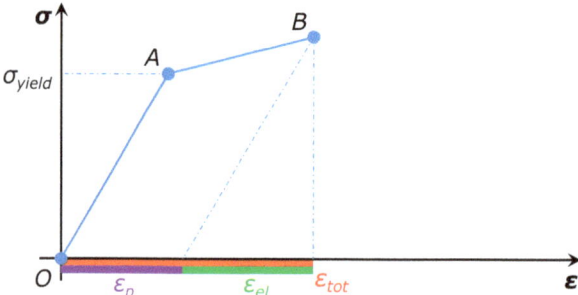

Figure 9. Bilinear plastic model.

ABAQUS TIE multipoint constraints were used to model the bonded connection between sub-components, as shown in Figure 10. These connections involved the composite skin and the composite omega stringers, the metal skin and the doubler, and the metal skin and the fillers.

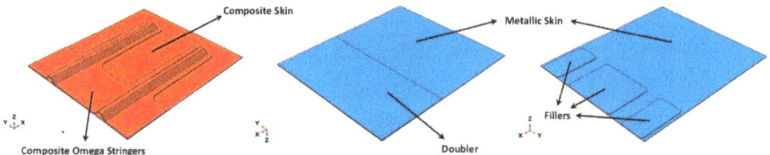

Figure 10. Bonded regions.

To numerically reproduce the fastener behaviour in fastened connections, two approaches were used:

- Model with *Tie*: TIE constraints: this was used to connect the surfaces of the fastened subcomponents;
- Model with *Fastener*: the fasteners were numerically simulated by means of ABAQUS fastener connectors. An elastic behaviour was supposed for the fastener connectors, which were placed in the locations of the experimental test-case, reported in Figure 11.

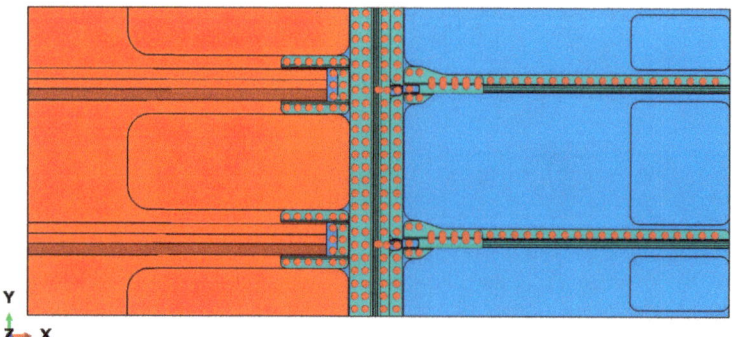

Figure 11. Fastener locations.

5. Numerical–Experimental Correlation

In this section, the numerical results, obtained by means of the previously described numerical model, are compared to the data recorded during the experimental test. In particular, the strains, obtained from the experimental test, are compared to the strains predicted at strain gauges locations for both the Tie and the Fastener numerical models.

Figures 12–18 report the numerical-experimental comparisons at strain gauge locations on the stringers (SG-1, SG-2, SG-5, and SG-6), on the composite skin (SG-3, SG-4, SG-7, SG-8, and SG-9), and on the aluminium skin (SG-10, SG-11, SG-12, SG-13, and SG-14). It is worth noting that, for this kind of test, a maximum deviation equal to ±5% can be observed in the experimental data [60,61].

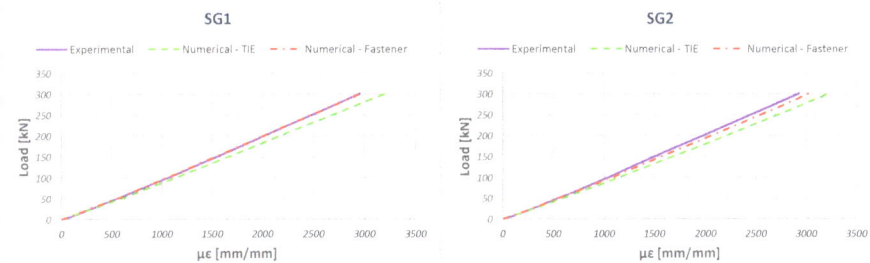

Figure 12. Numerical–experimental comparisons for SG-1 and SG-2: numerical TIE and Fasteners models.

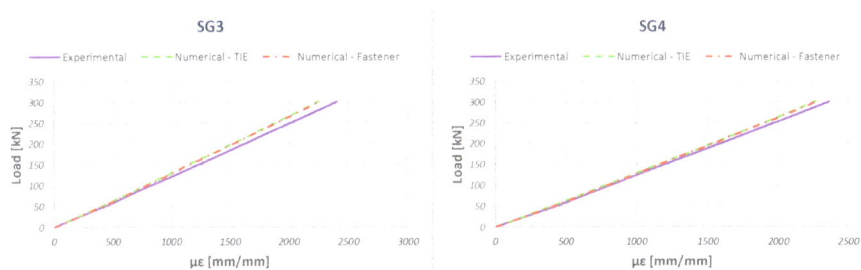

Figure 13. Numerical–experimental comparisons for SG-3 and SG-4: numerical TIE and Fasteners models.

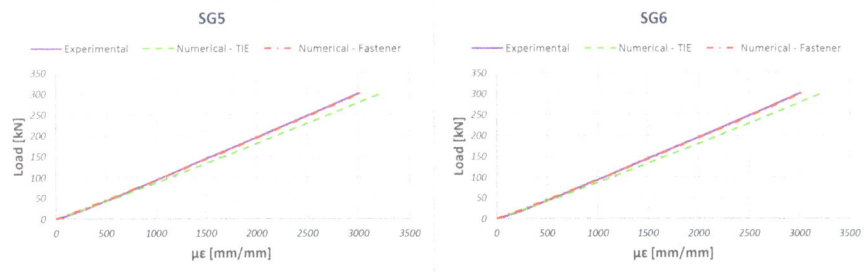

Figure 14. Numerical–experimental comparisons for SG-5 and SG-6: numerical TIE and Fasteners models.

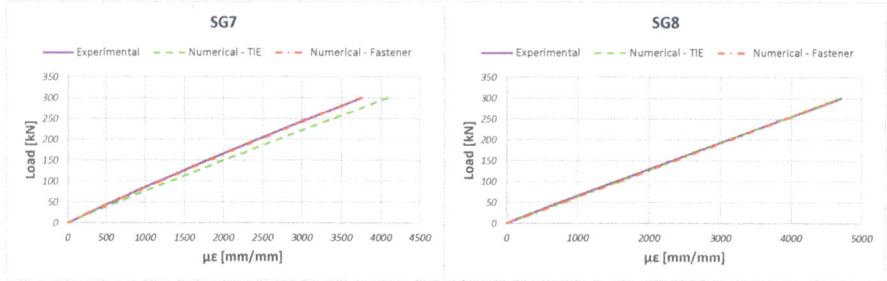

Figure 15. Numerical–experimental comparisons for SG-7 and SG-8: numerical TIE and Fasteners models.

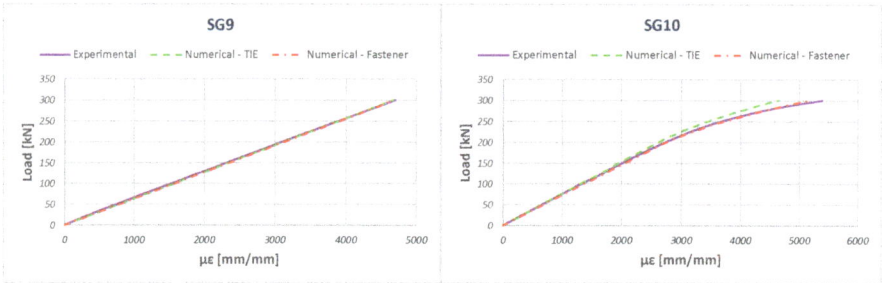

Figure 16. Numerical–experimental comparisons for SG-9 and SG-10: numerical TIE and Fasteners models.

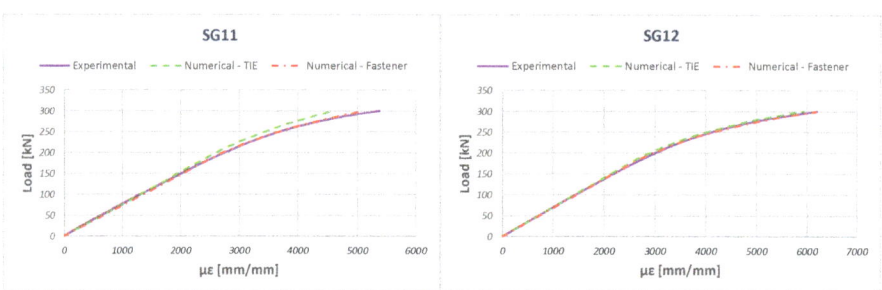

Figure 17. Numerical–experimental comparisons for SG-11 and SG-12: numerical TIE and Fasteners models.

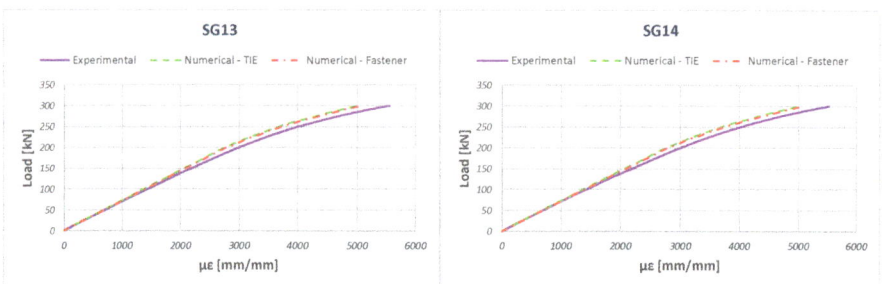

Figure 18. Numerical–experimental comparisons for SG-13 and SG-14: numerical TIE and Fasteners models.

According to Figures 13–18, good agreement is found between the experimentally measured strains and those predicted by the numerical TIE model; while an excellent agreement is found between the experiment and the numerical Fasteners model which uses fastener connectors to couple the structure components. Indeed, the fastener model is able to provide a more realistic deformation field of the structure resulting in a less stiff global behaviour when compared to the TIE model. This behaviour is confirmed by Figure 19, which introduces the comparison between the stiffness of the investigated numerical models. Finally, Figure 20 shows details of the displacements of the Tie and Fastener numerical models. From this figure, the capability of the Fastener-based numerical model to allow the relative displacement of all the structure subcomponents can be appreciated.

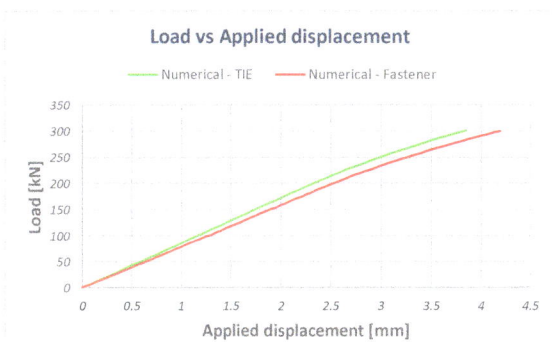

Figure 19. TIE and Fastener numerical models' stiffness comparisons.

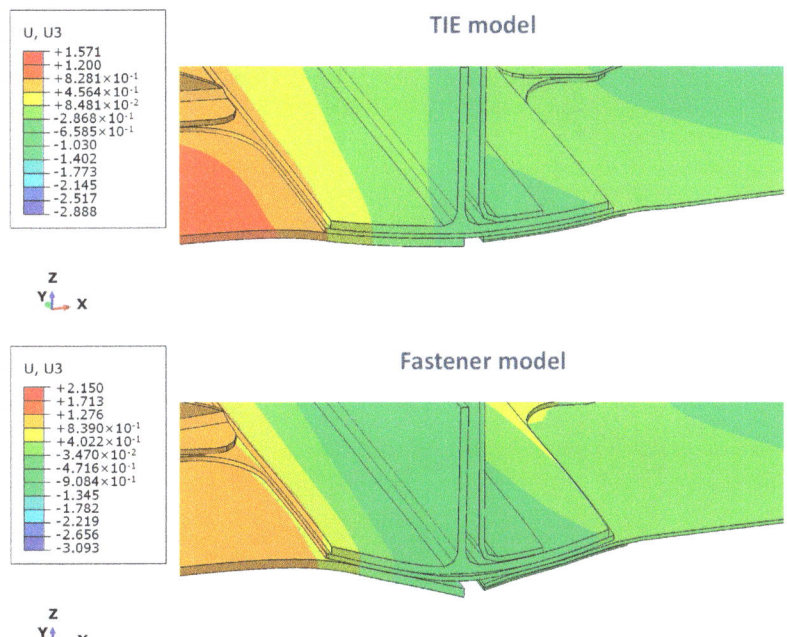

Figure 20. Out of plane displacements at 300 kN (10× deformation factor in the z-direction–units in mm).

As already remarked, Hashin's failure criteria were used to investigate the damage behaviour of the composite sub-components. Indeed, no damage occurred in composite sub-components if a tensile load of 300 kN was applied. This is confirmed by Figures 21 and 22, which introduce the values of the Hashin's failure criteria, respectively, for the fibres and for the matrix phases resulting from the Fasteners-based numerical model. Moreover, in Figures 21 and 22 the values of the Hashin's failure criteria induced by the stress concentration due to the fastener can be appreciated. Indeed, the different material characteristics of the CFRP and metal regions result in a complex stress field at their interface. The faster formulation adopted in this work allows the transfer of concentrated loads between the metal and CFRP regions. Despite the fact that this approach could lead to different predicted results at the interface, the results from a macroscopic point of view can be considered acceptable in terms of stress transfer between the CFRP and metal regions.

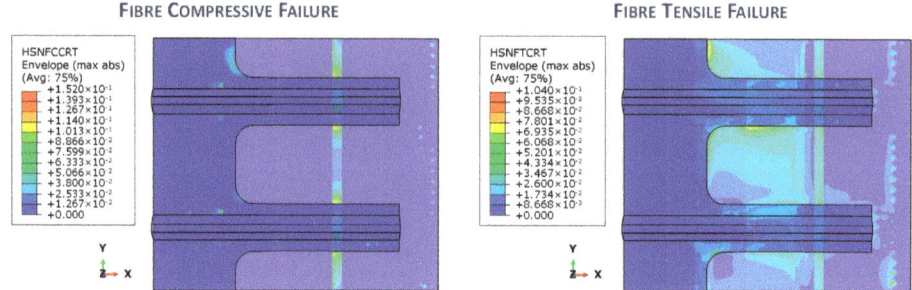

Figure 21. Hashin's failure criteria: fibre.

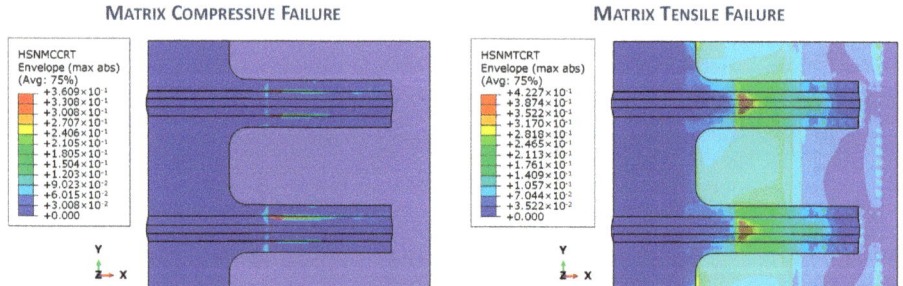

Figure 22. Hashin's failure criteria: matrix.

Finally, the predicted plastic deformations in the metallic region, when a tensile load of 300 kN is applied, are introduced in Figure 23. Indeed, the metallic plate experiences plastic deformations; however, as expected, an increase of the metallic skin thickness would restrain plasticity, as confirmed by the fact that the metal portion of the metal skin coupled with the doubler (see Figure 10) does not experience any plastic deformation.

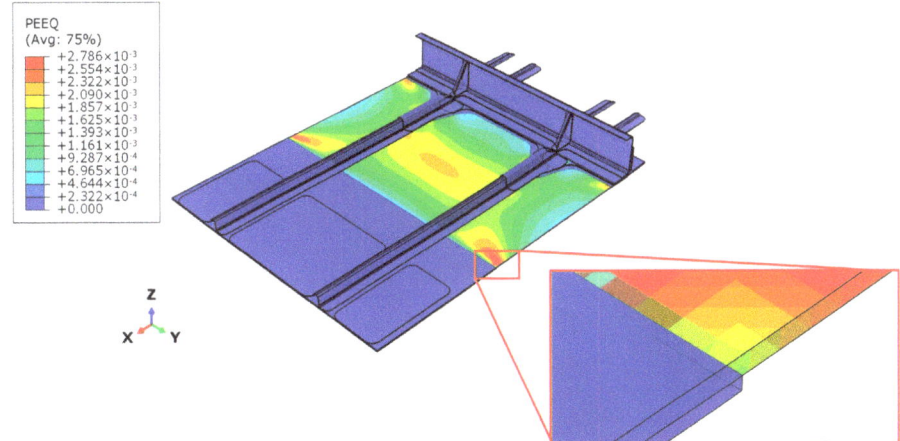

Figure 23. Plastic strains—load: 300 kN.

6. Sensitivity Analysis on the Inter-Laminar Damage Behaviour

In this section, the sensitivity analysis performed to assess the influence of the omega stringer joints on the damage behaviour of the composite components of the structure is summarized. An initial artificial 70 mm-long debonding was introduced between the interface of the composite skin and stringers. Then, the virtual crack closure technique (VCCT) [57] was adopted to numerically simulate the skin-stringer debonding growth in configurations with and without the omega stringer joints (see Figure 24). By comparing the results obtained on these configurations, in terms of skin-stringer debonding growth, for a tensile applied displacement equal to 2.5 mm (which is, actually, the displacement at collapse for the configuration without omega stringer joints) the effects of the omega stringer joints on debonding evolution were assessed.

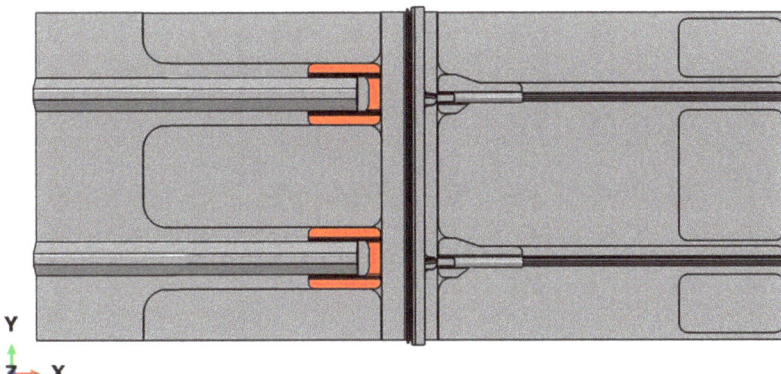

Figure 24. Omega stringer joint (in red).

Figure 25 reports the tensile load as a function of the applied displacement for both the investigated configurations (with and without omega stringer joints).

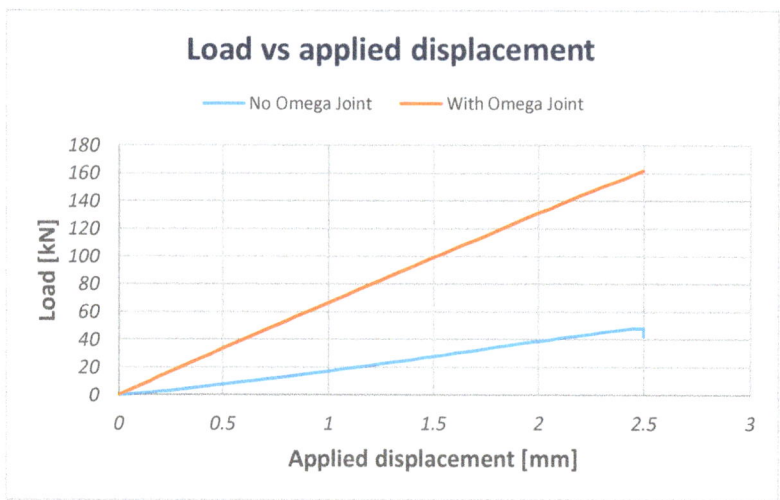

Figure 25. Load vs. applied displacement.

According to Figure 25, the maximum tensile load reached by the configuration without omega joints is 48 kN, which is considerably lower if compared to the 160 kN tensile load reached by the configuration with the omega joints. Indeed, in the model without omega stringer joints, the propagation of the skin-stringer debonding critically reduces the structure-carrying load capability, leading to a premature collapse. This trend is confirmed by Figure 26, which introduces the skin-stringer debonding propagation, in the configuration without omega joints, as the applied displacement increases.

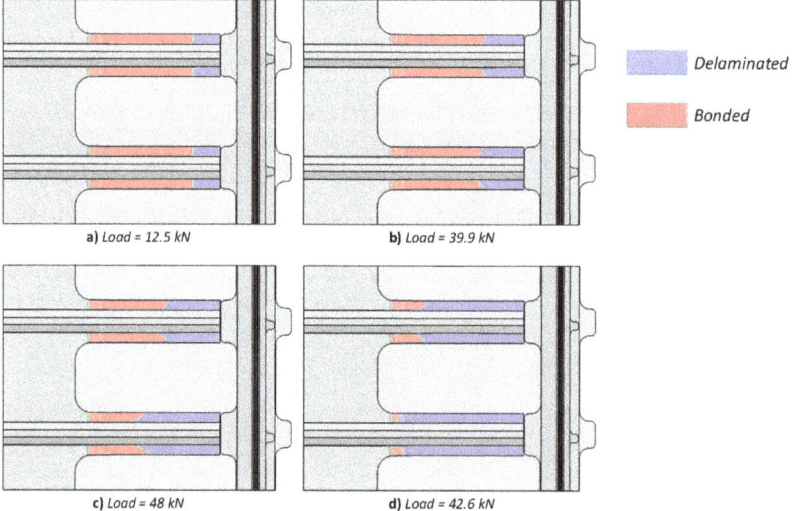

Figure 26. Skin-stringer debonding propagation. (**a**) skin-stringer debonding growth onset; (**b**) intermediate skin-stringer debonding evolution; (**c**) maximum attained load; (**d**) ultimate skin-stringer debonding.

The skin-stringer debonding growth onset occurs at a 12.5 kN load (Figure 26a). The skin-stringer debonding then propagates at both skin-stringer interfaces, up to the maximum applied tensile load of 48 kN (see Figure 26c) at which structural collapse takes place (see Figure 26d).

In contrast, the skin-stringer debonded area in the configuration with the omega joint (shown in Figure 27) does not increase, reaching a maxim applied tensile load of 160 kN for an applied displacement of 2.5 mm as confirmed by Figure 25.

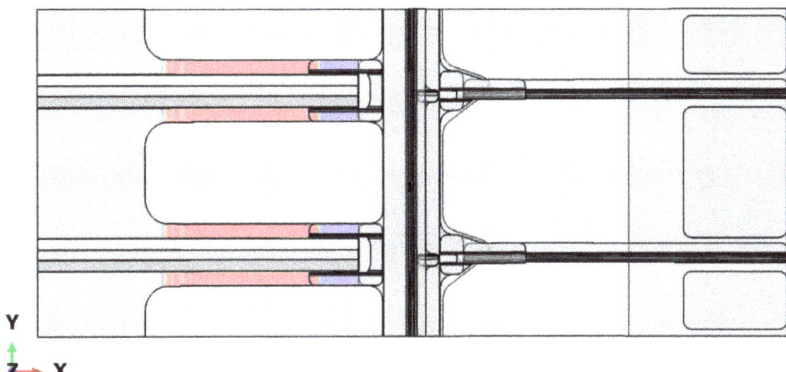

Figure 27. Omega joint configuration—delaminated area: load = 161 kN.

7. Conclusions

In this work, a numerical–experimental investigation into the mechanical behaviour of a fastened hybrid composite–metal stiffened panel has been presented. For numerical predictions, the fasteners have been simulated by using a tie approach and ABAQUS connectors. Both the numerical models take into account the damage behaviour of composite sub-components by Hashin failure criteria while plasticity has been considered for the metallic sub-components.

The numerical results have been correlated to data recorded from an experimental tensile test for a first preliminary validation of the adopted numerical approach. From this comparison, the fastener-based numerical model was revealed to be more accurate, if compared to the tie-based model, in predicting the experimental response in terms of deformations. Actually, tie constraints introduce excessive bonding between the sub-components without allowing the expected relative displacements between them. This preliminary numerical–experimental study demonstrated that the investigated hybrid panel does not experience any damage in composite sub-components up to a tensile load of 300 kN; while, at this loading level, some metallic sub-components experience extensive plastic deformation.

An additional sensitivity numerical analysis has been performed to assess the influence of omega stringer joints on the inter-laminar damage evolution in composite sub-components. Indeed, the virtual crack closure technique has been used to investigate the effect of the omega stringer joints on the evolution of an artificial debonding inserted between composite skin and stringers under tensile loading conditions. Numerical analysis demonstrated that the omega stringer joints are able to arrest the skin-stringer debonding growth avoiding a drastic reduction of the load-carrying capability of the panel.

Author Contributions: All authors equally contribute to this work. All authors have read and agreed to the published version of the manuscript.

Funding: This research received no external funding.

Conflicts of Interest: The authors declare no conflict of interest.

References

1. Sellitto, A.; Riccio, A.; Russo, A.; Zarrelli, M.; Toscano, C.; Lopresto, V. Compressive behaviour of a damaged omega stiffened panel: Damage detection and numerical analysis. *Compos. Struct.* **2019**, *209*, 300–316. [CrossRef]
2. Li, X.; Gao, W.; Liu, W. Post-buckling progressive damage of CFRP laminates with a large-sized elliptical cutout subjected to shear loading. *Compos. Struct.* **2015**, *128*, 313–321. [CrossRef]
3. Benedetti, I.; Gulizzi, V. A grain-scale model for high-cycle fatigue degradation in polycrystalline materials. *Int. J. Fatigue* **2018**, *116*, 90–105. [CrossRef]
4. Benedetti, I.; Nguyen, H.; Soler-Crespo, R.A.; Gao, W.; Mao, L.; Ghasemi, A.; Wen, J.; Nguyen, S.; Espinosa, H.D. Formulation and validation of a reduced order model of 2D materials exhibiting a two-phase microstructure as applied to graphene oxide. *J. Mech. Phys. Solids* **2018**, *112*, 66–88. [CrossRef]
5. Milazzo, A.; Benedetti, I.; Gulizzi, V. An extended Ritz formulation for buckling and post-buckling analysis of cracked multilayered plates. *Compos. Struct.* **2018**, *201*, 980–994. [CrossRef]
6. Riccio, A.; Linde, P.; Raimondo, A.; Buompane, A.; Sellitto, A. On the use of selective stitching in stiffened composite panels to prevent skin-stringer debonding. *Compos. Part B Eng.* **2017**, *124*, 64–75. [CrossRef]
7. Zhu, G.; Sun, G.; Liu, Q.; Li, G.; Li, Q. On crushing characteristics of different configurations of metal-composites hybrid tubes. *Compos. Struct.* **2017**, *175*, 58–69. [CrossRef]
8. Zhu, G.; Sun, G.; Yu, H.; Li, S.; Li, Q. Energy absorption of metal, composite and metal/composite hybrid structures under oblique crushing loading. *Int. J. Mech. Sci.* **2018**, *135*, 458–483. [CrossRef]
9. Kumar, P.; Rai, B. Delaminations of barely visible impact damage in CFRP laminates. *Compos. Struct.* **1993**, *23*, 313–318. [CrossRef]
10. Rogge, M.D.; Leckey, C.A.C. Characterization of impact damage in composite laminates using guided wavefield imaging and local wavenumber domain analysis. *Ultrasonics* **2013**, *53*, 1217–1226. [CrossRef]
11. Angelidis, N.; Irving, P.E. Detection of impact damage in CFRP laminates by means of electrical potential techniques. *Compos. Sci. Technol.* **2007**, *67*, 594–604. [CrossRef]
12. Romano, F.; Di Caprio, F.; Mercurio, U. Compression after impact analysis of composite panels and equivalent hole method. *Procedia Eng.* **2016**, *167*, 182–189. [CrossRef]
13. Borrelli, R.; Franchitti, S.; Di Caprio, F.; Mercurio, U.; Zallo, A. A repair criterion for impacted composite structures based on the prediction of the residual compressive strength. *Procedia Eng.* **2014**, *88*, 117–124. [CrossRef]
14. Borrelli, R.; Franchitti, S.; Di Caprio, F.; Romano, F.; Mercurio, U. A numerical procedure for the virtual compression after impact analysis. *Adv. Compos. Lett.* **2015**, *24*, 57–67. [CrossRef]
15. Perner, M.; Algermissen, S.; Keimer, R.; Monner, H.P. Avoiding defects in manufacturing processes: A review for automated CFRP production. *Robot. Comput.-Integr. Manuf.* **2016**, *38*, 82–92. [CrossRef]
16. Ma, L.; Soleimani, M. Hidden defect identification in carbon fibre reinforced polymer plates using magnetic induction tomography. *Meas. Sci. Technol.* **2014**, *25*, 055404. [CrossRef]
17. Hörrmann, S.; Adumitroaie, A.; Schagerl, M. The effect of ply folds as manufacturing defect on the fatigue life of CFRP materials. *Frattura ed Integrita Strutturale* **2016**, *10*, 76–81. [CrossRef]
18. Wang, P.; Lei, H.; Zhu, X.; Chen, H.; Wang, C.; Fang, D. Effect of manufacturing defect on mechanical performance of plain weave carbon/epoxy composite based on 3D geometrical reconstruction. *Compos. Struct.* **2018**, *199*, 38–52. [CrossRef]
19. Pramanik, A. Developments in the non-traditional machining of particle reinforced metal matrix composites. *Int. J. Mach. Tools Manuf.* **2014**, *86*, 44–61. [CrossRef]
20. Pramanik, A.; Basak, A.K.; Dong, Y.; Sarker, P.K.; Uddin, M.S.; Littlefair, G.; Dixit, A.R.; Chattopadhyaya, S. Joining of carbon fibre reinforced polymer (CFRP) composites and aluminium alloys—A review. *Compos. Part A Appl. Sci. Manuf.* **2017**, *101*, 1–29. [CrossRef]
21. Riccio, A.; Ricchiuto, R.; Di Caprio, F.; Sellitto, A.; Raimondo, A. Numerical investigation of constitutive material models on bonded joints in scarf repaired composite laminates. *Eng. Fract. Mech.* **2017**, *173*, 91–106. [CrossRef]
22. Rhee, K.Y.; Choi, N.-S.; Park, S.-J. Effect of plasma treatment of aluminum on the bonding characteristics of aluminum-CFRP composite joints. *J. Adhes. Sci. Technol.* **2002**, *16*, 1487–1500. [CrossRef]
23. Jumbo, F.; Ruiz, P.D.; Yu, Y.; Swallowe, G.M.; Ashcroft, I.A.; Huntley, J.M. Experimental and numerical investigation of mechanical and thermal residual strains in adhesively bonded joints. *Strain* **2007**, *43*, 319–331. [CrossRef]
24. Zhang, K.; Yang, Z.; Li, Y. A method for predicting the curing residual stress for CFRP/Al adhesive single-lap joints. *Int. J. Adhes. Adhes.* **2013**, *46*, 7–13. [CrossRef]

25. Ishii, K.; Imanaka, M.; Nakayama, H.; Kodama, H. Evaluation of the fatigue strength of adhesively bonded CFRP/metal single and single-step double-lap joints. *Compos. Sci. Technol.* **1999**, *59*, 1675–1683. [CrossRef]
26. Biscaia, H.; Cardoso, J.; Chastre, C. A finite element based analysis of double strap bonded joints with CFRP and aluminium. *Key Eng. Mater.* **2017**, *754*, 237–240. [CrossRef]
27. Ruiz, P.D.; Jumbo, F.; Huntley, J.M.; Ashcroft, I.A.; Swallowe, G.M. Experimental and numerical investigation of strain distributions within the adhesive layer in bonded joints. *Strain* **2011**, *47*, 88–104. [CrossRef]
28. Tinkloh, S.; Wu, T.; Tröster, T.; Niendorf, T. A micromechanical-based finite element simulation of process-induced residual stresses in metal-CFRP-hybrid structures. *Compos. Struct.* **2020**, *238*, 111926. [CrossRef]
29. Tsokanas, P.; Loutas, T.; Kotsinis, G.; Kostopoulos, V.; van den Brink, W.M.; Martin de la Escalera, F. On the fracture toughness of metal-composite adhesive joints with bending-extension coupling and residual thermal stresses effect. *Compos. Part B Eng.* **2020**, *185*, 107694. [CrossRef]
30. Xiao, Y.; Ishikawa, T. Bearing strength and failure behavior of bolted composite joints (part I: Experimental investigation). *Compos. Sci. Technol.* **2005**, *65*, 1022–1031. [CrossRef]
31. Xiao, Y.; Ishikawa, T. Bearing strength and failure behavior of bolted composite joints (part II: Modeling and simulation). *Compos. Sci. Technol.* **2005**, *65*, 1032–1043. [CrossRef]
32. Irisarri, F.-X.; Laurin, F.; Carrere, N.; Maire, J.-F. Progressive damage and failure of mechanically fastened joints in CFRP laminates-Part I: Refined Finite Element modelling of single-fastener joints. *Compos. Struct.* **2012**, *94*, 2269–2277. [CrossRef]
33. Irisarri, F.-X.; Laurin, F.; Carrere, N.; Maire, J.-F. Progressive damage and failure of mechanically fastened joints in CFRP laminates-Part II: Failure prediction of an industrial junction. *Compos. Struct.* **2012**, *94*, 2278–2284. [CrossRef]
34. Madukauwa-David, I.D.; Drissi-Habti, M. Numerical simulation of the mechanical behavior of a large smart composite platform under static loads. *Compos. Part B Eng.* **2016**, *88*, 19–25. [CrossRef]
35. Antony, S.; Drissi-Habti, M.; Raman, V. Numerical Analysis to Enhance Delamination Strength around Bolt Holes of Unidirectional Pultruded Large Smart Composite Platform. *Adv. Mater. Sci. Eng.* **2018**, *2018*, 3154904. [CrossRef]
36. Benezech, L.; Landon, Y.; Rubio, W. Study of manufacturing defects and tool geometry optimisation for multi-material stack drilling. *Adv. Mater. Res.* **2012**, *423*, 1–11. [CrossRef]
37. Zitoune, R.; Krishnaraj, V.; Sofiane Almabouacif, B.; Collombet, F.; Sima, M.; Jolin, A. Influence of machining parameters and new nano-coated tool on drilling performance of CFRP/Aluminium sandwich. *Compos. Part B Eng.* **2012**, *43*, 1480–1488. [CrossRef]
38. Breziner, L.; Hutapea, P. Influence of harsh environmental conditions on CFRP-aluminum single lap joints. *Aircr. Eng. Aerosp. Technol.* **2008**, *80*, 371–377. [CrossRef]
39. Coman, C.-D.; Pelin, G. Thermal effects on single-Lap, single-Bolt, hybrid metal—Composite joint stiffness. *Incas Bull.* **2018**, *10*, 75–88.
40. Ueda, M.; Miyake, S.; Hasegawa, H.; Hirano, Y. Instantaneous mechanical fastening of quasi-isotropic CFRP laminates by a self-piercing rivet. *Compos. Struct.* **2012**, *94*, 3388–3393. [CrossRef]
41. Hoang, N.-H.; Langseth, M.; Porcaro, R.; Hanssen, A.-G. The effect of the riveting process and aging on the mechanical behaviour of an aluminium self-piercing riveted connection. *Eur. J. Mech. A/Solids* **2011**, *30*, 619–630. [CrossRef]
42. Hoang, N.-H.; Porcaro, R.; Langseth, M.; Hanssen, A.-G. Self-piercing riveting connections using aluminium rivets. *Int. J. Solids Struct.* **2010**, *47*, 427–439. [CrossRef]
43. Pickin, C.G.; Young, K.; Tuersley, I. Joining of lightweight sandwich sheets to aluminium using self-pierce riveting. *Mater. Des.* **2007**, *28*, 2361–2365. [CrossRef]
44. Landgrebe, D.; Jäckel, M.; Niegsch, R. Influence of process induced damages on joint strength when self-pierce riveting carbon fiber reinforced plastics with aluminum. *Key Eng. Mater.* **2015**, *651–653*, 1493–1498. [CrossRef]
45. Rao, H.M.; Kang, J.; Huff, G.; Avery, K.; Su, X. Impact of rivet head height on the tensile and fatigue properties of lap shear self-pierced riveted CFRP to aluminum. *SAE Int. J. Mater. Manuf.* **2017**, *10*, 167–173. [CrossRef]
46. Kweon, J.-H.; Jung, J.-W.; Kim, T.-H.; Choi, J.-H.; Kim, D.-H. Failure of carbon composite-to-aluminum joints with combined mechanical fastening and adhesive bonding. *Compos. Struct.* **2006**, *75*, 192–198. [CrossRef]
47. Matsuzaki, R.; Shibata, M.; Todoroki, A. Improving performance of GFRP/aluminum single lap joints using bolted/co-cured hybrid method. *Compos. Part A Appl. Sci. Manuf.* **2008**, *39*, 154–163. [CrossRef]

48. Di Franco, G.; Zuccarello, B. Analysis and optimization of hybrid double lap aluminum-GFRP joints. *Compos. Struct.* **2014**, *116*, 682–693. [CrossRef]
49. Fu, M.; Mallick, P.K. Fatigue of hybrid (adhesive/bolted) joints in SRIM composites. *Int. J. Adhes. Adhes.* **2001**, *21*, 145–159. [CrossRef]
50. Di Franco, G.; Fratini, L.; Pasta, A. Analysis of the mechanical performance of hybrid (SPR/bonded) single-lap joints between CFRP panels and aluminum blanks. *Int. J. Adhes. Adhes.* **2013**, *41*, 24–32. [CrossRef]
51. Barut, A.; Madenci, E. Analysis of bolted-bonded composite single-lap joints under combined in-plane and transverse loading. *Compos. Struct.* **2009**, *88*, 579–594. [CrossRef]
52. Kapidžić, Z.; Nilsson, L.; Ansell, H. Finite element modeling of mechanically fastened composite-aluminum joints in aircraft structures. *Compos. Struct.* **2014**, *109*, 198–210. [CrossRef]
53. Hochard, C.; Payan, J.; Bordreuil, C. A progressive first ply failure model for woven ply CFRP laminates under static and fatigue loads. *Int. J. Fatigue* **2006**, *28*, 1270–1276. [CrossRef]
54. Guild, F.J.; Vrellos, N.; Drinkwater, B.W.; Balhi, N.; Ogin, S.L.; Smith, P.A. Intra-laminar cracking in CFRP laminates: Observations and modelling. *J. Mater. Sci.* **2006**, *41*, 6599–6609. [CrossRef]
55. Reiner, J.; Feser, T.; Schueler, D.; Waimer, M.; Vaziri, R. Comparison of two progressive damage models for studying the notched behavior of composite laminates under tension. *Compos. Struct.* **2019**, *207*, 385–396. [CrossRef]
56. Gong, W.; Chen, J.; Patterson, E.A. An experimental study of the behaviour of delaminations in composite panels subjected to bending. *Compos. Struct.* **2015**, *123*, 9–18. [CrossRef]
57. Riccio, A.; Russo, A.; Sellitto, A.; Raimondo, A. Development and application of a numerical procedure for the simulation of the "Fibre Bridging" phenomenon in composite structures. *Compos. Struct.* **2017**, *168*, 104–119. [CrossRef]
58. Lévêque, D.; Schieffer, A.; Mavel, A.; Maire, J.-F. Analysis of how thermal aging affects the long-term mechanical behavior and strength of polymer-matrix composites. *Compos. Sci. Technol.* **2005**, *65*, 395–401. [CrossRef]
59. De Luca, A.; Caputo, F. A review on analytical failure criteria for composite materials. *Aims Mater. Sci.* **2017**, *4*, 1165–1185. [CrossRef]
60. Khadyko, M.; Dumoulin, S.; Børvik, T.; Hopperstad, O.S. An experimental-numerical method to determine the work-hardening of anisotropic ductile materials at large strains. *Int. J. Mech. Sci.* **2014**, *88*, 25–36. [CrossRef]
61. Frodal, B.H.; Dæhli, L.E.B.; Børvik, T.; Hopperstad, O.S. Modelling and simulation of ductile failure in textured aluminium alloys subjected to compression-tension loading. *Int. J. Plast.* **2019**, *118*, 36–69. [CrossRef]

© 2020 by the authors. Licensee MDPI, Basel, Switzerland. This article is an open access article distributed under the terms and conditions of the Creative Commons Attribution (CC BY) license (http://creativecommons.org/licenses/by/4.0/).

Article

Impact Damage Resistance and Post-Impact Tolerance of Optimum Banana-Pseudo-Stem-Fiber-Reinforced Epoxy Sandwich Structures

Mohamad Zaki Hassan [1,*], S. M. Sapuan [2], Zainudin A. Rasid [3], Ariff Farhan Mohd Nor [1], Rozzeta Dolah [1] and Mohd Yusof Md Daud [1]

1. Razak Faculty of Technology and Informatics, Universiti Teknologi Malaysia, Jalan Sultan Yahya Petra, Kuala Lumpur 54100, Malaysia; ariffmdnor@yahoo.com (A.F.M.N.); rozzeta.kl@utm.my (R.D.); yusof.kl@utm.my (M.Y.M.D.)
2. Advanced Engineering Materials and Composites Research Centre, Department of Mechanical and Manufacturing Engineering, Universiti Putra Malaysia, Serdang 43400, Malaysia; sapuan@upm.edu.my
3. Malaysia-Japan International Institute of Technology, Universiti Teknologi Malaysia Jalan Sultan Yahya Petra, Kuala Lumpur 54100, Malaysia; arzainudin.kl@utm.my
* Correspondence: mzaki.kl@utm.my

Received: 25 December 2019; Accepted: 14 January 2020; Published: 18 January 2020

Abstract: Banana fiber has a high potential for use in fiber composite structures due to its promise as a polymer reinforcement. However, it has poor bonding characteristics with the matrixes due to hydrophobic–hydrophilic incompatibility, inconsistency in blending weight ratio, and fiber length instability. In this study, the optimal conditions for a banana/epoxy composite as determined previously were used to fabricate a sandwich structure where carbon/Kevlar twill plies acted as the skins. The structure was evaluated based on two experimental tests: low-velocity impact and compression after impact (CAI) tests. Here, the synthetic fiber including Kevlar, carbon, and glass sandwich structures were also tested for comparison purposes. In general, the results showed a low peak load and larger damage area in the optimal banana/epoxy structures. The impact damage area, as characterized by the dye penetration, increased with increasing impact energy. The optimal banana composite and synthetic fiber systems were proven to offer a similar residual strength and normalized strength when higher impact energies were applied. Delamination and fracture behavior were dominant in the optimal banana structures subjected to CAI testing. Finally, optimization of the compounding parameters of the optimal banana fibers improved the impact and CAI properties of the structure, making them comparable to those of synthetic sandwich composites.

Keywords: banana fiber; impact response; compression after impact; natural fiber

1. Introduction

The research interest into utilizing natural fibers as reinforcement in polymers has dramatically increased during the last decade. It has been claimed that they can replace their synthetic polymer counterparts. Natural fibers, such as banana, kenaf, sugar palm, pineapple leaf, and empty fruit bunch, are abundantly available in tropical countries, especially in Southeast Asia and Papua New Guinea. Among the natural fiber composites, banana has attracted significant interest since it is biodegradable, not a health hazard, of low abrasivity, cheap, and offers good sound absorption capabilities. It belongs to a subclass of monocotyledonous herbaceous flowering plants in the genus *Musa*. This tropical plant originated from Brazil and was widely consumed after the American Civil War. Roughly, 72.5 million tons of banana fruit are produced yearly throughout the world [1]. The most widely recognized banana that is consumed by humans is a member of the *Musa acuminata* species. For each ton of banana produced and harvested, around 100 kg of the fruit product is rejected and nearly 4 tons of biomass waste, including leaf, pseudo-stem, rotten fruit, peel, fruit-bunch-stem, and rhizome [2], are produced. In many nations, including Malaysia, the uses of banana fiber have been disregarded, despite research findings over the years [3].

Past studies have proven that the banana pseudo-stem is a promising fiber with a significant tensile strength [4] and stiffness [5]. Maleque et al. [6] mentioned that epoxy polymer reinforced by banana improved its mechanical strength by 90% and its impact strength by 40% compared to neat polymer. In addition, the optimal loading percentage of treated banana fiber incorporated into low-density polyethylene (LDPE) at a fiber loading of 20% offered the highest mechanical properties [7]. Further, the highest improvement of 4%, 11%, 14.5%, and 11.1% for Young's modulus and tensile, flexural, and impact strengths, respectively, was found for a 25% banana fiber loading that had been treated with alkali compared to values for neat LDPE [8]. In another study, Ahmed et al. [9] stated that 10% banana fiber loading treated with 5% alkaline suspension achieved the highest mechanical properties for acrylonitrile butadiene styrene. Venkateshwaran et al. [10] reported that an increase in banana fiber length of up to 15 mm and a 16% weight ratio increases the tensile strength and modulus of a banana/epoxy composite. However, in order to achieve higher mechanical properties, Udaya et al. [11] suggested an optimal fiber length and fiber weight ratio of 30 mm and 57%, respectively. However, the optimal findings for fiber treatment and fiber loading are still inconsistent. Higher loading and treatment of the banana fibers have led to poor interfacial adhesion and reduced mechanical properties.

A series of studies on the impact properties of a composite structure with banana fiber reinforcement have been conducted [12–16]. The impact strength of banana pseudo-stem unplasticized polyvinyl chloride composites were conducted by Zainudin and Sapuan [12] using the Izod impact test. It was found that fiber loading using banana fiber could possibly enhance the impact strength properties of the composites. Pavithran et al. [13] conducted the Charpy test in order to evaluate the effect of banana, sisal, and pineapple reinforcement on the fracture of coir/polyester composites and found that the sisal/polyester composites exhibited the highest value. The impact strengths of hybrid sisal, banana, coir, and sisal/banana/coir-fiber-reinforced epoxy were also compared by Balaji et al. [14]. Again sisal-fiber-reinforced epoxy offered a higher resistance to impact loading. In addition, Pothan et al. [15] examined the effect of fiber loading on a banana fiber/polyester composite following a low-velocity impact. They suggested that a fiber length from 30 to 40 mm and a fiber loading of 40% offered significantly higher impact resistance. Devireddy et al. [16] conducted impact response tests on banana, jute, and hybrid epoxy composites. It was mentioned that the hybrid composite offers an outstanding performance when compared to the individual natural fiber composites. Moreover, Narayana et al. [17] functionalized the hybrid nanocomposite made up of the nanoclay-reinforcement of either banana fiber, E-glass, or epoxy resin. The results showed that reinforcement using nanoclay inclusions was able to enhance the impact properties.

Energy absorption and compression residual strength capabilities are the main variables being analyzed in the impact testing. There have been several studies conducted on the low-velocity impact response and residual strength of composite structures. Dhakal et al. [18] studied the flexural strength after impact (FAI) of jute-reinforced unsaturated polyester composites. It was found that the FAI significantly decreased with the increase in test temperature and the damage assessment of the composites revealed delamination as the major failure mode. Ismail et al. [19] studied the post-impact behaviors of kenaf/glass hybrid composites with different weight proportions following a low-velocity impact test. The compression after impact (CAI) test demonstrated that the compression damage decreased as the impact energy increased. It was further discovered that the kenaf/glass hybrid composite with a 25% kenaf fiber weight ratio gave results that are comparable to those of a glass laminate composite. Mohd Nor et al. [20] examined the effect of nanofiller in a bamboo/epoxy composite to enhance the CAI properties. The addition of carbon nanotubes (CNTs) into the composite improved the compression post-impact response properties. The compressive residual strength test was also carried out on flax/polylactic acid (PLA) laminates [21]. The absorbed energy and normalized residual strength were analyzed and the main failure mode in a composite laminate was identified as being fiber failure. Numerous studies have explored the post-impact behavior of glass/epoxy composites, Kevlar/epoxy composites, and other synthetic types of composites. However, only limited findings on the CAI behavior of optimized banana fiber reinforced epoxy composites have been reported.

In order to improve the mechanical behavior of the natural fiber composites, many researchers have implemented a well-developed statistical approach including the Taguchi method and the response surface method (RSM). The optimal parameters, such as temperature, molding time, and volume fraction of kenaf-reinforced polyethylene composite, were determined using the Box–Behnken response surface method for ballistic protection, as reported by Akubue et al. [22]. Moreover, Yaghoobi and Fereidoon [23] evaluated the effect of the fiber load, fiber length, and compatibilizer content on the tensile strength and modulus using the Box–Behnken design. The results showed that the R^2 values and normal probability plots were in good agreement. Furthermore, Roslan et al. [24] investigated the mode I fracture toughness of optimized alkali-treated bamboo using the Box–Behnken method. It was suggested that this statistical analysis approach is highly suitable for optimizing the parameters for alkaline-treated bamboo fibers.

In previous work, the Box–Behnken method was used to determine the optimal parameters, including fiber length, fiber loading, and chemical treatment concentration, for a banana/epoxy composite [25]. However, there is a need to explore the behavior of this optimal fiber for sandwich structures to fill the knowledge gaps in this particular field of study. In this research, the optimized banana-fiber-reinforced epoxy composites were laminated with carbon/Kevlar twill woven skins to form banana epoxy sandwich structures. This current study focused on the low-velocity impact and compression after impact (CAI) response of these structures. The optimal behavior of banana fiber sandwich panels was compared to that of the synthetic fibers including Kevlar, carbon, and glass fibers. Prior to that, the tensile properties of neat epoxy-resin and optimal banana composite were also discussed. Further, details of the loading behavior, toughness, and damage evolution were obtained.

2. Materials and Methods

2.1. Materials

The banana pseudo-stem fiber from the *Musa acuminata* species [26] with a diameter range from 500 µm to 1 mm was supplied by Innovative Pultrusion Sdn Bhd, Seremban Negeri Sembilan, Malaysia. The EpoxAmite™ 100 epoxy resin and EpoxAmite™ 102 medium hardener [27] used as the base matrix were purchased from Kird Enterprise, Nilai Negeri Sembilan, Malaysia. This epoxy-resin system was mixed to a weight ratio of 10:2.9 g. The banana fibers were soaked in sodium hydroxide (NaOH) obtained from Orioner Hightech Sdn Bhd, Cyberjaya Selangor, Malaysia. In this study, a 2/2 twill weave carbon/Kevlar hybrid with a density of 210 g/m² was used for the skins. Carbon and Kevlar fiber tow was used as a comparison for the banana fiber. Those were purchased from EasyComposite Ltd., Stoke-on-Trent, UK [28]. Glass fibers were purchased from Alsey Kimia Sdn Bhd, Puchong Selangor, Malaysia. Detailed properties of all materials are given in Table 1.

Table 1. Properties of fibers, skins, and EpoxAmite™-102 hardener.

	Banana [26]	Kevlar [28]	Carbon [28]	Glass [28]	Carbon Kevlar Twill [28]	EpoxAmite™-102 Hardener [27]
Density (kg/m³)	1350	1340	1780	2600		1110
Flexural Strength (MPa)	52					84.25
Tensile Strength (MPa)	54	3260	4900	3450		56.4
Young's Modulus (GPa)	3.49	60–80	250	72–77		3.1
Elongation (%)		4.4	2	4.7		
Weight (g/m²)					210	
Weft					2(C)-1(K)	
Warp					2(C)-1(K)	

Table 1. Cont.

	Banana	Kevlar	Carbon	Glass	Carbon Kevlar Twill	EpoxAmite™-102 Hardener
	[26]	[28]	[28]	[28]	[28]	[27]
Cellulose (%)	63–64					-
Hemicellulose (%)	19					
Lignin (%)	5					-
Mixed viscosity (kg/ms)	-					0.65
Specific volume (m^3/kg)	-					9.03×10^{-4}

2.2. Fabrication of Composites

Initially, to eliminate any surface impurities, a long banana fiber, as shown in Figure 1a, was washed with deionized water and dried in a circulation oven Model H750CLAB200D16 (CMH Ltd., Lancing, UK) at 70 °C for 9 h. Then, the fibers were soaked in 5.45 wt.% sodium hydroxide (Figure 1b) for 5 h according to the optimal conditions suggested by the Box–Behnken design [25]. These fibers were ground down using a Cheso Model N3, (Cheso Machinery Pte. Ltd., Loyang Way, Singapore) crusher machine (Figure 1c). In order to obtain the 3.35-mm fiber length, a multi-stage sieve (Figure 1d)—model BS410/1986 (Endecots Ltd., London, UK)—fixed to a rotational shaker (Endecots Ltd., London, UK) (Figure 1e) was used. The speed of this shaker was maintained at 275 rpm for 45 min. The short banana pseudo-stem fibers were utilized as the reinforcement, as illustrated in Figure 1f. Similar processes were repeated for the Kevlar, carbon, and glass fiber tows, except for the alkaline treatment. These synthetic fibers were chopped using a carbon fiber shear cutter Model 3670C-8 (EasyComposite Ltd., Stoke-on-Trent, UK) and sieved using a rotational shaker. Then, the epoxy-resin matrix with fibers was gradually mixed at 29.86 wt.% of the fiber loading.

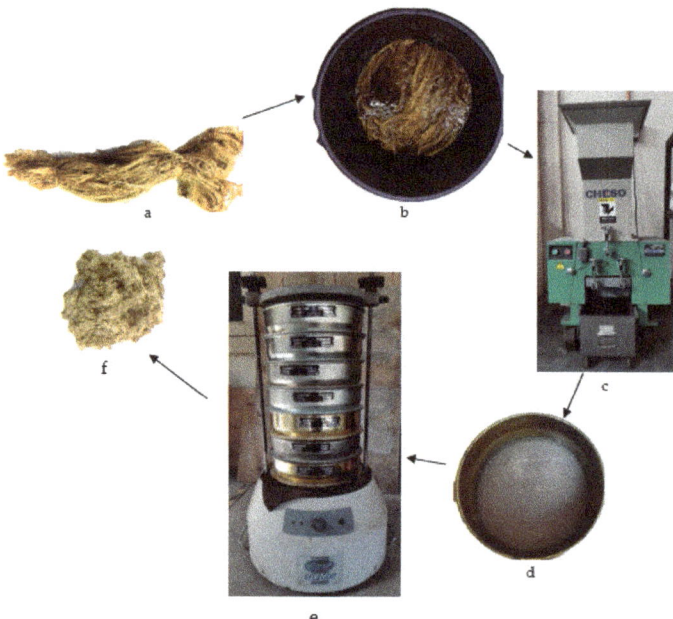

Figure 1. Photos of (**a**) long banana fibers, (**b**) chemically retted fibers, (**c**) fiber chopping machine, (**d**) 3.35-mm brass frame sieve, (**e**) rotational shaker, and (**f**) short banana fibers.

2.3. Tensile and Sandwich Structure Preparation

The tensile test specimens of banana epoxy composites were fabricated using the mold shown in Figure 2. A dog bone specimen was manufactured following ASTM D638 [29].

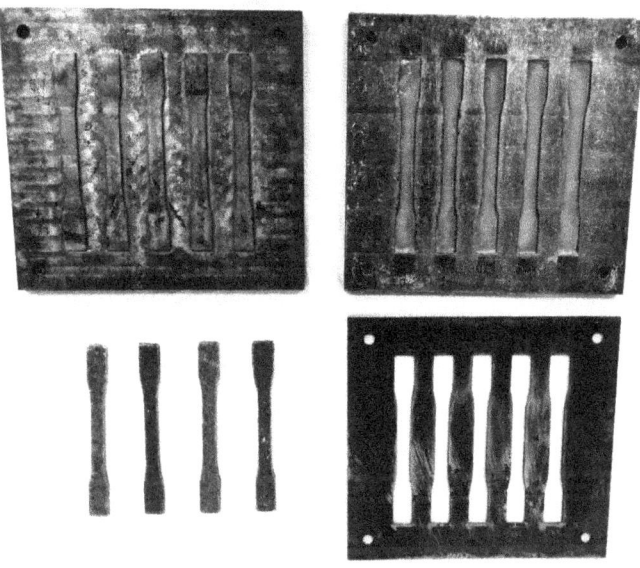

Figure 2. Photo of tensile specimen fabrication.

In order to fabricate the sandwich structure, a banana and synthetic fiber epoxy composite was layered with carbon/Kevlar plies. Here, these 2/2 woven hybrid skins were initially cut to a size of 300 × 300 mm before being placed into picture frame molds. Prior to that, those fibers were gradually mixed with epoxy-resin paste. Further, the sandwich structures were prepared using a hot press machine. The specimens were then heated to 70 °C for one hour under a pressure of 1 bar before leaving them to cure overnight. The samples were inspected visually before being sectioned into 150 × 100 mm samples. In this study, at least five samples for each configuration were examined. The configuration of this structure is shown in Figure 3.

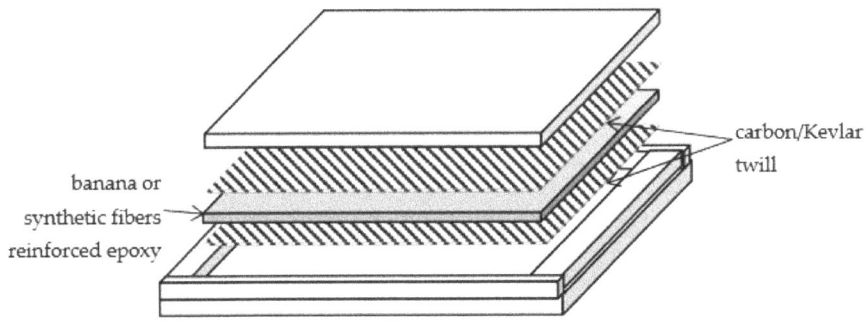

Figure 3. Schematic of the fiber-reinforced epoxy sandwich structure fabrication.

2.4. Tensile Test Properties

The tensile test was conducted using a Universal Testing Machine, (Shimadzu AGX-S, Kyoto, Japan). This table stand tensile machine was fixed with a 10 kN load cell. Testing was obtained at a crosshead speed of 1 mm/min. Tests were conducted for all five samples and their average values were used as the final result.

2.5. Low-Velocity Impact Test

The low-velocity impact properties of the sandwich structures were evaluated using a floor-standing impact tower—CEAST 9340, (Instron, Pianezza TO, Italy)—according to the ASTM D3763 standard [30]. The tests were investigated at energy levels from 5 J to 20 J using a hemispherical steel indenter with a 12.7-mm diameter. Then, the load–displacement traces were recorded, followed by calculating the energy absorbed from the area under the curves.

2.6. Dye Penetrant Application

The damage area subjected to the impact was easily located using a dye penetrant that complied with the ASTM E1417 standard [31]. This nondestructive testing utilized a Spotcheck SKL-SP2 kit supplied by Kird Enterprise, Nilai Negeri Sembilan, Malaysia. Initially, all samples were cleaned using a solvent to remove dirt, sand, and grease. The aerosol red-color penetrant was sprayed on the surface and the samples were left for 10 min. Then, the excess red penetrant was gently cleaned, and finally, the well-shaken developer was applied to the impacted area to increase the visibility of the damaged region. The red spots that remained visible on the sample's surface were due to the damaged area affected by the impact loading.

2.7. Damage Area Measurement

Specimen damage for an area of the impacted sandwich structure was characterized using ImageJ version Java 1.8.0_172 (National Institutes of Health (NIH), Maryland, US and Laboratory for Optical and Computational Instrumentation (LOCI), University of Wisconsin, US) software. Initially, to convert the color-scanned photo to grayscale, the Image/Type/8-bit command was used. Then, a straight line was drawn from edge-to-edge of the photo as a known measurement distance. The Analyze/Set Scale/Known Distance/Unit of Measurement (mm)/Global command was chosen to set the scale parameter. Moreover, the damage area color was inverted using the Image/Adjust/Threshold manual setting. Finally, the calculated area outlines were measured using a rectangular selection tool called the Analyze/Analyze Particle function.

2.8. The Compression after Impact Test

The residual compressive strengths of the post-impacted specimens were evaluated using the CAI test setup. The specimens were fully clamped using the anti-buckling Boeing CAI test fixture according to the ASTM D7137-17 standard [32]. An in-plane compression load was applied at a crosshead displacement rate of 1.25 mm/min until the specimen failed. A Shimadzu AGX universal testing machine fitted with a 300 kN load cell was used to obtain the load–displacement traces. The compressive residual strength of the materials was characterized using the ultimate load prior to failure over the cross-sectional area of the specimen.

3. Results and Discussion

3.1. Tensile Properties of Composites

Figure 4 illustrates the stress–strain curves of the neat epoxy system, banana fiber composite, and optimal banana reinforced epoxy composites. It can be seen that the maximum stress and tensile modulus of the optimal banana fiber reinforced epoxy composite were increased by 66% and 22%, respectively, compared to the neat epoxy resin. It is suggested that the properly optimized the fiber blending condition improved the load-bearing capabilities between the fibers and matrixes. A similar finding was also reported by Yaghoobi and Fereidoon [23]. In this figure, the untreated, 0.25-mm length, and mixed at 50 wt.% fiber loading of a banana-fiber-reinforced epoxy composite was also included for comparison. Interestingly, these composites unfolded at lower tensile strength and modulus values than the virgin epoxy-resin. In addition, the tensile modulus of this unoptimized composite was found to be 792 MPa, which was lower than that of the optimal banana epoxy composite with 1628 MPa. It was demonstrated that adding inappropriate natural fiber "debris" in the polymeric matrixes decreased the material properties of the structure.

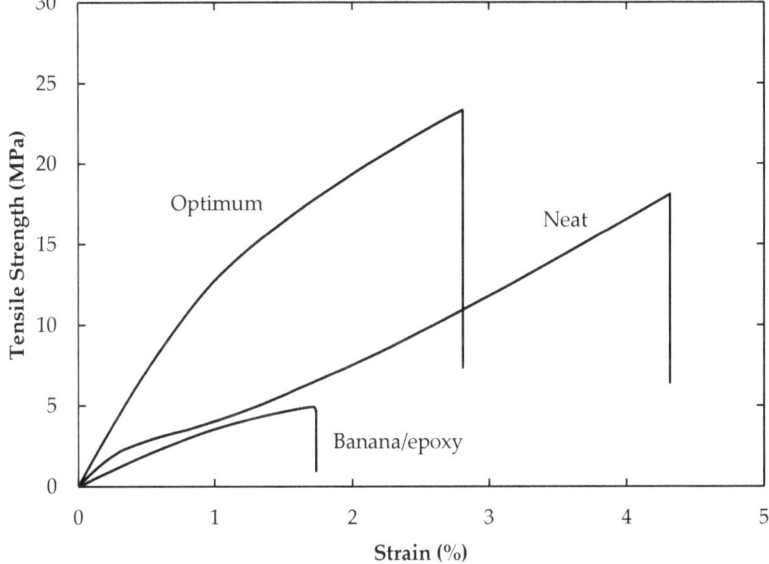

Figure 4. Typical stress–strain curve for neat, optimal banana/epoxy composite, and banana/epoxy composite.

In addition, the tensile modulus of this impair composite was obtained to be 792 MPa, which was evidently lower than that of the optimum banana epoxy composite with at 1628 MPa. It can be demonstrated that adding inappropriate "debris" of natural fiber in the polymeric matrixes decreased the material properties of the structure. The average and standard deviation of tensile properties of the composites are illustrated in Table 2.

Table 2. The average and standard deviation values of the modulus of elasticity, tensile strength, and percentage elongation of the virgin epoxy and natural fiber composites.

Material	Tensile Strength (MPa)	Modulus of Elasticity (MPa)	Elongation (%)
Neat Epoxy	19.18 ± 2.6	980 ± 1.8	4.21 ± 0.8
Banana/Epoxy	8.23 ± 1.4	792 ± 2.6	1.72 ± 1.3
Optimum Banana/Epoxy	23.30 ± 2.3	1628 ± 4.8	2.81 ± 2.4

3.2. Low-Velocity Impact Response of Sandwich Structures

Figure 5 illustrates typical load–displacement traces for the low-velocity impact response at four different impact energy levels correspond to the optimal banana, Kevlar, glass, and carbon epoxy sandwich structures. All traces initially exhibited a similar pattern with the force increasing in an almost linear manner before reaching a peak load. This was associated with the plastic deformation of the sandwich structures directly beneath the hemispherical impactor. Figure 5a presents typical load versus displacement traces for the impact loading of the optimal banana epoxy sandwich structures. It can be seen from the figure that increasing the incident impact energy increased the effective slope of the curve up to 3 mm of displacement, suggesting a rise in the contact stiffness of the system. Beyond this point, a non-linear trace pattern was observed, resulting in the initiation of unstable crack propagation and fiber fracturing. A very drastic fall in the contact load was apparent in the trace for the 20 J impact energy level due to sub-critical propagation and brittle failure of the core. Continued loading often resulted in microcracking and instabilities, with this being most evident in the 15 J impact energy trace where very small load jumps were apparent in the trace at displacements above 3 mm. In addition, the enclosed area within the loading and unloading curves was a measure of the energy absorbed due to the damage in the laminates. Then, the impactor was rebounding since the force and deflection values decreased in the unloading phase. At all the energy levels used in this study, the impactor did not fully perforate the sandwich structures and a portion of the impact energy was conserved as elastic energy. This energy was then transferred back to the steel impactor, causing it to rebound. He et al. [33] suggested that if the core structure was a highly brittle material, the rapid rebound of the skin under impact can result in debonding at the interface between the adjacent layers of the face sheet and the core. Figure 5b–d reveals the typical load displacement traces for Kevlar, glass, and carbon epoxy sandwich structures after being subjected to 5, 10, 15, and 20 J impact energies. From the figure, it is evident that the Kevlar/epoxy composite has a characteristically higher contact stiffness than that of the other composite structures; however, the carbon/epoxy structure offered less "resting" displacement during the unloading phase. It can be suggested that this has resulted in less of a dent depth for the carbon epoxy structures.

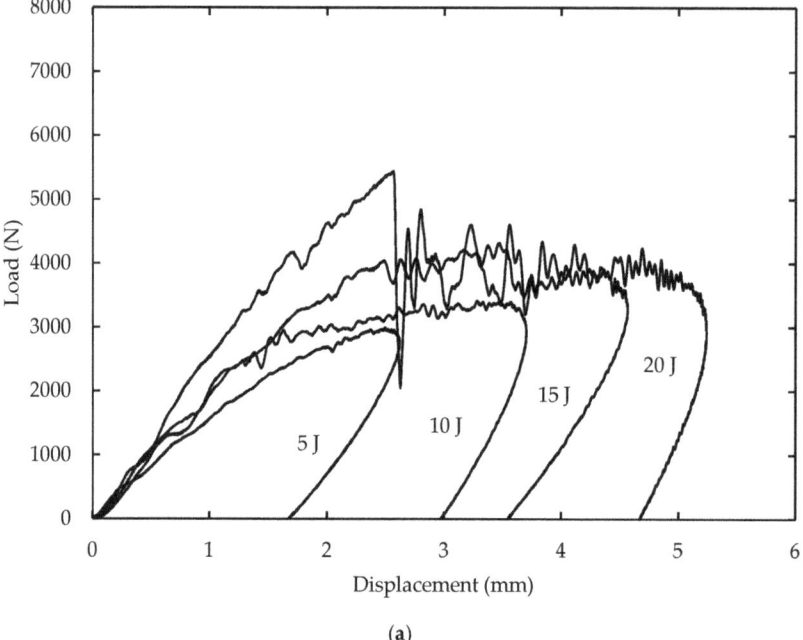

(a)

Figure 5. Cont.

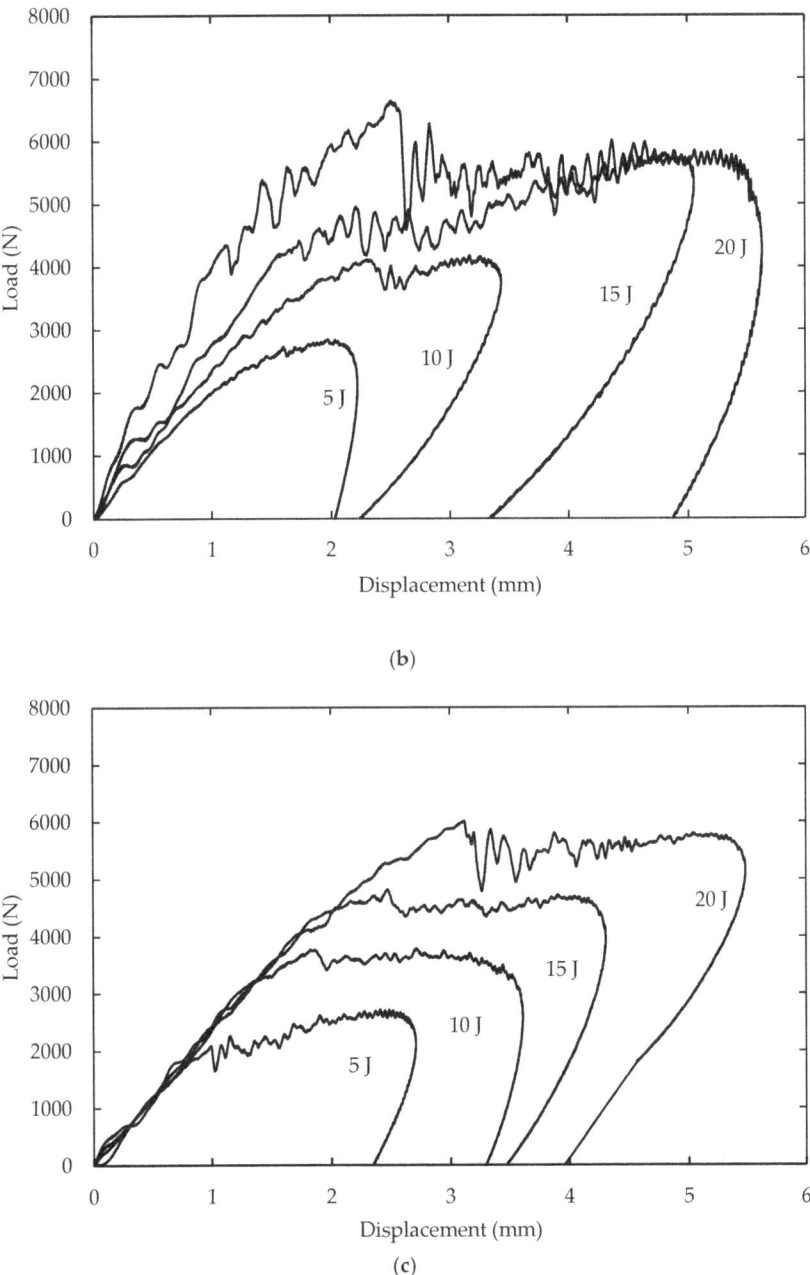

(b)

(c)

Figure 5. *Cont.*

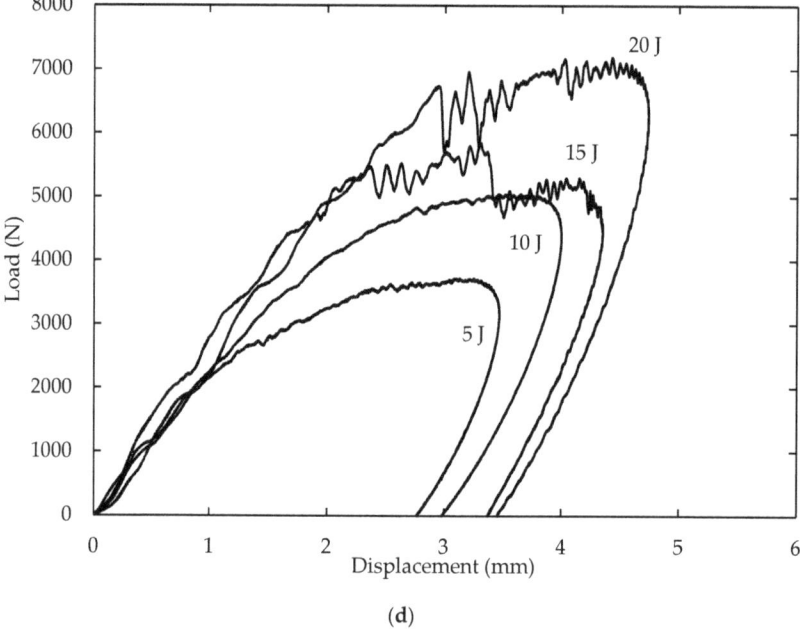

Figure 5. Typical load–displacement traces of (**a**) banana, (**b**) Kevlar, (**c**) glass, and (**d**) carbon/epoxy sandwich structures.

The variations in the peak force at different impact energy levels for composite structures are shown in Figure 6. It can be observed that the peak impact force was increased with increases in the energy level. A similar observation was also reported by Olsson et al. [34]. In addition, the optimal banana/epoxy composite offered a lower peak force than those of the synthetic/epoxy structures. Joseph et al. [35] claimed that fiber pullout and fiber compatibility were the major contributors to the toughness of the composite. Thus, synthetic fibers offered a significantly higher peak load, suggesting higher bonding capabilities with the matrix due to the surface smoothness and consistency of the cross-section. In the case of natural fibers, such a mechanism was not favored due to the mechanical interlocking between fibers and the matrices. In this study, to increase the compatibility between the banana fiber and the matrix, the chemical treatment was promoted; however, the peak force still remained unimpressive.

The peak displacement indicates the maximum deformation of the sandwich panels, which results in a significant area of damage after subjection to different energy levels. Figure 7 shows representative maximum dent depth against energy level traces for different core composites. In general, the maximum displacement steadily increased with rising impact energy. It can be seen that carbon fiber had the highest dent depth at low energy levels and it was the lowest at higher impact energies. This was in part due to the high in-plane tensile properties of the material. Properties of the core, including rigidity and brittleness, also influenced this behavior. In a recent study by Chen and Hodgkinson [36], it was found that at high peak displacement, the uppermost skins of the specimens exhibited more evidence of splitting and delamination of the surface ply. Further, comparing the maximum displacement values for optimum banana and synthetic structures, it can be seen that the values vary between those structures. This helps to explain why the differences between the behavior of core structures were within the scattering; as a result, peak displacement is another independent parameter of sandwich panels.

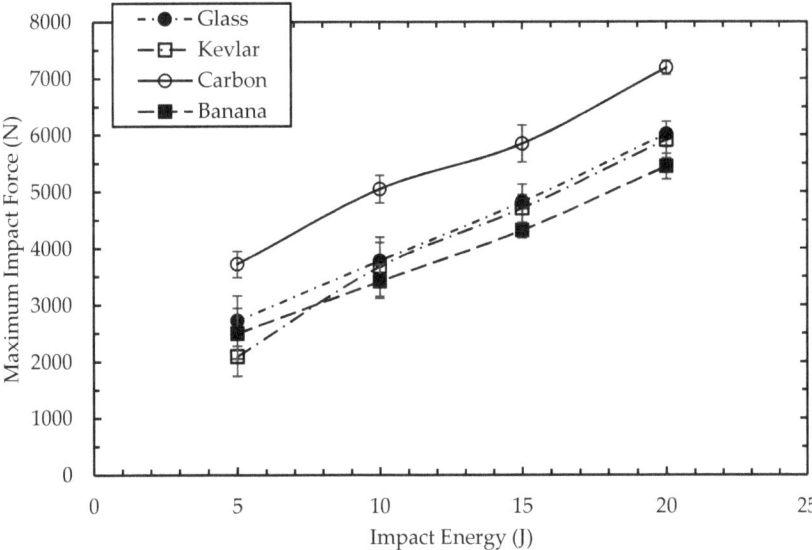

Figure 6. The variation of the maximum impact force at different impact energy levels for composite structures.

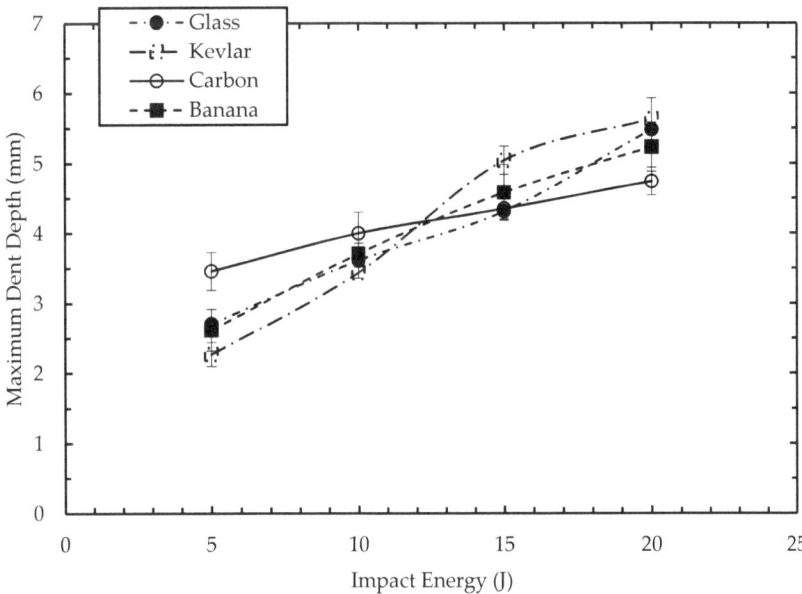

Figure 7. Variation of the maximum dent depth against energy level traces for composite structures.

The total impact energy is a combination of the energy absorbed and the elastic energy loss of the impacted structures [37]. If the percentage of the energy absorbed is high, the energy converted to elastic energy lost by the impactor is low. As a result, the impact energy is fully transferred to the structures at the point of maximum displacement [38]. In order to evaluate the impact response and resistance of the composite structures, the energy absorbed against impact energy traces are shown in Figure 8. The absorbed energy was measured by calculating the area under the force–displacement traces. An observation of the figure suggests that the absorbed energy tended to increase with the incident impact energy, although there did appear to be scattering in the findings, particularly between the optimal banana/epoxy and carbon/epoxy composites. The lower energy absorption means that there was not much energy lost due to failure. Thus, each failure mechanism, including matrix cracks, interlayer failures, delamination, and fiber breakage, absorbed a fraction of the impact energy. The variation of energy absorbed depends on the skin thickness [39], the mechanical properties of fibers and matrices [40], the density of the core [41], and the impactor head [42].

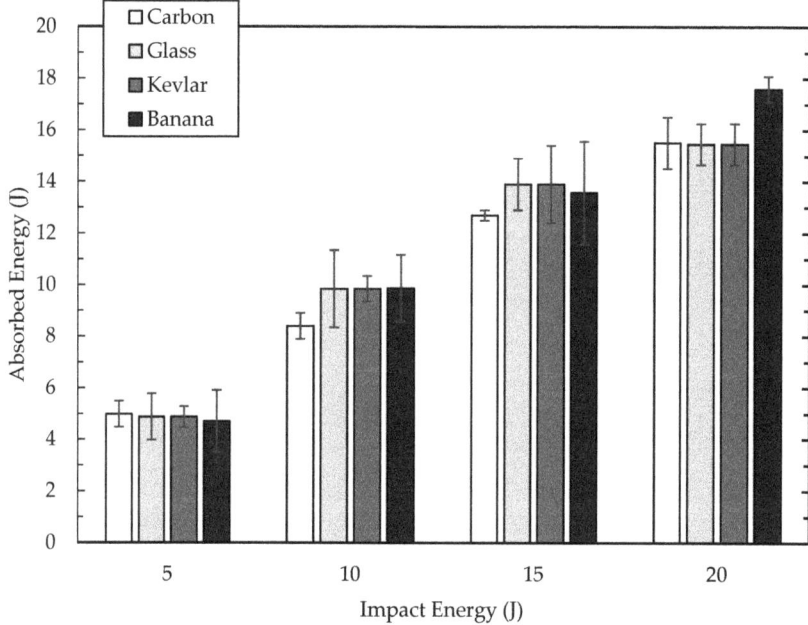

Figure 8. Typical absorbed energies at various incident energy levels.

The evolution of the damage area with impact energy for the optimum banana/epoxy sandwich structure is presented in Figure 9. The sample data was elucidated using a dye penetrant inspection beginning at the uppermost impacted face and followed by measurement of the area using ImageJ software. According to the load–displacement curves (Figure 5), the onset of damage, essentially matrix cracking, was at approximately 3000 N. As the impact energy increased, greater propagation of the damage area in the transverse direction of the specimen was obtained. As can be seen from Figure 9a, a closely peanut shape of visible marks, just like in plain laminates [43], was observed, which was dominated by matrix cracking of the skin. Further loading often resulted in delamination and fiber breakage, which was revealed by the changes of dye color, as shown in Figure 9b–d, which illustrate a similar shape for the dented areas. It can be suggested that this damage mainly corresponds to delamination of the skin and brittle fracture of the core. Here, a small permanent indentation (residual indentation) was created on the uppermost skin, which added to the fracture resulting from the composite core around the impact site. If the core layer is a brittle material, the fast rebound of the face sheet under the impact will result in debonding at the interface between the face sheet and the core. Selver et al. [44] mentioned that larger dent depths in the structure may be attributed to impact energy that was absorbed by a smaller area, which generates a greater plasticity of the composite, stiffness degradation, and more localized damage under the hemispherical steel impactor.

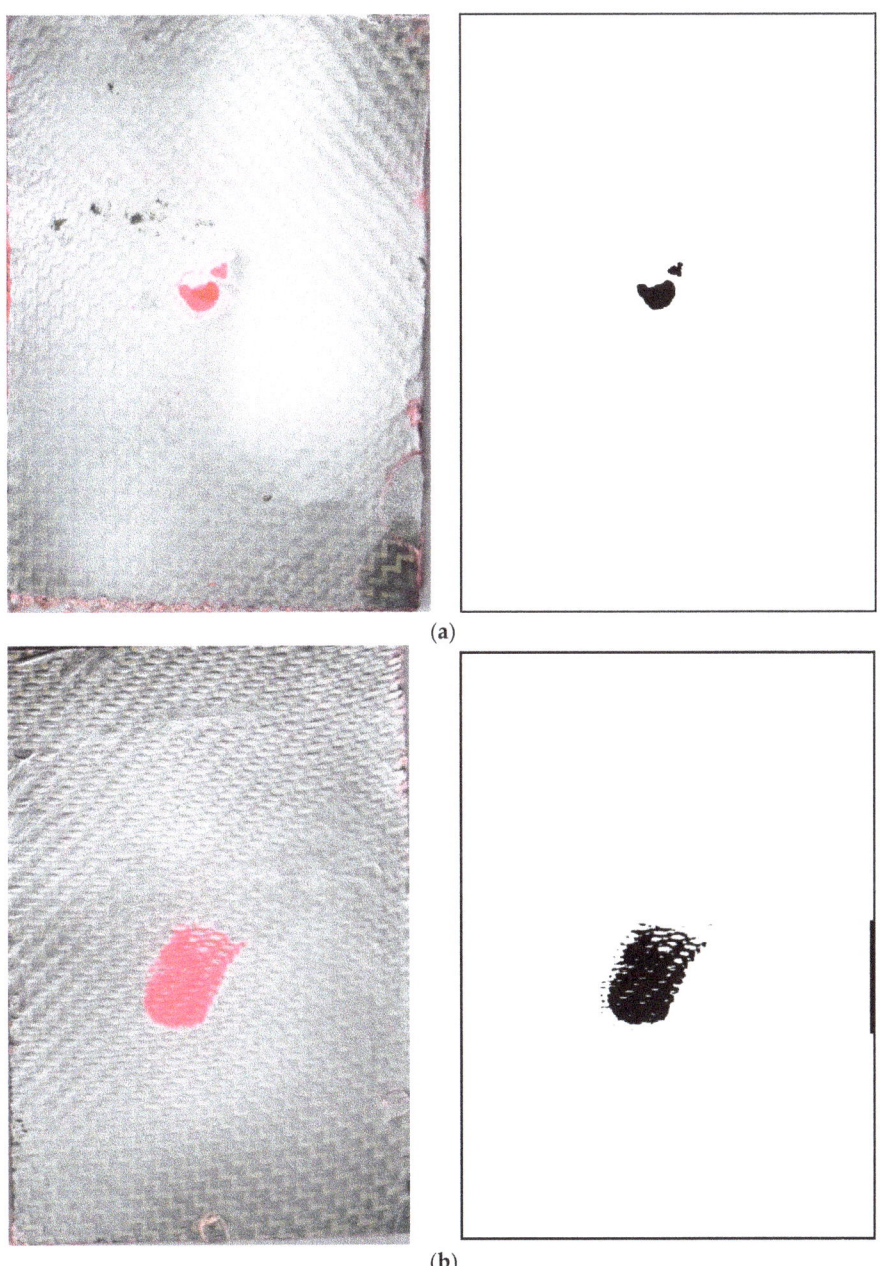

(a)

(b)

Figure 9. *Cont.*

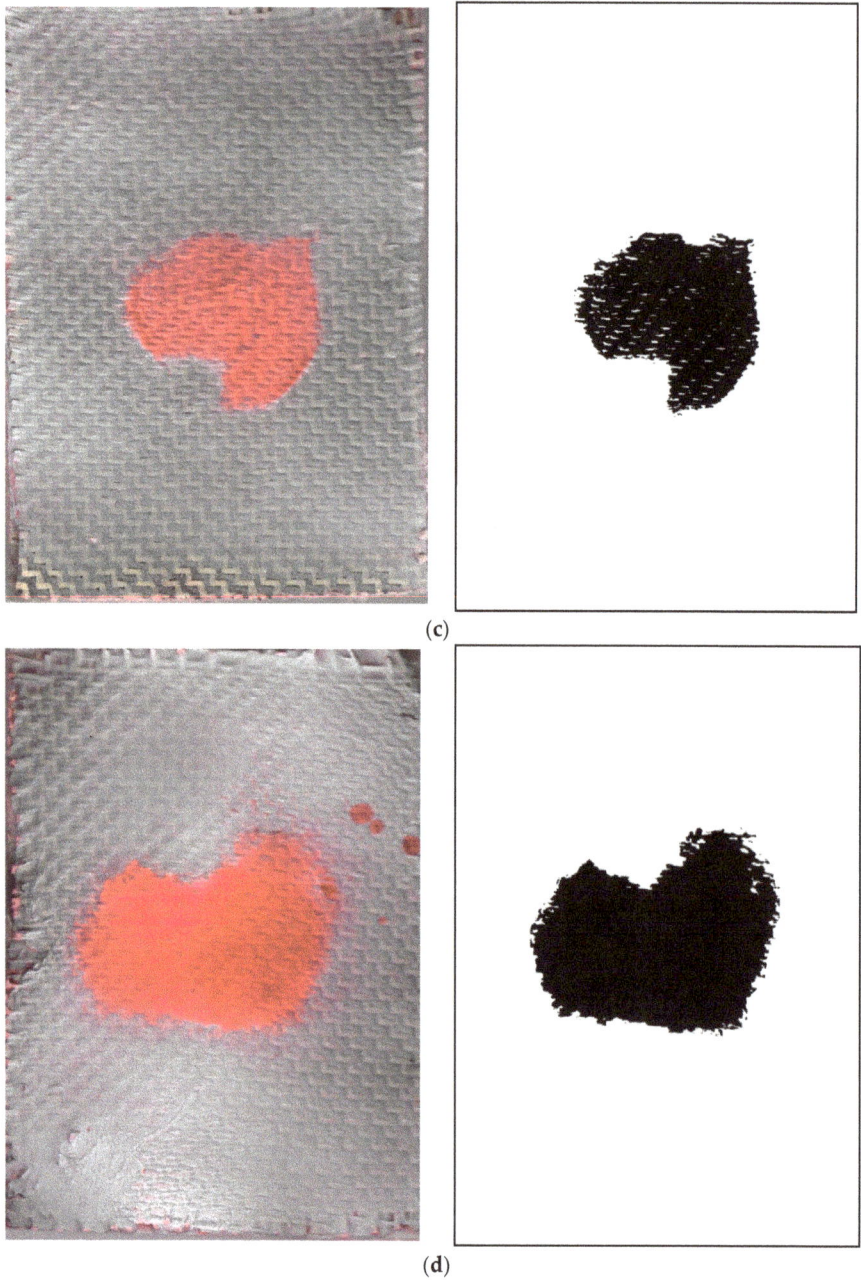

Figure 9. Damage area of impacted specimens of the banana/epoxy sandwich structure with respect to different impact energy levels: (**a**) 5 J, (**b**) 10 J, (**c**) 15 J, and (**d**) 20 J.

Figure 10 shows the front- and back-face damaged areas versus the nominal impact energies. The areas were calculated by post-processing the results of the dye penetrant test, as shown in Figure 9. Many works [45,46] reported that the relationship between the in-plane damaged area and the incident impact energy of the impactor is linear. As can been seen from Figure 10a, in the impact process for sandwich structures where the energy level exceeded 10 J, after the damage to the top skin and the core, there was an amount of residual energy, which was associated with the damage to the bottom face sheet. A pronounced in-plane damage area for the back face of the optimal banana structures is shown in Figure 10b, suggesting a weak core material leading to serious failure of the bottom ply.

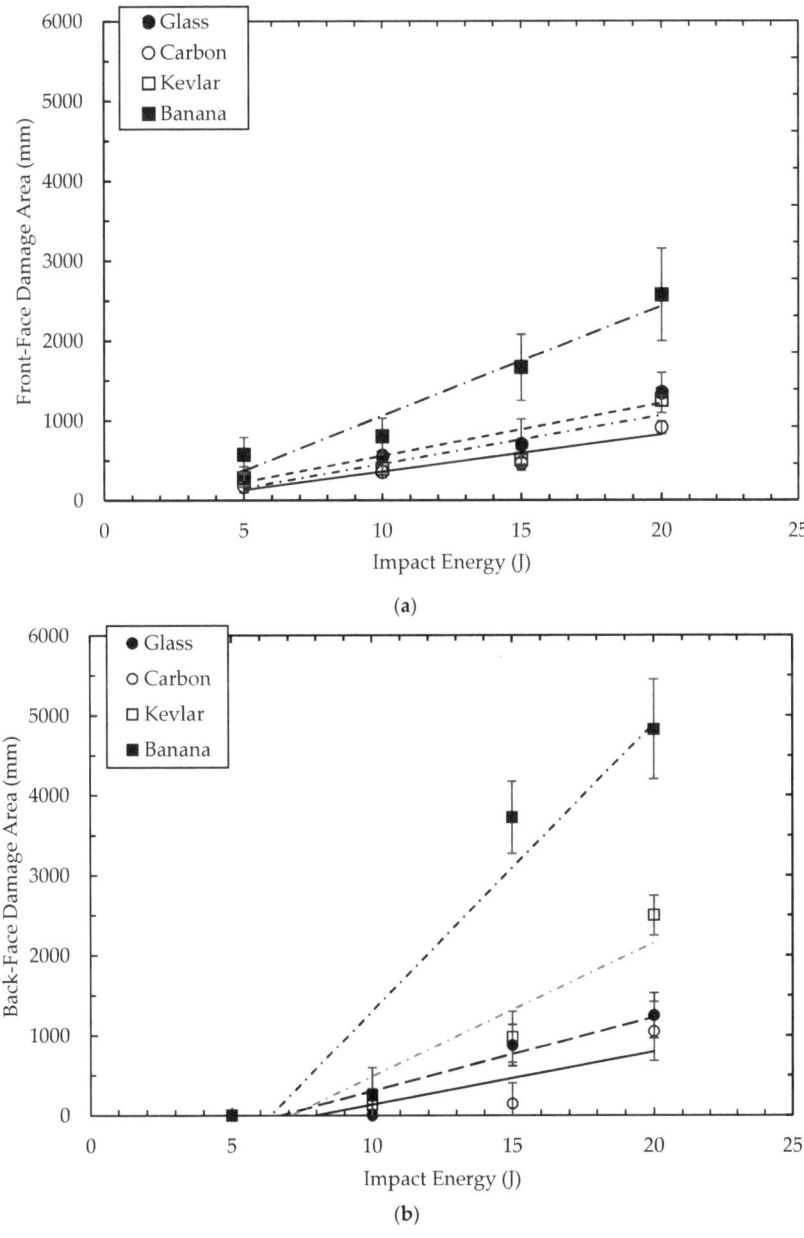

Figure 10. Nominal (**a**) front- and (**b**) back-face damaged areas versus incident impact energies.

3.3. Compression after Impact (CAI) of Structure Properties

Figure 11 shows the variations of CAI strength performance with the displacement for the optimal banana, carbon, Kevlar, and glass sandwich structures following an impact energy of 10 J. For all sandwich panels, the structures exhibited an initial linear CAI strength before reaching a maximum value, following which the strength gradually fell, an effect that was also reported previously [47]. The figure shows that an increase in peak stress resulted in a decrease in the displacement, highlighting the presence of the stronger and stiffer core structures. The carbon/epoxy based sandwich structure was seen to offer a greater residual strength than those of other systems. Furthermore, the findings also show that the CAI strength of the natural-fiber-reinforced epoxy was still lower than its synthetic fiber counterparts. The differences in residual stress were recorded at 22%, 63%, and 136% between banana and Kevlar, glass, and carbon respectively.

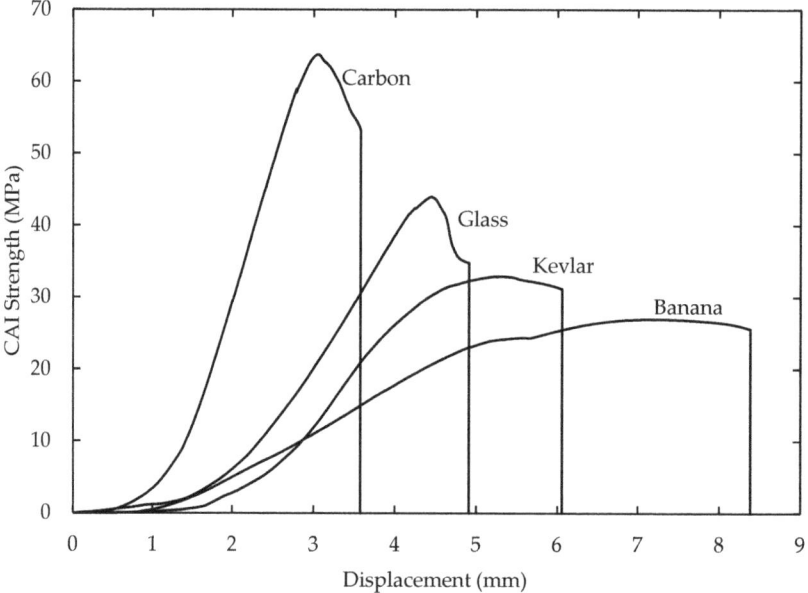

Figure 11. The variation of compression after impact (CAI) strength with the displacement for four different sandwich structures subjected to an energy impact of 10 J.

Figure 12 illustrates the average CAI strength versus impact energy traces of the sandwich structures, including those of unimpacted specimens. As observed, the compression strengths decreased with the increase in impact energy. The carbon fiber sandwich structure exhibited a massive residual strength under all impact energies with the optimal banana composite system providing the least. The outstanding residual strength performance of the carbon/epoxy sandwich structure was due to the high resistance to shear cracking between the carbon and epoxy of the core. The use of the surface treatment and fiber conditions of the banana pseudo-stem fiber in order to increase the surface bonding between the fiber and matrix was still insufficient and failed to offer high resistance to CAI strength compared to synthetic fibers. With respect to CAI behavior in the sandwich structures experiments, core breaking and skin buckling were identified as the major failure modes of the sandwich structures [48]. Castanié et al. [49] mentioned that matrix cracking and delamination between the face sheet and the core were found to be the most common damage mechanisms of the sandwich structure under residual compression strength.

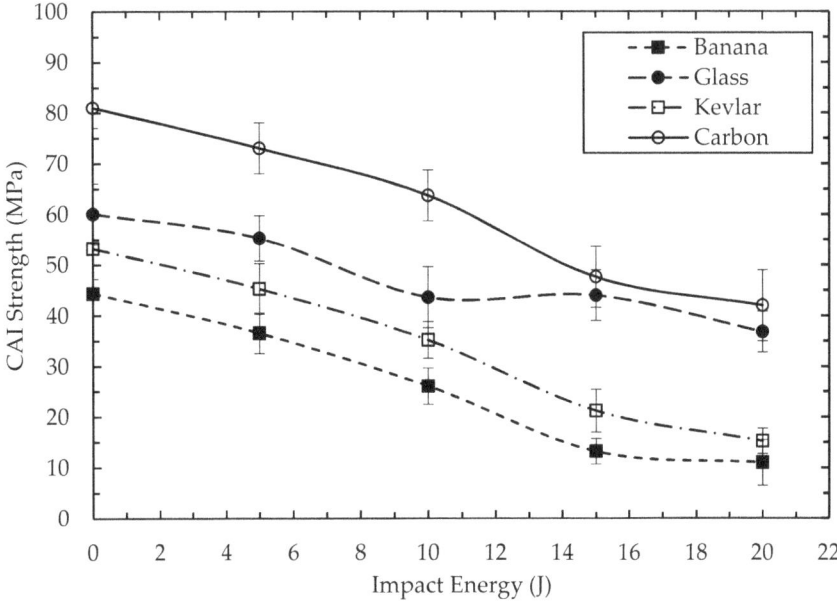

Figure 12. The average CAI strength versus impact energy traces of the sandwich structures.

In order to compare the properties of impact damage on the strength of the material, the ratio between the CAI strength of the specimens damaged by a given impact and the unimpacted sample was measured. Figure 13 presents the normalized residual strength as a function of the incident impact energy. The optimal banana/epoxy sandwich structure showed the greatest reduction in normalized strength; however, the lowest drop in the normalized residual strength was obtained in the carbon/epoxy core structures. In addition, the normalized CAI strength gradually decreased with the increasing incident impact energies. Similar findings were reported by Wang et al. [50]. At an impact energy of 15 J, the normalized reductions of this strength were: 41% in the carbon/epoxy structure, 43% in the glass/epoxy structure, 60% in the Kevlar/epoxy structure, and 70% in the optimum banana/epoxy structure. The lowest reduction in normalized value, especially of the synthetic fiber structure, may be due to the stiffer core, making the structure more stable and with a high resistance to impact. As a result, synthetic fiber/epoxy systems were more prone to buckling during CAI testing.

Figure 14 shows the CAI setup conditions of the optimum banana/epoxy sandwich structures subjected to a 15 J impact energy. The major delamination face was observed perpendicular to the loading direction. As can be seen in the figure, the failure form of the outer skin was more severe at the point of impact. In addition, the debonding phenomenon occurred between the core and the outer surface extended rapidly from the middle of the damaged area to both sides of the edge. All failure modes for the testing sample were similarly observed and were associated with the compression shear failure and local buckling at the point of impact [51].

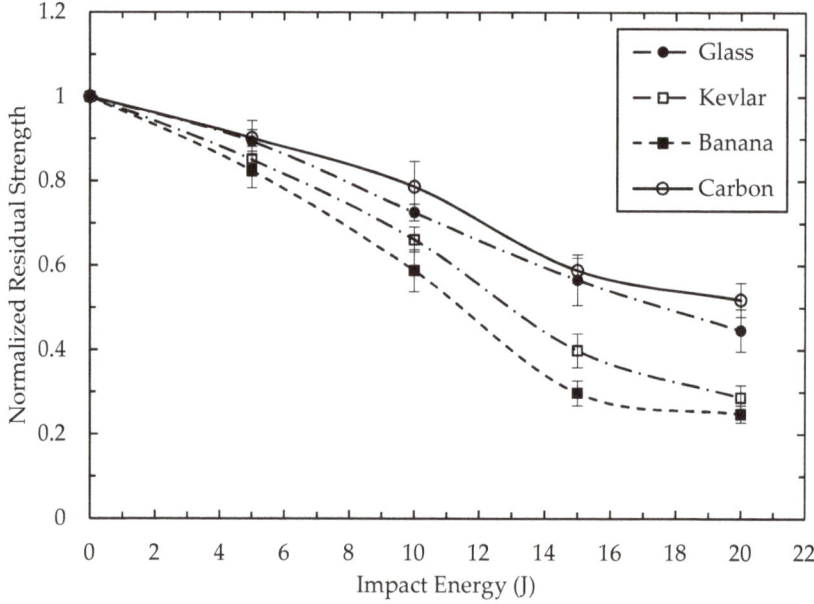

Figure 13. Normalized residual strength as a function of incident impact energy.

Figure 14. The CAI testing condition of the banana-epoxy sandwich structures subjected to an impact energy of 15 J.

Typical failure modes of the optimal banana, glass, carbon, and Kevlar epoxy sandwich structures subjected to compression loading are illustrated in Figure 15. It can be seen that the optimal banana epoxy core largely fractured from the middle position and the uppermost skin was folded (Figure 15a). From the side view of the failure specimen, compression shear cracks located at the impacted zone and delamination of the outer face sheet were extensively observed for the glass/epoxy system (Figure 15b). In addition, shear matrix cracking toward the direction of CAI loading was also obtained for the Kevlar structure, as shown in Figure 15c. This damage occurred due to stress concentration at the impact point of loading and the failure mode propagating throughout the area. The difference in the material properties between the high stiffness of the skin and the soft core contributed to the catastrophic failure of the sandwich structure during CAI testing [52]. Less core buckling and face sheet microbuckling of the carbon/epoxy structures was observed (Figure 15d), suggesting a high stiffness of the core and resistance to the axial compressive stress within the face sheets. V-shaped shear cracking was clearly apparent in this sandwich panel.

Figure 15. Cross-section of failure modes for the CAI testing of the (**a**) banana, (**b**) glass, (**c**) carbon, and (**d**) Kevlar epoxy sandwich structures.

4. Conclusions

The low-velocity impact test and residual compression behaviors for the optimized fiber loading, fiber length, and alkaline treatment content of the banana-fiber reinforced sandwich structures were obtained. In this study, glass/epoxy, carbon/epoxy, and Kevlar/epoxy composites were also tested for comparison. This work contributes several key findings, as follows:

i. The maximum stress and tensile modulus of the optimal banana-fiber-reinforced epoxy composite were increased up to 90% and 22%, respectively, compared to the epoxy-resin system.
ii. The effect of optimizing the compounding parameters of banana-reinforced epoxy was significantly comparable in terms of impact, damage tolerance, and residual impact due to higher bonding between fibers and matrices.

iii. The low-velocity impact response of the optimal banana/epoxy composite gave a lower peak load. However, a significantly large damage area, higher energy absorption, and greater dent depth were obtained.

iv. The banana fiber sandwich structure recorded a lower CAI resistance and a greater reduction in normalized strength than those using synthetic fibers.

v. Delamination and core fracture were mainly observed in the optimal banana/epoxy structure. However, under the CAI testing matrix, shear cracking dominated in the Kevlar, carbon, and glass composites.

Finally, the results depicted significantly lower values for the impact resistance and residual damage tolerance of the optimum banana composite structures. Further investigation through matrix modification by adding nanofillers to natural fiber composites may yield interesting discoveries.

Author Contributions: Formal analysis, Z.A.R.; Funding acquisition, R.D. and M.Y.M.D.; Investigation, M.Z.H. and A.F.M.N.; Supervision, S.M.S. All authors have read and agreed to the published version of the manuscript.

Funding: This research was funded by Universiti Teknologi Malaysia under "Geran Universiti Penyelidik" (GUP) Tier 2 Scheme Q.K.130000.2656.15J85, Tier 2 Scheme Q.K.130000.2656.17J66, and UTMER Scheme Q.K.130000.2656.18J24.

Acknowledgments: The authors are thankful to the Ministry of Education Malaysia and Universiti Putra Malaysia for supporting this work through a Visiting Scholar (Post-Doctoral) scholarship.

Conflicts of Interest: The authors declare no conflict of interest.

References

1. Aziz, N.A.; Ho, L.H.; Azahari, B.; Bhat, R.; Cheng, L.H.; Ibrahim, M.N.M. Chemical and functional properties of the native banana (*Musaacuminata × balbisiana Colla cv.* Awak) pseudo-stem and pseudo-stem tender core flours. *Food Chem.* **2011**, *128*, 748–753. [CrossRef]
2. Subagyo, A.; Chafidz, A. Banana Pseudo-Stem Fiber: Preparation, Characteristics, and Applications. *IntechOpen* **2018**. [CrossRef]
3. Abdullah, N.; Sulaiman, F.; Miskam, M.A.; Taib, R.M. Characterization of banana (*Musa* spp.) pseudo-stem and fruit-bunch-stem as a potential renewable energy resource. *Int. J. Biol. Vet. Agric. Food. Eng.* **2014**, *8*, 712.
4. Hassan, M.Z.S.S.; Rasid, Z.A. Thermal Degradation and Mechanical Behavior of Banana Pseudo-Stem Reinforced Composite. *Int. J. Recent Technol. Eng.* **2019**, *8*, 4.
5. Sapuan, M.S.; Leenie, A.; Harimi, M.; Beng, Y.K. Mechanical properties of woven banana fibre reinforced epoxy composites. *Mater. Des.* **2006**, *27*, 689–693. [CrossRef]
6. Maleque, M.; Belal, F.; Sapuan, S. Mechanical properties study of pseudo-stem banana fiber reinforced epoxy composite. *Arab. J. Sci. Eng.* **2007**, *32*, 359–364.
7. Ogunsile, B.O.; Oladeji, T.G. Utilization of banana stalk fiber as reinforcement in low density polyethylene composite. *Matéria* **2016**, *21*, 953–963. [CrossRef]
8. Prasad, N.; Agarwal, V.K.; Sinha, S. Banana fiber reinforced low-density polyethylene composites: Effect of chemical treatment and compatibilizer addition. *Iran. Polym. J.* **2016**, *25*, 229–241. [CrossRef]
9. Ahmed, M.S.; Attia, T.; El-Wahab, A.A.A.; El-Gamsy, R.; El-latif, M.H.A. Effect of adding banana pseudo stem on the mechanical properties of ABS composites. *J. Al-Azhar Univ. Eng. Sect.* **2018**, *13*, 1090–1098. [CrossRef]
10. Venkateshwaran, N.; ElayaPerumal, A.; Jagatheeshwaran, M. Effect of fiber length and fiber content on mechanical properties of banana fiber/epoxy composite. *J. Reinf. Plast. Compos.* **2011**, *30*, 1621–1627. [CrossRef]
11. Udaya Kiran, C.; Ramachandra Reddy, G.; Dabade, B.; Rajesham, S. Tensile properties of sun hemp, banana and sisal fiber reinforced polyester composites. *J. Reinf. Plas. Compos.* **2007**, *26*, 1043–1050. [CrossRef]
12. Zainudin, E.; Sapuan, S. Impact strength and hardness properties of banana pseudo-stem filled unplastisized PVC composites *Multidiscip. Model. Mater. Struct.* **2009**, *5*, 277–282. [CrossRef]
13. Pavithran, C.; Mukherjee, P.; Brahmakumar, M.; Damodaran, A. Impact properties of natural fibre composites. *J. Mater. Sci. Lett.* **1987**, *6*, 882–884. [CrossRef]
14. Balaji, A.; Sivaramakrishnan, K.; Karthikeyan, B.; Purushothaman, R.; Swaminathan, J.; Kannan, S.; Udhayasankar, R.; Madieen, A.H. Study on mechanical and morphological properties of sisal/banana/coir fiber-reinforced hybrid polymer composites. *J. Braz. Soc. Mech. Sci. Eng.* **2019**, *41*, 386. [CrossRef]

15. Pothan, L.A.; Thomas, S.; Neelakantan, N. Short banana fiber reinforced polyester composites: Mechanical, failure and aging characteristics. *J. Reinf. Plas. Compos.* **1997**, *16*, 744–765. [CrossRef]
16. Devireddy, S.B.R.; Biswas, S. Physical and thermal properties of unidirectional banana–jute hybrid fiber-reinforced epoxy composites. *J. Reinf. Plas. Compos.* **2016**, *35*, 1157–1172. [CrossRef]
17. Narayana, K.S.; Suman, K.; Ravindra, A. Experimental Investigation on Mechanical Characterization of Nanoclay-Reinforced Banana Fiber/E-Glass/Epoxy Resin Hybrid Nanocomposite. In *Recent Advances in Material Sciences*; Springer: Singapore, 2019; pp. 609–625.
18. Dhakal, H.; Arumugam, V.; Aswinraj, A.; Santulli, C.; Zhang, Z.; Lopez-Arraiza, A. Influence of temperature and impact velocity on the impact response of jute/UP composites. *Polym. Test.* **2014**, *35*, 10–19. [CrossRef]
19. Ismail, M.F.; Sultan, M.T.; Hamdan, A.; Shah, A.U.; Jawaid, M. Low velocity impact behaviour and post-impact characteristics of kenaf/glass hybrid composites with various weight ratios. *J. Mater. Res. Technol.* **2019**, *8*, 2662–2673. [CrossRef]
20. Nor, A.F.M.; Sultan, M.T.H.; Jawaid, M.; Azmi, A.M.R.; Shah, A.U.M. Analysing impact properties of CNT filled bamboo/glass hybrid nanocomposites through drop-weight impact testing, UWPI and compression-after-impact behaviour. *Compos. Part B Eng.* **2019**, *168*, 166–174. [CrossRef]
21. Rubio-López, A.; Artero-Guerrero, J.; Pernas-Sánchez, J.; Santiuste, C. Compression after impact of flax/PLA biodegradable composites. *Polym. Test.* **2017**, *59*, 127–135. [CrossRef]
22. Akubue, P.C.; Igbokwe, P.K.; Nwabanne, J.T. Production of kenaf fibre reinforced polyethylene composite for ballistic protection. *Int. J. Sci. Eng. Res.* **2015**, *6*, 7.
23. Yaghoobi, H.; Fereidoon, A. Modeling and optimization of tensile strength and modulus of polypropylene/kenaf fiber biocomposites using Box–Behnken response surface method. *Polym. Compos.* **2018**, *39*, E463–E479. [CrossRef]
24. Roslan, S.A.; Hassan, M.Z.; Rasid, Z.A.; Bani, N.A.; Sarip, S.; Daud, M.Y.M.; Muhammad-Sukki, F. Mode I Fracture Toughness of Optimized Alkali-Treated Bambusa Vulgaris Bamboo by Box-Behnken Design. In *Advances in Material Sciences and Engineering*; Springer: Singapore, 2020; pp. 565–575.
25. Hassan, M.Z.; Sapuan, S.; Roslan, S.A.; Sarip, S. Optimization of tensile behavior of banana pseudo-stem (*Musa acuminate*) fiber reinforced epoxy composites using response surface methodology. *J. Mater. Res. Technol.* **2019**, *8*, 3517–3528. [CrossRef]
26. Venkateshwaran, N.; Elayaperumal, A.; Sathiya, G. Prediction of tensile properties of hybrid-natural fiber composites. *Compos. Part B Eng.* **2012**, *43*, 793–796. [CrossRef]
27. EpoxAmite™. Available online: https://www.smooth-on.com/product-line/epoxamite/ (accessed on 17 January 2020).
28. Materials, Equipment and Supplies for Advanced Composites. Available online: https://www.easycomposites.co.uk (accessed on 17 January 2020).
29. ASTM International. *ASTM D638-14, Standard Test Method for Tensile Properties of Plastics*; ASTM International: West Conshohocken, PA, USA, 2014.
30. ASTM International. *ASTM D3763-18, Standard Test Method for High Speed Puncture Properties of Plastics Using Load and Displacement Sensors*; ASTM International: West Conshohocken, PA, USA, 2018.
31. ASTM International. *ASTM E1417/E1417M-16, Standard Practice for Liquid Penetrant Testing*; ASTM International: West Conshohocken, PA, USA, 2016.
32. ASTM International. *ASTM D7137/D7137M-17, Standard Test Method for Compressive Residual Strength Properties of Damaged Polymer Matrix Composite Plates*; ASTM International: West Conshohocken, PA, USA, 2017.
33. He, W.; Liu, J.; Wang, S.; Xie, D. Low-velocity impact response and post-impact flexural behaviour of composite sandwich structures with corrugated cores. *Compos. Struc.* **2018**, *189*, 37–53. [CrossRef]
34. Olsson, R.; Block, T.B. Criteria for skin rupture and core shear cracking induced by impact on sandwich panels. *Compos. Struc.* **2015**, *125*, 81–87. [CrossRef]
35. Joseph, S.; Sreekala, M.; Oommen, Z.; Koshy, P.; Thomas, S. A comparison of the mechanical properties of phenol formaldehyde composites reinforced with banana fibres and glass fibres. *Compos. Sci. Technol.* **2002**, *62*, 1857–1868. [CrossRef]
36. Chen, F.; Hodgkinson, J. Impact behaviour of composites with different fibre architecture. *Proc. Inst. Mech. Eng. Part G J. Aerosp. Eng.* **2009**, *223*, 1009–1017. [CrossRef]
37. Foo, C.C.; Seah, L.K.; Chai, G.B. Chai Low-velocity impact failure of aluminium honeycomb sandwich panels. *Compos. Struc.* **2008**, *85*, 20. [CrossRef]

38. Safri, S.; Sultan, M.; Cardona, F. Impact damage evaluation of Glass-Fiber Reinforced Polymer (GFRP) using the drop test rig–an experimental based approach. *ARPN J. Eng. Appl. Sci.* **2015**, *10*, 9916–9928.
39. Hassan, M.; Umer, R.; Balawi, S.; Cantwell, W. The impact response of environmental-friendly sandwich structures. *J. Compos. Mater.* **2014**, *48*, 3083–3090. [CrossRef]
40. Liu, T.; Hou, S.; Nguyen, X.; Han, X. Energy absorption characteristics of sandwich structures with composite sheets and bio coconut core. *Compos. Part B Eng.* **2017**, *114*, 328–338. [CrossRef]
41. Hassan, M.; Cantwell, W. The influence of core properties on the perforation resistance of sandwich structures—An experimental study. *Compos. Part B Eng.* **2012**, *43*, 3231–3238. [CrossRef]
42. Hassan, M.; Cantwell, W. Strain rate effects in the mechanical properties of polymer foams. *Int. J. Polym. Technol.* **2011**, *3*, 27–34.
43. Maierhofer, C.; Krankenhagen, R.; Röllig, M. Application of thermographic testing for the characterization of impact damage during and after impact load. *Compos. Part B Eng.* **2019**, *173*, 106899. [CrossRef]
44. Selver, E.; Potluri, P.; Hogg, P.; Soutis, C. Impact damage tolerance of thermoset composites reinforced with hybrid commingled yarns. *Compos. Part B Eng.* **2016**, *91*, 522–538. [CrossRef]
45. Zhang, T.; Yan, Y.; Li, J. Experiments and numerical simulations of low-velocity impact of sandwich composite panels. *Polym. Compos.* **2017**, *38*, 646–656. [CrossRef]
46. Giannopoulos, I.K.; Theotokoglou, E.E.; Zhang, X. Impact damage and CAI strength of a woven CFRP material with fire retardant properties. *Compos. Part B Eng.* **2016**, *91*, 8–17. [CrossRef]
47. Caminero, M.; García-Moreno, I.; Rodríguez, G. Experimental study of the influence of thickness and ply-stacking sequence on the compression after impact strength of carbon fibre reinforced epoxy laminates. *Polym. Test.* **2018**, *66*, 360–370. [CrossRef]
48. Yang, B.; Wang, Z.; Zhou, L.; Zhang, J.; Tong, L.; Liang, W. Study on the low-velocity impact response and CAI behavior of foam-filled sandwich panels with hybrid facesheet. *Compos. Struct.* **2015**, *132*, 1129–1140. [CrossRef]
49. Castanié, B.; Aminanda, Y.; Bouvet, C.; Barrau, J.J. Core crush criterion to determine the strength of sandwich composite structures subjected to compression after impact. *Compos. Struct.* **2008**, *86*, 243–250. [CrossRef]
50. Wang, J.; Chen, B.; Wang, H.; Waas, A.M. Experimental study on the compression-after-impact behavior of foam-core sandwich panels. *J. Sandw. Struct. Mater.* **2015**, *17*, 446–465. [CrossRef]
51. Sanchez-Saez, S.; Barbero, E.; Zaera, R.; Navarro, C. Compression after impact of thin composite laminates. *Compos. Sci. Technol.* **2005**, *65*, 1911–1919. [CrossRef]
52. Bai, R.; Guo, J.; Lei, Z.; Liu, D.; Ma, Y.; Yan, C. Compression after impact behavior of composite foam-core sandwich panels. *Compos. Struct.* **2019**, *225*. [CrossRef]

© 2020 by the authors. Licensee MDPI, Basel, Switzerland. This article is an open access article distributed under the terms and conditions of the Creative Commons Attribution (CC BY) license (http://creativecommons.org/licenses/by/4.0/).

Article

Investigation on the Triaxial Mechanical Characteristics of Cement-Treated Subgrade Soil Admixed with Polypropylene Fiber

Wei Wang [1], Chen Zhang [1], Jia Guo [2,*], Na Li [1,*], Yuan Li [1], Hang Zhou [1] and Yong Liu [3]

1. School of Civil Engineering, Shaoxing University, Shaoxing 312000, China; wellswang@usx.edu.cn (W.W.); zhch2513620@126.com (C.Z.); liyuan3596@163.com (Y.L.); zhouhang1119@126.com (H.Z.)
2. Department of Civil Engineering and Mechanics, Faculty of Engineering, China University of Geosciences, Wuhan 430074, China
3. State Key Lab Water Resources & Hydropower Engineering School, Wuhan University, Wuhan 430072, China; liuy203@whu.edu.cn
* Correspondence: jiaguo@cug.edu.cn (J.G.); lina@usx.edu.cn (N.L.)

Received: 15 September 2019; Accepted: 25 October 2019; Published: 27 October 2019

Abstract: In order to evaluate the improvement effect of fiber on the brittle failure of cement-treated subgrade soil, a series of triaxial unconsolidated undrained (UU) tests were carried out on samples of polypropylene fiber-cement-treated subgrade soil (PCS) with polypropylene fiber mass content of 0‰, 2‰, 4‰, 6‰, and 10‰. The results showed that, (1) the deviatoric stress-axial strain curve of PCS samples were all strain-softening curves. (2) For the same fiber mass content, the peak stress, residual stress, and strain at peak stress of PCS samples gradually increases with the increase in the confining pressure, while their brittleness index gradually decreases. (3) With the increase in confining pressure, compared with that of the 0‰ PCS sample, the increase in peak stress, residual stress, and strain at peak stress of 6‰ PCS sample were in the ranges of 24%–29%, 87%–110%, and 85%–120%, respectively. The decrease in the brittleness index and failure angle was 52%–79% and 16%, while the cohesion and internal friction angle increased by 25.9% and 7.4%, respectively. The results of this study indicate that it is feasible to modify cement subgrade soil with an appropriate amount of polypropylene fiber to mitigate its brittle failure.

Keywords: fiber-cement-treated subgrade soil; mechanical properties; triaxial test; brittleness index; failure angle

1. Introduction

Soft soil subgrades with high compressibility and low strength are often encountered in engineering practice. Generally, the mechanical properties of these subgrades cannot meet the requirements of practical engineering conditions. It is necessary to make ground improvements to avoid serious damage to the roadbed, pavement, and even the upper road structures [1–4]. As a typical example of the chemical improvement method, cement soil is widely applied to various roadbed treatment projects due to its advantages of good integrity, strong water stability, and low cost [5–9]. However, a large number of tests show that, due to the defects of low tensile strength, large brittleness, and poor deformability, cement soil is not suitable for every kind of project [10–13]. In response to the above problems, many researchers have proposed a variety of improved methods based on different materials [14–20], and found that the use of fiber reinforcement technology as a method for improving the mechanical properties performed well in terms of soil improvement [21,22].

In recent years, studies [23–25] show that adding some fibers into the cemented soil can improve the tensile strength, toughness, bearing capacity of soil, and brittle failure to some extent. Tran et al. [26]

conducted unconfined tests on cement soil samples with different fiber mass contents; their results showed that the modification of a soil sample was optimum with a fiber mass content of 3‰– 5‰. Ayeldeen et al. [27] carried out unconfined compressive tests on cemented soft soil samples with different fiber mass contents and found that the compressive strength first increases and then decreases as the fiber mass content increases in samples. When the fiber mass content was around 5‰, the compressive strength reached its maximum value. Yang et al. [28] conducted unconfined tests on cement loess samples with different polypropylene (PP) fiber mass contents and found that fibers have the best effect on the compressive strength and ductility in samples with 4‰ fiber mass content. Estabragh et al. [29] performed unconfined tests on PP fiber cement soft soils with different lengths and contents and pointed out that the compressive strength and tensile strength of cement soft soils increased with the increase in fiber mass content within a certain range, while the fiber length had little effect. By using the direct tensile test and split test results, Xiao and Liu [30] investigated the effect of randomness in orientation of fibers on the tensile strength of a sample. In their study, two types of fibers were considered, namely, PP and polyvinyl alcohol (PVA) fibers. A prediction model was proposed for the tensile strength of cement-admixed clays. The above literature review suggests that fiber reinforcement is a good soil improvement measure, but few have considered the effect of confining pressure. In reality, the ground improvement level is often buried below the ground level. Sometimes the buried depth is as high as 30 m or even deeper. In such circumstances, the confining pressure is likely to affect the mechanical behavior of the cemented soils.

In this study, a series of triaxial unconsolidated undrained (UU) tests was carried out under different confining pressures. The brittleness index was introduced as the criterion for identifying brittle failure. The modification effects of different confining pressures and fiber mass contents on cement subgrade soil are discussed to provide a reference for the application of polypropylene fiber-cement-treated subgrade soil (PCS) in practical engineering problems.

2. Experimental Scheme

2.1. Test Material and Equipment

The soil was taken from Shaoxing City, Zhejiang Province. The specific physical properties are shown in Table 1. PO32.5 Portland cement was used for the test. Polypropylene (PP) fiber of 6-mm length was used, which had chemical stability, strong gripping force with the surface of cement subgrade soil, and high tensile strength. The main technical indexes are shown in Table 2, and the bulk fiber is shown in Figure 1. The full-automated triaxial test apparatus used was TKA-TTS-3S.

Table 1. Physical property indexes of subgrade soil.

Density (g/cm^3)	Pore Ratio	Water Content (%)	Liquid Limit (%)	Plastic Limit (%)	Liquidity Index	Plastic Index
1.65	1.64	30.0	46.2	26.4	1.7	19.8

Table 2. Main technical indexes of polypropylene (PP) fiber.

Fiber Type	Diameter (μm)	Length (mm)	Tensile Strength (MPa)	Modulus Elasticity (GPa)	Stretch Limits (%)
Bunchy monofilament	18–48	6	>358	>3.50	>15

Figure 1. Polypropylene fibers used in this study.

2.2. Test Scheme

In this study, pavement soils were used. According to the practical engineering background, the test was divided into 5 groups with 4 samples in each group. Water content, cement content, fiber mass content, confining pressure σ, and curing time of the experimented samples were designed as shown in Table 3.

Table 3. Composition and testing condition.

Water Content (%)	Cement Content (%)	Fiber Mass Content (‰)	Confining Pressure (kPa)	Curing Time (day)
30	20	0, 2, 4, 6, 10	100, 200, 300, 400	7

2.3. Sample Preparation

According to the Chinese National Geotechnical Test Standard (GB/T 50123-1999) and the designed test scheme [31], the sample preparation process for triaxial UU tests can be divided into the following steps:

1. Subgrade soil was placed in a 105 °C constant temperature oven, baked for 24 h, and pulverized after drying.
2. The crushed subgrade soil was passed through a sieve with an aperture size of 2 mm to remove large particles and impurities such as stones.
3. Appropriate quantities of subgrade soil, cement, PP fiber, and water were weighed; after mixing evenly, according to the geotechnical test specification, a three-valve saturator was used to make triaxial samples of height H = 80 mm and diameter D = 39.1 mm.
4. After the sample was prepared, all triaxial samples were placed in the constant temperature (20 °C) and humidity (95%) curing box for standard curing for 7 days.

3. Test Results and Analysis

3.1. Stress–Strain Curve

Figure 2 shows the curves between deviatoric stress q and axial strain ε curve (hereinafter referred to as "stress–strain curve") of PCS samples. It can be seen that the stress–strain curves of the samples are all softening curves. The axial strain ε was considered within 10%, which is often the maximum strain level in practice.

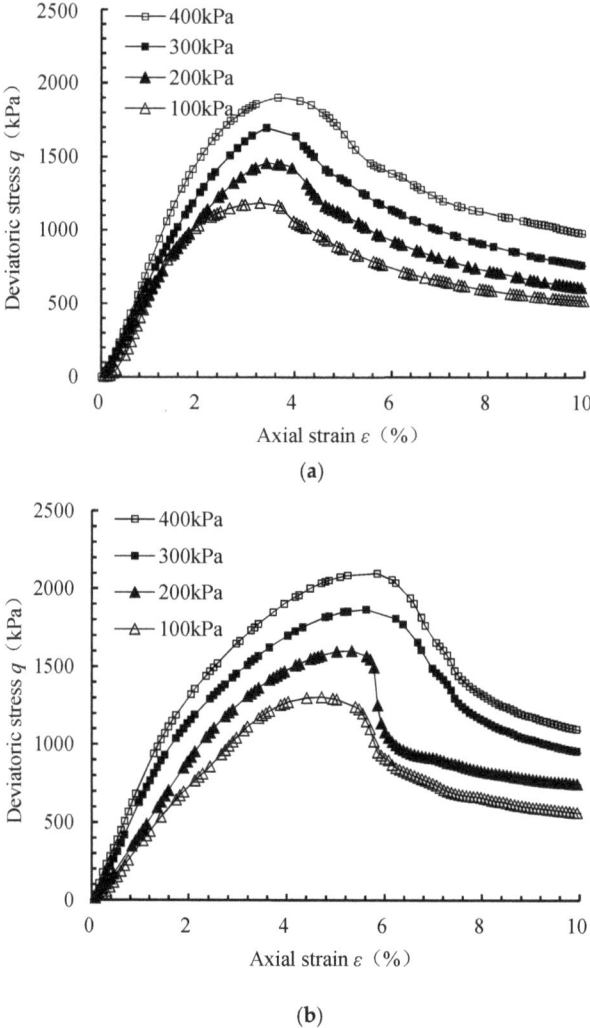

Figure 2. Cont.

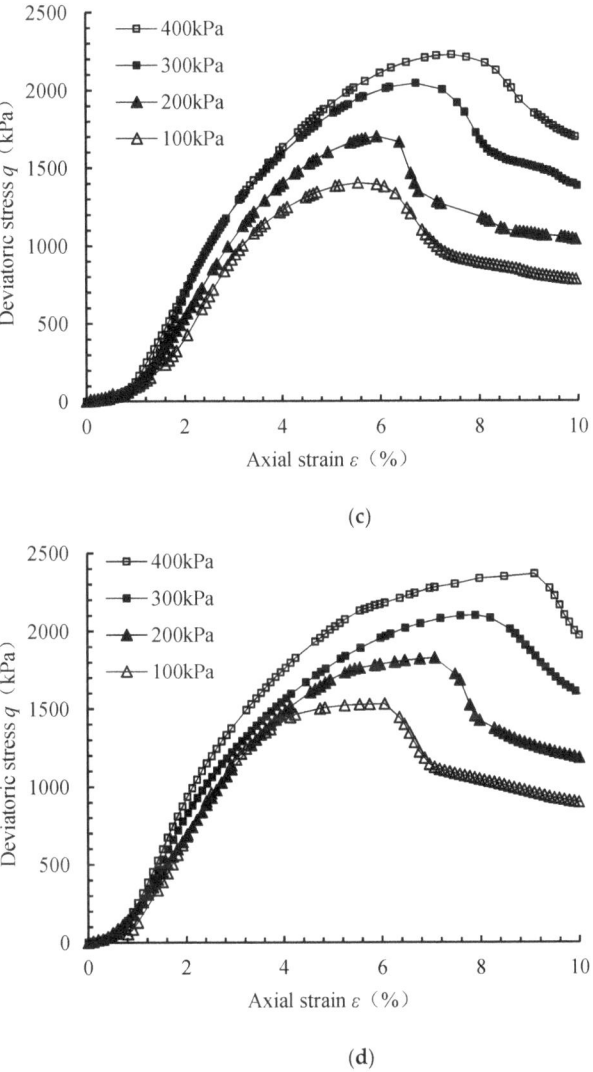

(c)

(d)

Figure 2. *Cont.*

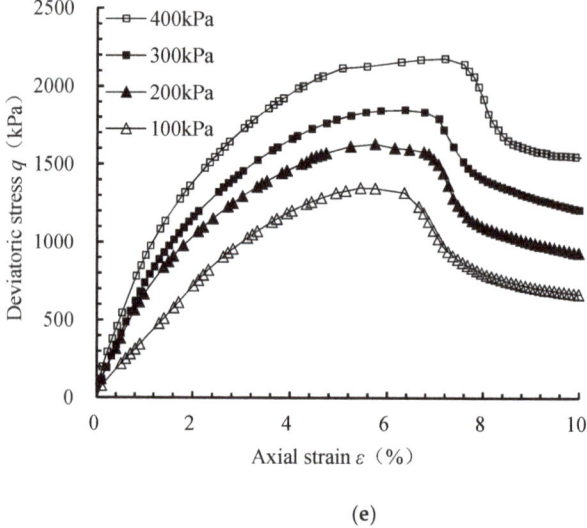

(e)

Figure 2. Deviatoric stress-axial strain curve with various fiber mass contents; (**a**) 0‰ fiber mass content; (**b**) 2‰ fiber mass content; (**c**) 4‰ fiber mass content; (**d**) 6‰ fiber mass content; (**e**) 10‰ fiber mass content.

3.2. Peak Stress and Strain at Peak Stress

The peak stress q_{max} and strain at peak stress ε_{qmax} diagrams of PCS samples can be obtained from the relevant data in Figure 2, as shown in Figures 3 and 4.

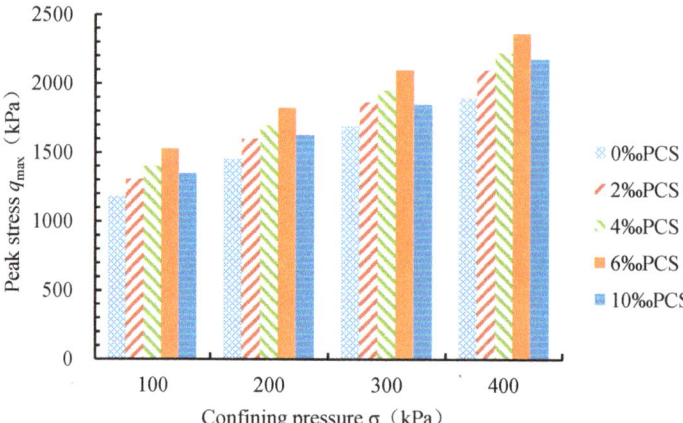

Figure 3. Peak deviatoric stress of PCS (fiber-cement-treated subgrade soil with polypropylene (PP) fiber).

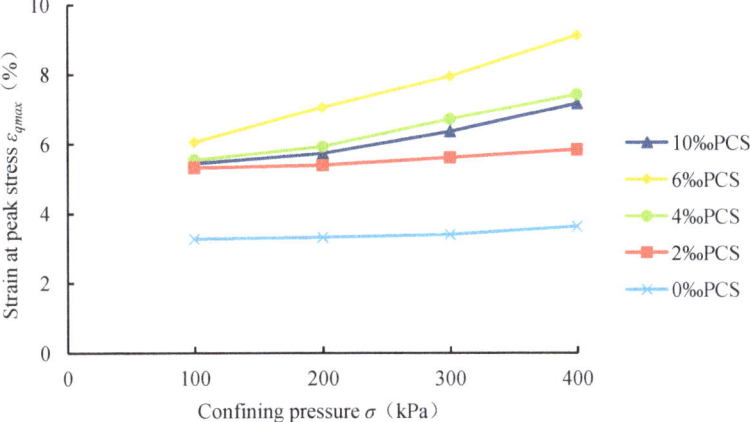

Figure 4. Strain at peak stress of PCS.

It can be seen from Figures 3 and 4 that the peak stress and strain at peak stress of PCS samples gradually increase with the increase in confining pressure given a constant fiber mass content. In addition, under the same confining pressure, the peak stress and strain at peak stress of PCS samples first increase and then decrease with the increase in fiber mass content. When the fiber mass content is 6‰, the peak stress and strain at peak stress reach the maximum. Compared with the 0‰ PCS sample, with the change in confining pressure, the peak stress increases from 24% to 29% and the strain at peak stress increases from 85% to 120%.

The above results show that the modification effect of fiber increases with the increase in confining pressure; among them, the ductility is significantly improved.

3.3. Stress at 10% Axial Strain

It is evident that after reaching peak stress, the stress reduces gradually with greater strain, which means that, after the peak stress, the sample still has some strength to bear the internal force and it is helpful to the ductile failure mold. The stress value with 10% axial strain is referred to as $q_{0.1}$. The $q_{0.1}$ diagram of PCS samples can be obtained from the relevant data in Figure 2, as shown in Figure 5.

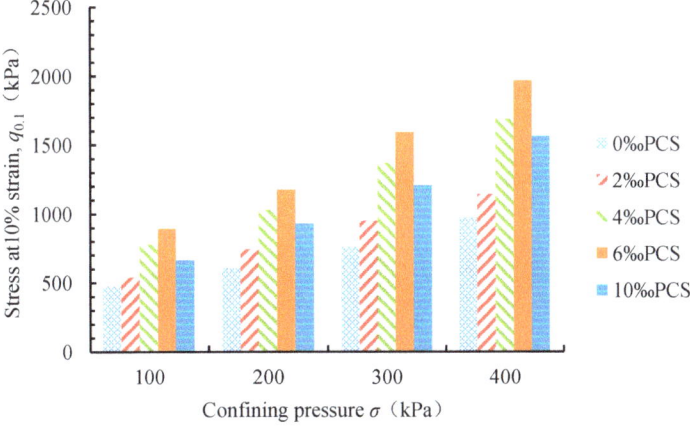

Figure 5. Stress at 10% axial strain of PCS.

It can be seen from Figure 5 that when the fiber mass content is fixed, $q_{0.1}$ in PCS samples increases with the increase in confining pressure. In addition, under the same confining pressure, with the increase in fiber mass content, $q_{0.1}$ of PCS samples first increases and then decreases. When the fiber mass content is 6‰, the residual stress reaches the maximum; compared with 0‰ PCS sample, $q_{0.1}$ increases from 87% to 110% with the change in confining pressure. It can be seen that with the addition of fiber, the ability of the sample to resist destruction is greatly improved.

3.4. Strength Curve

In order to draw the strength envelope of PCS samples and obtain their strength parameters, the peak of q is used as the failure point, where $q = \sigma_1 - \sigma_3$; taking the normal stress as the abscissa, the shear stress as the ordinate, $(\sigma_1 + \sigma_3)/2$ as the center, and $(\sigma_1 - \sigma_3)/2$ as the radius on the abscissa, the ultimate stress diagram is drawn on the τ–σ stress plan graph, as shown in Figure 6; and the envelope of the ultimate stress circle under different confining pressures is drawn as shown in Figure 7; the strength parameters c and φ of the samples are obtained from the strength envelope, as shown in Table 4.

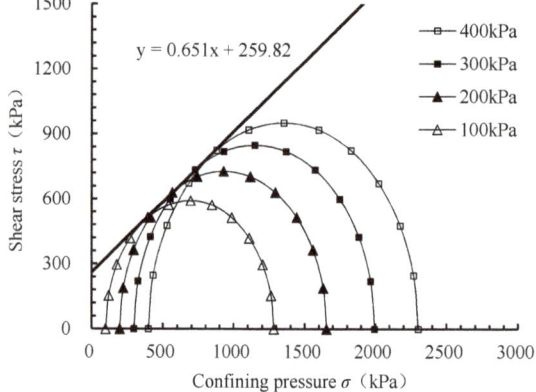

(a)

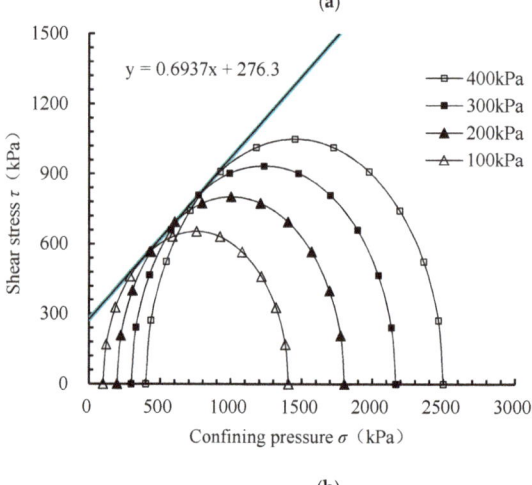

(b)

Figure 6. Cont.

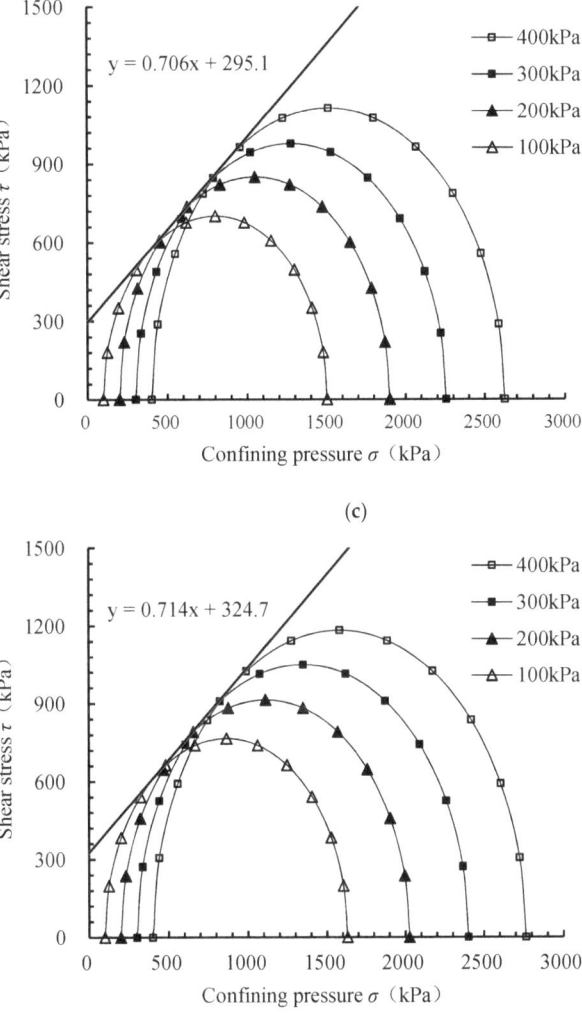

Figure 6. Cont.

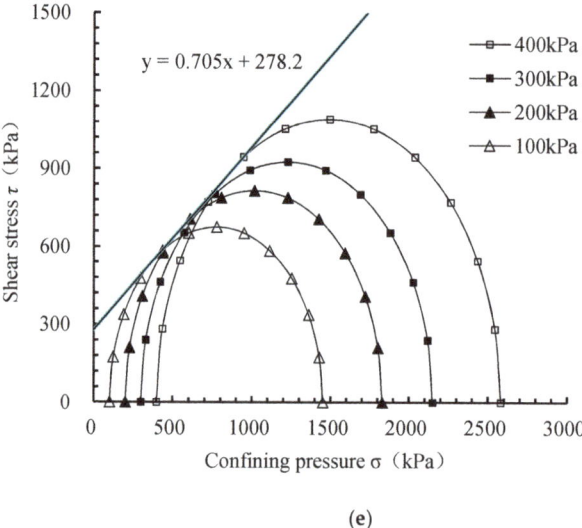

(e)

Figure 6. Mole envelope diagram; (**a**) 0‰ PCS; (**b**) 2‰ PCS; (**c**) 4‰ PCS; (**d**) 6‰ PCS; (**e**) 10‰ PCS.

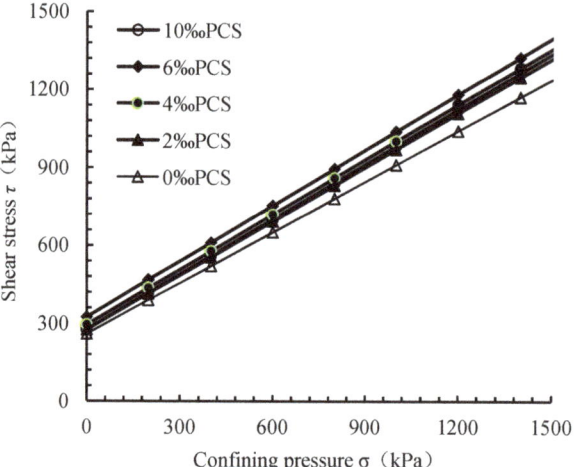

Figure 7. Strength envelope of PCS.

Table 4. Strength parameters.

Fiber Mass Content (‰)	Strength Equation	c (kPa)	ϕ (°)
0	$\tau = 0.651X + 259.8$	259.8	33.1
2	$\tau = 0.694X + 276.3$	276.3	34.8
4	$\tau = 0.706X + 295.1$	295.1	35.2
6	$\tau = 0.714X + 324.7$	324.7	35.5
10	$\tau = 0.705X + 278.2$	278.2	35.2

The relationship between the cohesion and the internal friction angle of PCS samples and fiber mass content can be obtained by referring to Figures 6 and 7, and Table 4, as shown in Figures 8 and 9.

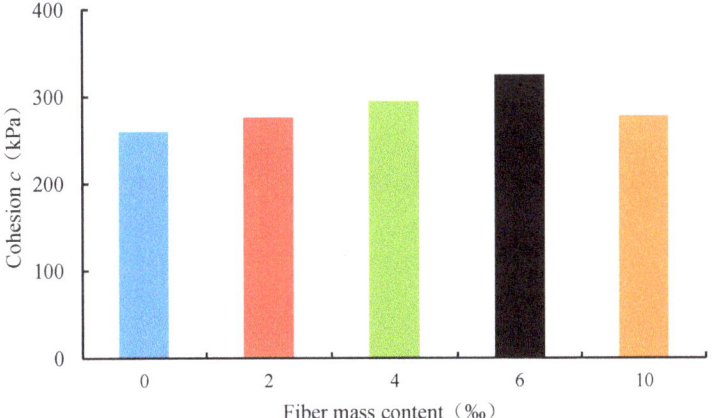

Figure 8. Cohesion of samples under different fiber mass contents.

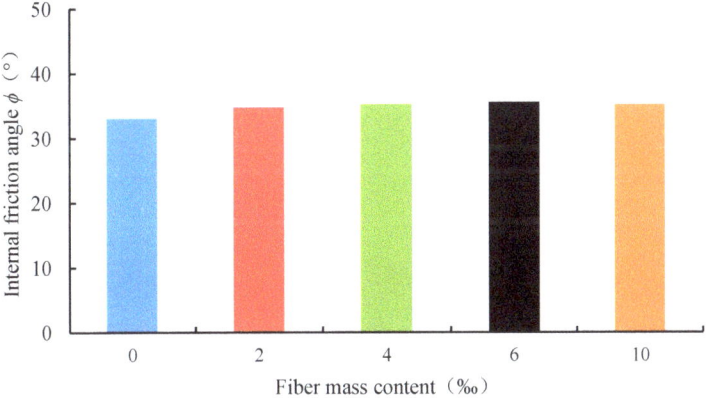

Figure 9. Internal friction angle of samples under different fiber mass contents.

It can be seen from Table 4, together with Figures 8 and 9 that the cohesion c and internal friction angle ϕ of 0‰ PCS sample are 259.8 kPa and 33.1°, respectively. Their values for 6‰ PCS sample are 324.7 kPa and 35.5°, respectively. In other words, when the fiber mass content increases from 0‰ to 6‰, the sample's cohesion c and internal friction angle ϕ increase 25.9% and 7.4%, respectively. However, further increase in fiber mass content lead to a decrease in c and ϕ. When the fiber mass content is 6‰, c and ϕ reach their maximum values, indicating that the shear strength is enhanced. The existence of an optimal fiber mass content may be justified by considering the effect of fibers in a PCS sample. During the cement hydration process, the fibers were combined with the surrounding matrix (cement-treated soil); as a result, a fiber can offer tensile strength in the sample if the cement slurry can coat each fiber well. This condition can be fulfilled if the fiber content is low. With the increase of fiber content, the cement slurry could be relatively insufficient to combine each fiber with its surrounding matrix; in addition, the overlaps among fibers are even likely to undermine the cement hydration process. As such, the strength parameters of the PCS sample have a decreasing tendency when the fiber content is too large. For instance, the strength parameters decrease when the fiber mass content increases from 6‰ to 10‰. In this study, the optimal fiber mass content was found to be 6‰ as indicated in Figures 8 and 9. A similar phenomenon was reported in reference [32], which examined the strength parameters of cement-tailings with fibers.

4. Sample Failure Characteristics

4.1. Brittleness Index

In order to study the brittleness characteristics of the samples during failure, Consoli et al. [33] proposed the brittleness index as the evaluation criteria, as shown in Equation (1):

$$I = q_{max}/q_{0.1} - 1 \tag{1}$$

where q_{max} and $q_{0.1}$ represent the peak stress and stress at 10% axial strain, respectively, and I represents the brittleness index. The brittleness index of PCS samples can be obtained by substituting the relevant data in Figure 2 into Equation (1), as shown in Figure 10. The figure indicates that:

1. When the fiber mass content is fixed, with the increase in confining pressure, the brittleness index of PCS samples decreases and the ductility increases. Under a confining pressure of 100 kPa, the brittleness index is the highest and the brittle failure is obvious. The brittleness index reaches its lowest value when the confining pressure is 400 kPa, which suggest the best mitigation effect of brittle failure.

2. Under the same confining pressure, with the increase in fiber mass content, the brittleness index of PCS samples first decreases and then increases. When the fiber mass content is 6‰, the brittleness index reaches the minimum. Compared with 0‰ PCS sample, the brittleness index decreases from 52% to 79% with the increase in confining pressure, indicating that the incorporation of fiber can improve the brittle failure of the sample, and the modification effect is enhanced with the increase in confining pressure.

The main reason is that, with the change of confining pressure, there will be an interfacial interaction force between the fiber and the soil particles. When the PCS sample is subjected to an external force, the fiber will bear a part of the external force through the interfacial interaction, whereby improving the brittle failure of the sample.

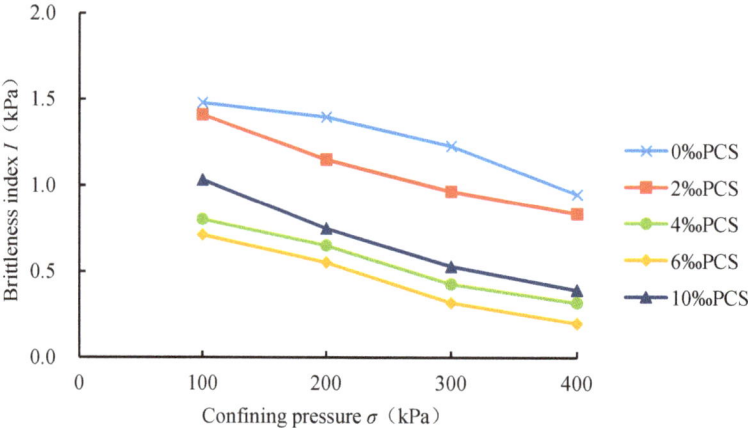

Figure 10. Brittleness index against confining pressure.

4.2. Failure Angle of Sample

Fiber content plays an important role in the PCS's failure model. Due to the confining pressure having no significant influence on the failure angle, in order to investigate the effect of fiber on the failure angle of cement subgrade soil samples, PCS samples with typical failure behavior [15] under 300 kPa confining pressure were selected, as shown in Figure 11. According to the failure pattern of these samples, the failure section diagram was drawn as shown in Figure 12. The dimensions are specified in mm, the oblique line is the failure surface of the sample, and the angle between the failure surface and the horizontal direction is determined as the failure angle. The relationship between the fiber mass content and the angle can be obtained from the data in Figure 12, as shown in Figure 13.

Figure 11. PCS sample failure. (**a**) 0‰ PCS; (**b**) 2‰ PCS; (**c**) 4‰ PCS; (**d**) 6‰ PCS; (**e**) 10‰ PCS.

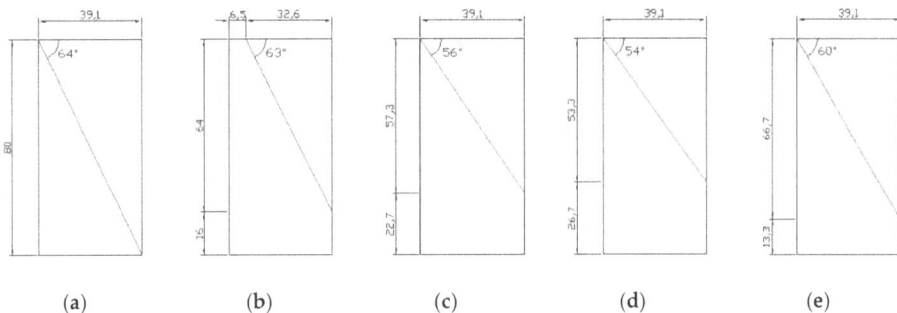

Figure 12. Profile map of PCS sample failure. (**a**) 0‰ PCS; (**b**) 2‰ PCS; (**c**) 4‰ PCS; (**d**) 6‰ PCS; (**e**) 10‰ PCS.

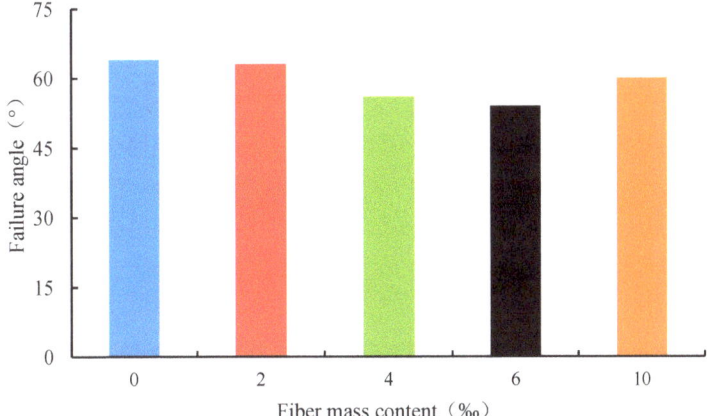

Figure 13. Fiber content and failure angle.

Figure 13 indicates that, under the same confining pressure, the failure angle of PCS samples first decreases and then increases as the fiber mass content increases. When the fiber mass content is 6‰, the failure angle reaches its minimal value. This value is 16% less than that of a 0‰ PCS sample, indicating that the incorporation of fiber can improve the brittle fracture pattern of a sample to some extent. When the fiber mass content further increases to 10‰, the failure angle is still less than that of a 0‰ PCS sample, and the modification is significantly improved with the increase in fiber mass content. The main reason is that when the fiber mass content is too large, the fibers are likely to overlap with each other in the soil; as a result, the soil cannot be well compacted. Hence, the modification effect is limited. In addition, when the fiber mass content is within a certain range, the fiber can better fill the voids inside the PCS sample, a more stable structure can be formed between the fiber and soil particles, thereby improving the failure form of the sample. On the contrary, when the fiber mass content is too high, the fibers readily overlap with each other in the soil. In this regard, neither can the soil be well compacted, nor can the fiber be well connected. As such, the failure angle of the sample is improved and the modification effect decreases.

5. Conclusions

1. The deviatoric stress-axial strain curves of PCS samples are all strain-softening curves.

2. Under the same confining pressure, with the increase in fiber mass content, the peak stress, strain at peak stress, and residual stress of PCS samples first increase and then decrease. When the fiber mass content is 6‰, the modification effect is optimum. Compared with the 0‰ PCS sample, with the change in confining pressure, the peak stress increases from 24% to 29%, the strain at peak stress increases from 85% to 120%, and the residual stress increases from 87% to 110%. With the increase in confining pressure, the modification effect of fiber is gradually improved. Among the tests, the ductility is significantly improved.

3. The incorporation of fiber can increase the cohesion c and the internal friction angle ϕ of PCS samples. When the fiber mass content is 6‰, the cohesion c and the internal friction angle ϕ reach the maximum value. Compared with the 0‰ PCS sample, the cohesion c increases by 25.9% and the internal friction angle ϕ increases by 7.4%. It can be seen that the incorporation of fiber improves the cohesion c and the internal friction angle ϕ of the sample, thereby improving the shear strength.

4. When the fiber mass content is fixed, with the increase in confining pressure, the brittleness index of the PCS sample decreases and the ductility increases. Under a confining pressure of 100 kPa, the brittleness index is the highest and the brittle failure is obvious. Under a confining pressure of 400 kPa, the brittleness index reached its lowest value and the brittleness failure shows the most mitigation. When the fiber mass content is 6‰, the brittleness index reaches the minimum. Compared with the 0‰ PCS sample, with the increase in confining pressure, the brittleness index decreases from 52% to 79%. The incorporation of fiber can improve the brittle failure behavior of the sample, and the modification effect is enhanced with the increase in the confining pressure.

5. Under the same confining pressure, with the increase in fiber mass content, the failure angle of PCS samples first decreases and then increases. The PCS sample with 6‰ fiber mass content exhibits the minimum failure angle, providing solid evidence that the incorporation of fiber mitigates the brittle fracture form of the sample to some extent.

As a limitation, the current study only considered small soil specimens. In reality, the situation is more complicated, as there will be impurities such as stones and organic matter. The effect of impurities in a specimen and the effect of specimen size were not considered in this study, but which is worthy of future study.

Author Contributions: The authors confirm contribution to the paper as follow: W.W. and N.L. proposed the idea and wrote the paper; Y.L. (Yong Liu) revised the manuscript; C.Z., Y.L. (Yuan Li), and H.Z. conducted the tests and analyzed the data; J.G. reviewed the results and approved the final version of the manuscript.

Funding: This research was funded by the National Natural Science Foundation of China (Grant numbers [41772311]), the Zhejiang Provincial Natural Science Foundation of China (Grant number [LY17E080016]), the Open Research Fund of State Key Laboratory of Geomechanics and Geotechnical Engineering, Institute of Rock and Soil Mechanics, Chinese Academy of Science (Grant number [Z017013]), the Scientific Research Projects of Zhejiang Department of Housing and Urban and Rural Construction (Grant number [2017K179]), and the International Scientific and Technological Cooperation Projects of Shaoxing University (Grant number [2019LGGH1007]).

Conflicts of Interest: The authors declare no conflict of interest.

References

1. Pan, Y.T.; Liu, Y.; Hu, J.; Sun, M.M.; Wang, W. Probabilistic investigations on the watertightness of jet-grouted ground considering geometric imperfections in diameter and position. *Can. Geotech. J.* **2017**, *54*, 1447–1459. [CrossRef]
2. Liu, Y.; Lee, F.H.; Quek, S.T.; Chen, J.E.; Yi, J.T. Effect of spatial variation of strength and modulus on the lateral compression response of cement-admixed clay slab. *Géotechnique* **2015**, *65*, 851–865. [CrossRef]
3. Wang, W.; Li, N.; Zhang, F.; Zhou, A.Z.; Chi, S. Experimental and mathematical investigations on unconfined compressive behavior of coastal soft soil under complicated freezing processes. *Pol. Marit. Res.* **2016**, *23*, 112–116. [CrossRef]
4. Li, N.; Zhu, Q.Y.; Wang, W.; Song, F.; An, D.A.; Yan, H.R. Compression characteristics and microscopic mechanism of coastal soil modified with cement and fly ash. *Materials* **2019**, *12*, 3182. [CrossRef] [PubMed]
5. Yao, K.; Li, N.; Chen, D.H.; Wang, W. Generalized hyperbolic formula capturing curing period effect on strength and stiffness of cemented clay. *Constr. Build. Mater.* **2019**, *199*, 63–71. [CrossRef]
6. Liu, Y.; He, L.Q.; Jiang, Y.J.; Sun, M.M.; Chen, E.J.; Lee, F.H. Effect of in situ water content variation on the spatial variation of strength of deep cement-mixed clay. *Géotechnique* **2019**, *69*, 391–405. [CrossRef]
7. Xing, H.F.; Xiong, F.; Zhou, F. Improvement for the strength of salt-rich soft soil reinforced by cement. *Marine Georesour. Geotechnol.* **2017**, *36*, 38–42.
8. Choobbasti, A.J.; Kutanaei, S.S. Microstructure characteristics of cement-stabilized sandy soil using nanosilica. *J. Rock. Mech. Geotech. Eng.* **2017**, *9*, 981–988. [CrossRef]
9. Liu, Y.; Jiang, Y.; Xiao, H.; Lee, F.H. Determination of representative strength of deep cement-mixed clay from core strength data. *Géotechnique* **2017**, *67*, 350–3645. [CrossRef]
10. Anagnostopoulos, C.A. Strength properties of an epoxy resin and cement-stabilized silty clay soil. *Appl. Clay Sci.* **2015**, *114*, 517–529. [CrossRef]
11. Yao, K.; Xiao, H.; Chen, D.H.; Liu, Y. A direct assessment for the stiffness development of artificially cemented clay. *Géotechnique* **2019**, *69*, 741–747. [CrossRef]
12. Jia, L.; Zhao, F.L.; Guo, J.; Yao, K. Properties and reaction mechanisms of magnesium phosphate cement mixed with ferroaluminate cement. *Materials* **2019**, *12*, 2561. [CrossRef] [PubMed]
13. Mengue, E.; Mroueh, H.; Lancelot, L. Physicochemical and consolidation properties of compacted lateritic soil treated with cement. *Soils Found.* **2017**, *57*, 60–79. [CrossRef]
14. Yao, K.; Wang, W.; Li, N.; Zhang, C.; Wang, L.X. Investigation on strength and microstructure characteristics of Nano-MgO admixed with cemented soft soil. *Constr. Build. Mater.* **2019**, *206*, 160–168. [CrossRef]
15. Xiao, H.W.; Wang, W.; Goh, S.H. Effectiveness study for fly ash cement improved marine clay. *Constr. Build. Mater.* **2017**, *157*, 1053–1064. [CrossRef]
16. Yao, K.; An, D.L.; Wang, W.; Li, N.; Zhang, C.; Zhou, A.Z. Effect of nano-mgo on mechanical performance of cement stabilized silty clay. *Marine Georesour. Geotechnol.* **2019**. [CrossRef]
17. Wang, W.; Zhang, C.; Li, N.; Tao, F.F.; Yao, K. Characterisation of nano magnesia-cement-reinforced seashore soft soil by direct-shear test. *Marine Georesour. Geotechnol.* **2019**, *37*, 989–998. [CrossRef]
18. Jiang, P.; Qiu, L.Q.; Li, N.; Wang, W.; Zhou, A.Z.; Xiao, J.P. Shearing performance of lime reinforced iron tailing powder based on energy dissipation. *Adv. Civ. Eng.* **2018**. [CrossRef]
19. Lenoir, T.; Preteseille, M.; Ricordel, S. Contribution of the fiber reinforcement on the fatigue behavior of two cement-modified soils. *Int. J. Fatigue* **2016**, *93*, 71–81. [CrossRef]
20. Wang, W.; Li, Y.; Yao, K.; Li, N.; Zhou, A.Z.; Zhang, C. Strength properties of nano-MgO and cement stabilized coastal silty clay subjected to sulfuric acid attack. *Marine Georesour. Geotechnol.* **2019**. [CrossRef]
21. Zerrouk, A.; Lamri, B.; Vipulanandan, C. Performance evaluation of human hair fiber reinforcement on lime or cement stabilized clayey-sand. *Key Eng. Mater.* **2015**, *668*, 207–217. [CrossRef]
22. Gowthaman, S.; Nakashima, K.; Kawasaki, S. A state-of-the-Art review on soil reinforcement technology using natural plant fiber materials: Past findings present trends and future directions. *Materials* **2018**, *11*, 553. [CrossRef] [PubMed]

23. Chen, M.; Shen, S.L.; Wu, H.N. Laboratory evaluation on the effectiveness of polypropylene fibers on the strength of fiber-reinforced and cement-stabilized Shanghai soft clay. *Geotext. Geomembr.* **2015**, *43*, 515–523. [CrossRef]
24. Anggraini, V.; Asadi, A.; Syamsir, A. Three point bending flexural strength of cement treated tropical marine soil reinforced by lime treated natural fiber. *Measurement* **2017**, *111*, 158–166. [CrossRef]
25. Sharma, R.K. Laboratory study on stabilization of clayey soil with cement kiln dust and fiber. *Geotech. Geol. Eng.* **2017**, *35*, 2291–2302. [CrossRef]
26. Tran, K.Q.; Satomi, T.; Takahashi, H. Improvement of mechanical behavior of cemented soil reinforced with waste cornsilk fibers. *Constr. Build. Mater.* **2018**, *178*, 204–210. [CrossRef]
27. Ayeldeen, M.; Kitazume, M. Using fiber and liquid polymer to improve the behaviour of cement-stabilized soft clay. *Geotext. Geomembr.* **2017**, *45*, 592–602. [CrossRef]
28. Yang, B.H.; Weng, X.Z.; Liu, J.Z. Strength characteristics of modified polypropylene fiber and cement-reinforced loess. *J. Cent. South Univ.* **2017**, *24*, 560–568. [CrossRef]
29. Estabragh, A.R.; Ranjbari, S.; Javadi, A.A. Properties of clay soil and soil cement reinforced with polypropylene fibers. *ACI Mater. J.* **2017**, *114*, 195–205. [CrossRef]
30. Xiao, H.; Liu, Y. A prediction model for the tensile strength of cement-admixed clay with randomly orientated fibres. *Eur. J. Environ. Civ. Eng.* **2018**, *22*, 1131–1145. [CrossRef]
31. GB/T 50123-1999, Standard for Soil Test Method. China Planning Press, 1999. Available online: http://www.zys168.net/Upload/DownLoad/DownLoadFile/20107191920846.pdf (accessed on 26 October 2019). (In Chinese)
32. Xue, G.; Yilmaz, E.; Song, W.; Gao, S. Mechanical, flexural and microstructural properties of cement-tailings matrix composites: Effects of fiber type and dosage. *Compos. Part B* **2019**, 131–142. [CrossRef]
33. Consoli, N.C.; Pedro, D.M.; Ulbrich, L.A. Influence of fiber and cement addition on behavior of sandy soil. *J. Geotech. Geoenviron. Eng.* **1998**, *124*, 1211–1214. [CrossRef]

© 2019 by the authors. Licensee MDPI, Basel, Switzerland. This article is an open access article distributed under the terms and conditions of the Creative Commons Attribution (CC BY) license (http://creativecommons.org/licenses/by/4.0/).

Article

Shear Failure Mode and Concrete Edge Breakout Resistance of Cast-In-Place Anchors in Steel Fiber-Reinforced Normal Strength Concrete

Jong-Han Lee [1], Eunsoo Choi [2] and Baik-Soon Cho [3],*

1. Department of Civil Engineering, Inha University, Incheon 22212, Korea; jh.lee@inha.ac.kr
2. Department of Civil Engineering, Hongik University, Seoul 04066, Korea; eunsoochoi@hongik.ac.kr
3. Department of Civil and Urban Engineering, CTRC, Inje University, Gimhae 50834, Korea
* Correspondence: civcho@inje.ac.kr

Received: 25 August 2020; Accepted: 28 September 2020; Published: 1 October 2020

Abstract: Concrete edge failure of a single anchor in concrete is strongly dependent on the tensile performance of the concrete, which can be greatly improved by the addition of steel fibers. This study investigated the effect of steel fibers on the shear failure mode and edge breakout resistance of anchors installed in steel fiber-reinforced concrete (SFRC) with fiber volume percentages of 0.33, 0.67, and 1.00%. The anchor used in the study was 30 mm in diameter, with an edge distance of 75 mm and embedment depth of 240 mm. In addition to the anchor specimens, beam specimens were prepared to assess the relationship between the tensile performance of SFRC beams and the shear resistance of SFRC anchors. The ultimate flexural strength of the beam and the breakout shear resistance of the anchor increased almost linearly with increasing volume fractions of fiber. Therefore, based on the ACI 318 design equation, a term was proposed using the ultimate flexural strength of concrete instead of the compressive strength to determine the concrete breakout shear resistance of an anchor in the SFRC. The calculated shear resistance of anchors in both the plain concrete and SFRC were in good agreement with the measurements. In addition to the load capacity of the SFRC anchors, the energy absorption capacity showed a linear increase with that of the SFRC beam.

Keywords: anchor; shear behavior; concrete edge breakout resistance; ultimate flexural strength; energy absorption capacity; steel fiber

1. Introduction

Concrete anchors are commonly used to support structural members and equipment in civil and industrial structures, including power plants. In addition, anchor systems are used to connect new structural members for strengthening and retrofitting existing structures [1]. In particular, large-scale structural members, facilities, and equipment are connected using pre-installed cast-in-place (CIP) anchors [2]. The failure of anchors installed in concrete is mainly dependent on the strength of the steel anchor and concrete. Concrete failure causes sudden destruction of an anchor system, which can directly affect the proper performance of structures and human safety. Thus, the evaluation of concrete fracture strength in the anchor system is essential for the stability and durability of a structure.

Studies on the concrete edge breakout strength of anchor bolts in concrete have mainly focused on anchors installed in non-reinforced plain concrete. For single and multiple anchors in plain concrete, design equations were developed based on the experimental results and discussions of previous researchers [3–5] for the concrete breakout strength of anchors using the concrete capacity design (CCD) method. This method assumes that the angle of the breakout cone shape is 35 degrees. The CCD method has been adopted in the current design standards [6–9] to determine the concrete breakout capacity of anchors installed in concrete. Olalusi and Spyridis [10] derived a statistical model from the database of anchor tests for the concrete breakout capacity of single anchors in shear, and they showed that the CCD model was highly scattered and biased. The CCD method is a semi-empirical design method largely dependent on the test data. For limited data of concrete cone failure, Bokor and Sharma [11] evaluated load and displacement behavior of anchor groups in tension as the basis for development of concrete cone failure of anchorage with various aspects. In addition, Pürgstaller et al. [12] applied concrete anchors to nonstructural components and investigated the hysteresis shear behavior of the anchors. Dengg et al. [13] assessed the applicability of anchor systems using tunnel excavation material.

Concrete has very weak tensile capacity, so steel bars are used to support the tensile force generated in the concrete. The ACI 318M-08 [8] has proposed modification factors to increase the concrete breakout strength of anchors installed in concrete reinforced with steel bars. However, experimental studies on the effect of the reinforcement on the resistance of anchors are still limited. Moreover, when using steel bars in concrete, careful attention should be paid to the placement and corrosion of the bars [14]. Thus, many studies have also been carried out to improve the tensile capacity and ductility of cement-based materials using discrete short-length fibers, such as steel, carbon, textile, and natural fibers. In particular, steel fibers with great strength and ductility have been actively studied. For the application of steel fibers in cement-based materials, pullout tests have been performed to evaluate the bond and slip mechanisms between the steel fibers and cement-based matrix, as well as the effects of the fibers on bridging and arresting crack propagation and opening [15–17]. The addition of steel fibers enhances flexural capacity in concrete beams, and the increase in the flexural capacity is proportional to the fiber content [18,19]. In addition to the flexural behavior of steel fiber-reinforced concrete (SFRC), Narayanan and Darwish [20], Sharma [21], and Amin and Foster [22] applied steel fibers to increase the shear strength of concrete beams. Khuntia et al. [23] assessed the increase in the shear strength of concrete with steel fibers in the normal and high-strength concrete matrix. Furthermore, the influence of the concrete strength on the strength and ductility of the SFRC concrete has been studied. Mansur et al. [24] indicated the improvement of strength with steel fibers in high-strength concrete, and Holschemacher et al. [25] investigated the effect of fiber type and content in high-strength concrete. Recently, Lee [26] evaluated the influence of the matrix strength of concrete on the post-cracking residual strength of SFRC. With the contribution of previous studies to the application of steel fibers in concrete materials and structures, SFRC is currently used reliably in practical applications, such as industrial slabs, pavements, tunnel shotcrete, and precast tunnel segments [27–30].

However, the application of SFRC to anchor systems in concrete is very limited. Most previous studies have focused on anchors in non-reinforced plain concrete, and some studies have been concerned with anchor systems in steel bar-reinforced concrete. Thus, the current design equations only provide the concrete resistance of anchors installed in plain concrete or steel bar-reinforced concrete. Recently, for an anchor in SFRC, Nilforoush et al. [31] evaluated the breakout concrete capacity of an anchor bolt in SFRC members under tensile load and showed a great increase in the tensile resistance and toughness of the anchor. Tóth and Boker [32] investigated the behavior of anchorages in SFRC and showed increases in concrete breakout capacity and displacement due to the presence of fibers. Mahrenholtz et al. [33] performed experiments on anchor channel bolt systems in plain and fiber-reinforced concrete to develop the basis of the design rules for fasteners installed in FRC. Lee et al. [34] performed shear tests of SFRC anchors to assess the relationship between the mechanical properties of SFRC and the shear resistance of anchors embedded in SFRC. Then, they proposed a modified design equation that can calculate the concrete breakout strength of anchors in plain concrete and SFRC using the equivalent flexural strength ratio.

SFRC exhibits similar compressive strength to plain concrete but a significant difference in tensile capacity. Thus, this study aims to expand knowledge on anchors in SFRC and replace the compressive strength of concrete with the tensile capacity, which can be more simply and generally employed for concrete anchors than the equivalent flexural strength ratio. For this, this study performed shear tests for anchors 30 mm in diameter with an edge distance of 75 mm and embedment depth of

240 mm. The effect of steel fibers on the shear failure mode and breakout resistance of the anchors was investigated. The fiber volume percentages were changed from zero to 1.00%, and the failure mode was investigated to determine the cracking resistance, ultimate shear resistance, and energy absorption capacity of the anchors.

The increases in the concrete breakout strength and ductility of the anchor system are strongly associated with the tensile performance of the concrete. Thus, beam specimens were also prepared to assess the relationship between the tensile performance of SFRC beams and the shear resistance of SFRC anchors. The ultimate flexural strength of the beam and breakout shear resistance of the anchor increased almost linearly with increasing fiber volume fractions. Thus, based on the ACI 318 design equation, a term is proposed using the ultimate flexural strength of concrete instead of the compressive strength to determine the concrete breakout shear resistance of the anchors. In addition, this study proposes a relationship between the energy absorption capacity of the SFRC anchors and that of the SFRC beams. The results of the proposed equation showed good agreement with the experimental values.

2. Experiments

2.1. SFRC Material

The concrete used in the study was provided by a ready-mixed concrete company, and its compressive strength was designed as 27 MPa. The water-to-cement ratio was 0.54, and the slump value was 150 mm. The mix proportion of the concrete is given in Table 1. All of the materials except the steel fibers were mixed in a ready-mixed concrete truck for around 30 min until reaching the site where specimens were manufactured. The steel fibers were then added into the truck at the site and mixed with the concrete mixture. The steel fibers used in the study had hooked ends that are commonly used in practice. The length and diameter of the fibers were 60 and 0.75 mm, respectively, and thus the aspect ratio was 80. The ultimate tensile strength of the fiber was 1100 MPa.

Table 1. Mix proportion of concrete (kg/m^3).

Cement	Fine Aggregate	Coarse Aggregate	Fly Ash	Superplasticizer
279	931	929	31	1.17

The compressive strength of concrete was measured using cylinder specimens with a diameter of 10 mm and length of 20 mm. The average compressive strength of the non-reinforced plain concrete measured from four cylinder specimens was approximately 28 MPa. The average compressive strengths of the SFRC with fiber volume percentages of 0.33, 0.67, and 1.00%, measured from four cylinders for each fiber volume fraction, were in the range of 26 to 28 MPa. Thus, the measured compressive strength showed very good agreement with the design strength of 27 MPa. Little difference was found between the non-reinforced and SFRC, which means that there was little influence of the steel fibers on the compressive strength of the concrete.

2.2. Preparation of the Anchor and Beam Specimens

The concrete breakout failure of an anchor installed in concrete is strongly associated with the tensile performance of concrete. Steel fibers mixed in concrete have a great influence on the generation and growth of concrete cracks. Thus, to investigate the effect of steel fibers on the concrete edge breakout failure of an anchor, the anchor specimens of SFRC with fiber volume percentages of 0.33, 0.67, and 1.00%, which correspond to approximately 26, 54, and 80 kg/m^3, respectively, were designed, as shown in Table 2. The steel anchor used in the study was an M30-S45C with a diameter of 30 mm and yield strength of 450 MPa. One steel anchor was installed in the center of each edge of a concrete block. Thus, four anchor specimens per test variable were prepared in one concrete block. One concrete block without steel fibers was also prepared, and a total of 16 anchor specimens were fabricated (four anchors in one concrete block per test variable).

Table 2. Specimen names and numbers according to fiber volume fractions.

Name of Specimens	Number of Concrete Blocks	Number of Anchor Specimens	Fiber Volume Percentages, v_f (%)
V000			0.00
V033	1	4	0.33
V067			0.67
V100			1.00

To induce the concrete edge breakout failure of an anchor, the edge distance of the anchor was defined as 75 mm (=2.5 times the diameter of the anchor). The embedded depth of the anchor in concrete was 240 mm, which was 8 times the diameter of the anchor. Thus, the concrete block was designed to have a square top surface with a length of 500 mm so that the inclined cracks generated on the surface of the block due to the shear load could not affect the other edge faces. The height of the block was defined as 680 mm in consideration of the embedded depth of 240 mm of the anchor. Figure 1 shows the dimensions of the concrete block and anchor specimens. The concrete blocks with steel anchors were cured in air for around 90 days.

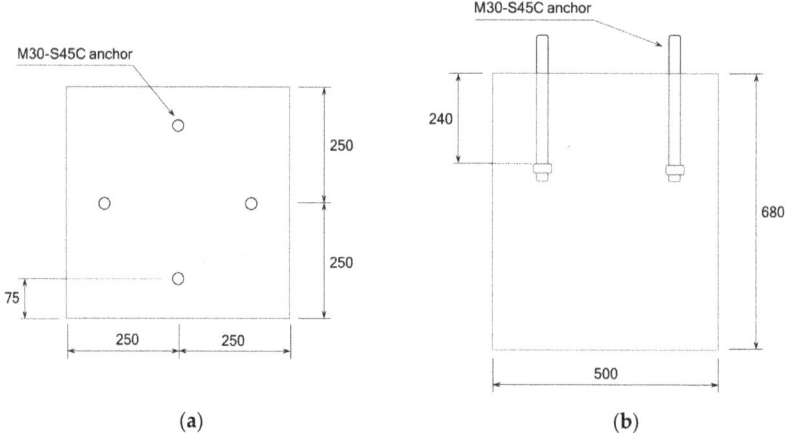

Figure 1. Dimensions of the anchor specimen in concrete: (**a**) top surface; (**b**) side surface (units: mm).

Beam specimens were also manufactured to assess the tensile capacity of the SFRC according to the ASTM C 1609 standard [35]. The dimensions of the beam specimens were 150 × 150 × 500 mm³. The beam specimens were cast at the same time as the anchor specimens using the same concrete mixture and fiber volume fractions. Control beam specimens without steel fibers were also prepared. The same number of beam specimens as anchor specimens was prepared to ensure the reliability of the experiments (four specimens per fiber volume fraction). Thus, 16 beam specimens were manufactured in total. After de-molding, the beam specimens were cured in air under the same conditions as the anchor specimens.

2.3. Installation and Measurement of the Specimens

The concrete anchor blocks were supported using steel angles at the four corners of the front and back sides at the bottom of the block, as shown in Figure 2a. To prevent the reaction force of the steel angles from affecting the shear behavior of the anchor, the contact width and height of the steel angle with the anchor block were limited to 100 and 400 mm, respectively. Four steel angles were fixed to holes in the laboratory floor using high-strength bolts and nuts. Rubber plates were installed between the concrete surface and the steel angle to minimize the influence of frictional force and stress concentration on the anchor block.

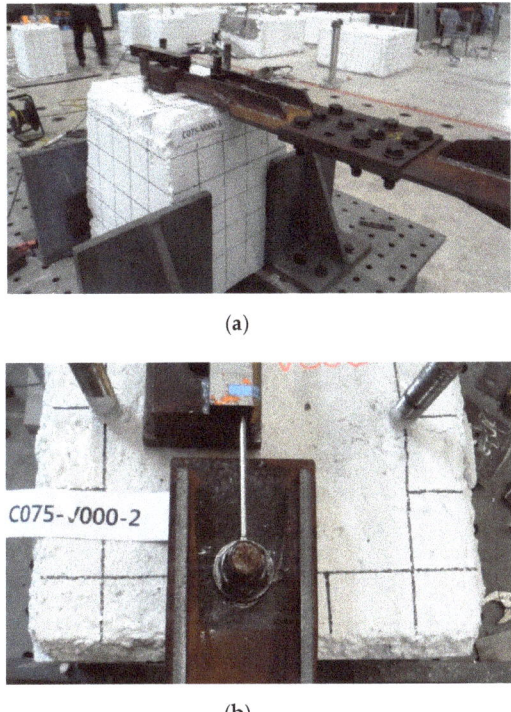

Figure 2. Test setup and instrumentation of the anchor specimen: (**a**) shear test setup; (**b**) linear variable differential transformer (LVDT) installed on the anchor.

To apply a monotonic shear load to the anchor installed in the concrete block, a loading plate 150 mm wide and 20 mm thick with a hole at the end of the plate was connected to the anchor bolt, as shown in Figure 2b. To avoid contact between the loading plate and the top surface of the anchor, a washer with a thickness of 2 mm was inserted into the anchor bolt before the loading plate was installed. Then, a nut was used to loosely connect the loading plate and anchor bolt to prevent the loading plate from moving during the experiment and to maintain a pin connection. The other end of the loading plate was connected to an actuator with a capacity of 200 kN. The actuator was installed on a fixed frame to maintain the horizontal condition during the test. The applied load was monitored with a load cell installed at the actuator. The displacement of the anchor was recorded using a linear variable differential transformer (LVDT) installed on the back of the anchor, as shown in Figure 2b. The test was continued until concrete breakout failure completely occurred.

The beam specimen was installed on a steel roller under a simply supported condition. The distance between the supports was 450 mm, and the cast top surface was used as the side surface of the beam. To distribute the supporting force evenly through the width of the beam, a rubber plate which was 30 mm in width and 3 mm in thickness was placed at the contact portion between the steel roller and beam. Two vertical loads were then applied at a rate of 0.2 mm/min on the top surface of the beam to generate a pure moment between the two loads. A rubber plate was also installed between the point of the loading and the top surface of the beam to prevent local cracks in the concrete at the loading points and to allow the uniform distribution of the load.

The bending test was continued for around 15 min until the deflection of the span reached 1/150 of the span length at mid-span. The vertical deflection was measured using LVDTs installed at the front and back sides of the beam at mid-span. Gopalaratnam et al. [36] reported that the deflection measured at the middle of the beam could be approximately twice the real deflection due to additional deflections caused by the elastic and inelastic behavior of the loading device and the slip of the specimen. Thus, a lateral frame was fabricated and attached to the beam according to the recommendation of ASTM C 1609 [35] to exclude the additional deflection and to measure deflections at mid-span.

3. Bending Test Results and Discussion

3.1. Tensile Strength before and after Cracking

According to the ASTM C 1609 [35], the stress f can be calculated by

$$f = \frac{PL}{bd^2} \tag{1}$$

where P is the applied load, L is the span length, and b and d are the width and depth of the beam, respectively. Figure 3 shows the relationship between the cracking stress f_{cr} and ultimate stress f_u and the steel fiber content. The f_{cr} corresponds to the first peak stress defined in the ASTM C 1609 [35] and the stress at the limit of proportionality defined in the BS EN 14,651 [37]. f_{cr} is equal to f_u in the non-reinforced plain concrete beam, which exhibited no strength recovery greater than f_{cr} after cracking. The linear lines in Figure 3 were obtained from the linear regression analysis of the stress with respect to the steel fiber volume percentages from zero to 1.00%. f_{cr} is almost constant, while f_u increases in proportion to the steel fiber content. f_u was approximately 7.40 MPa for the reinforced concrete beam with a fiber volume percentage of 1.00%, which was 2.21 times that of the non-reinforced plain beam.

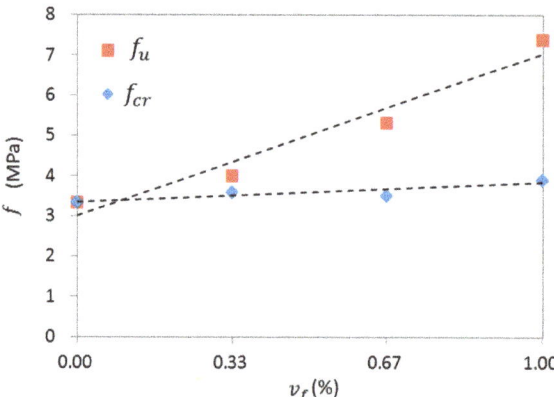

Figure 3. Cracking and ultimate strengths with increasing fiber volume percentages from zero to 1.00%.

The residual tensile strength is defined as the stress at the deflection of L/600 (=0.75 mm) and L/150 (=3.0 mm) according to the ASTM C 1609 [35]. JSCE SF-4 [38] defines the residual strength as the mean stress from the beginning (=zero deflection) to the deflection of L/150. Lee et al. [14] excluded the uncracked region to calculate the residual strength up to a certain deflection. In this study, the method proposed by Lee et al. [14] is used to define the residual strengths of f_{600}^{eq} and f_{150}^{eq}, which are the mean stresses from the crack initiation to the deflection of L/600 and L/150, respectively.

Figure 4 shows f_{600}^{eq} and f_{150}^{eq} with respect to the steel fiber volume fractions. The linear lines shown in Figure 4 were obtained from the linear regression analysis, except for the residual strength of the non-reinforced concrete beam, which was zero. The results of the linear regression analysis show that f_{600}^{eq} and f_{150}^{eq} increase in proportion to the content of steel fiber. f_{150}^{eq} is slightly greater than f_{600}^{eq}, which means that the steel fibers play an effective role in increasing the residual strength. The f_{150}^{eq} of the reinforced beams with fiber volume percentages of 0.33, 0.67, and 1.00% was 3.96, 4.83, and 6.66 MPa, which are approximately 110, 136, and 170% of f_{cr}, respectively.

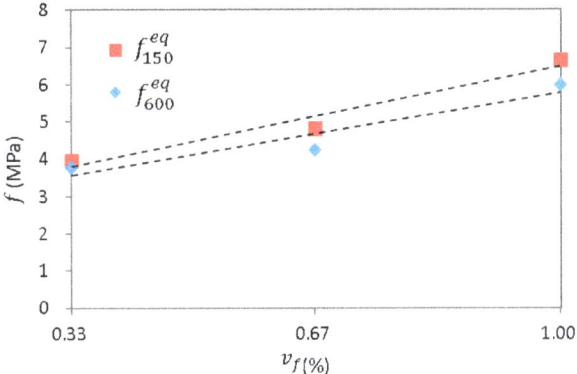

Figure 4. Residual strengths f_{600}^{eq} and f_{150}^{eq} with increasing fiber volume percentages from 0.33 to 1.00%.

3.2. Energy Absorption Capacity

The energy absorption was calculated from the area of the stress and deflection curves obtained from the bending tests. Figure 5 shows the relationship between the deflection and energy absorption of the beams. In the plain concrete beam, the residual strength was very small after cracking, and the increase in the energy absorption capacity was minimal with increases in deflection. However, the energy absorption capacity of the SFRC beams increased almost linearly with increasing fiber content.

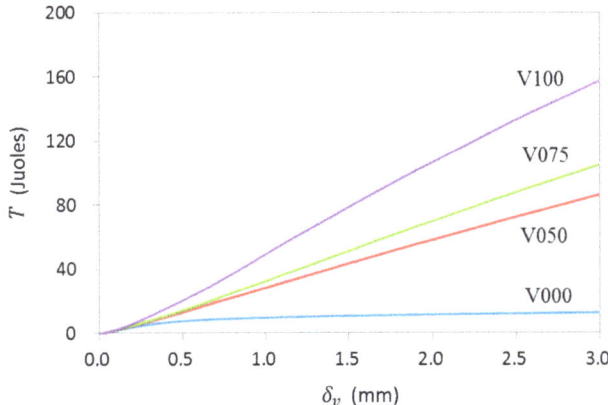

Figure 5. Energy absorption and deflection curves of the plain concrete and steel fiber-reinforced concrete (SFRC) beams obtained from the bending tests.

Figure 6 shows the energy absorption capacities of T_{600} and T_{150} with increasing fiber volume fractions. T_{600} and T_{150} are the energy absorption capacities at deflections of $L/600$ (=0.75 mm) and $L/150$ (=0.0 mm), respectively. The linear lines obtained from the regression analysis with respect to the fiber volume percentages from zero to 1.00% show that T_{600} and T_{150} increase proportionally to the increase in the steel fiber content. The slope of the regression line of T_{150} is greater than that of T_{600}, which means that the steel fibers effectively maintained greater residual strength than the cracking strength until the deflection of $L/600$. For the concrete beam with a fiber volume percentage of 1.00%, T_{600} and T_{150} were 31.02 and 143.72 J, which correspond to 3.47 and 11.25 times the values of the non-reinforced plain beams, respectively.

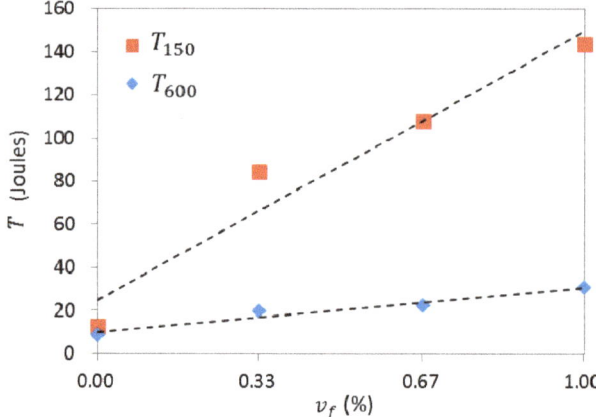

Figure 6. Energy absorption capacities T_{600} and T_{150} with increasing fiber volume percentages from zero to 1.00%.

4. Shear Test Results and Discussion

4.1. Test Results and Failure Mode

Figure 7 shows the load and displacement curves measured in the shear tests of the anchors installed in the plain and SFRC. The load and displacement curve can be divided into pre-cracking and post-cracking regions based on a point when a concrete crack occurs on the top surface of the concrete block. Before the crack occurs, the tensile stress generated in the concrete block is below the cracking strength of the concrete. Thus, all of the materials including concrete linearly and elastically resist the external shear loads, and the load and displacement curve of the anchor in the SFRC is similar to that in the non-reinforced plain concrete, as shown in Figure 7.

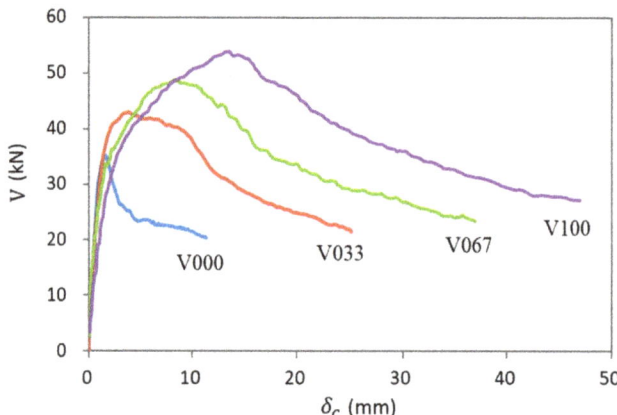

Figure 7. Typical load and displacement curves measured from the shear tests of anchors in the plain concrete and SFRC.

When a crack occurred on the top surface of the concrete block, the slope of the load and displacement curve decreased due to the strength reduction of the concrete. Cracks initiated almost simultaneously on the left and right sides of the steel anchor on the top surface of the concrete block. The left and right cracks occurred at an angle on the top surface and proceeded to the edge of the block. The inclined cracks reduced the stiffness of the anchor block, which resulted in the lower slope of the load and displacement curve. The load, at which the inclined cracks occurred on the top surface of the anchor block, is defined as the cracking shear load of the anchor, V_{cr}.

While the inclined crack propagated to the edge of the top surface, another new vertical crack was generated from the top to the bottom on the front side of the anchor block. The vertical crack proceeded to a depth a little longer than the embedded depth of 240 mm. Then, the anchor reached its ultimate shear load, V_u. The inclined crack, which reached the edge of the block, was directed to the end tip of the vertical crack and thus formed into a semi-circular crack shape on the front surface of the block. Table 3 summarizes the cracking and ultimate shear loads, V_{cr} and V_u, and the displacement at the maximum shear load, $d_{c,u}$ for the anchor specimens in the plain and SFRC.

Table 3. Summary of the cracking and ultimate shear loads and the displacement at the ultimate load obtained from the shear tests of anchors.

Name of Specimen	V_{cr} (kN)		V_u (kN)	$d_{c,u}$ (mm)
V000	29.6		38.0	2.55
	30.3		35.2	2.17
	40.5		45.8	1.52
	23.6		26.4	1.66
	Mean	31.0	36.3	1.98
	(Std.)	(7.01)	(8.01)	(0.48)
V033	35.2		42.9	3.78
	29.5		38.2	3.81
	35.3		41.8	4.06
	28.1		30.5	2.76
	Mean	32.0	38.3	3.60
	(Std.)	(3.78)	(5.63)	(0.58)
V067	34.8		48.7	8.58
	34.6		55.9	9.56
	33.4		48.7	8.64
	29.8		43.6	7.59
	Mean	33.1	49.2	8.59
	(Std.)	(2.31)	(5.05)	(0.80)
V100	37.4		52.1	12.27
	30.7		59.4	9.72
	28.8		53.8	13.76
	30.7		43.7	9.56
	Mean	31.9	52.2	11.33
	(Std.)	(3.77)	(6.48)	(2.04)

Figure 8a shows a typical example of crack distributions of anchors in the plain concrete upon concrete breakout failure. The steel fibers mixed in concrete changed the crack distribution and shape of the concrete anchor under shear load. The anchors in SFRC showed a great improvement of the shear load and displacement resistance after cracking, while the strength of the anchor in the plain concrete quickly decreased and reached the final brittle fracture. This is because the steel fibers distributed in the concrete control the growth of the cracks and greatly improve the tensile capacity of the concrete block.

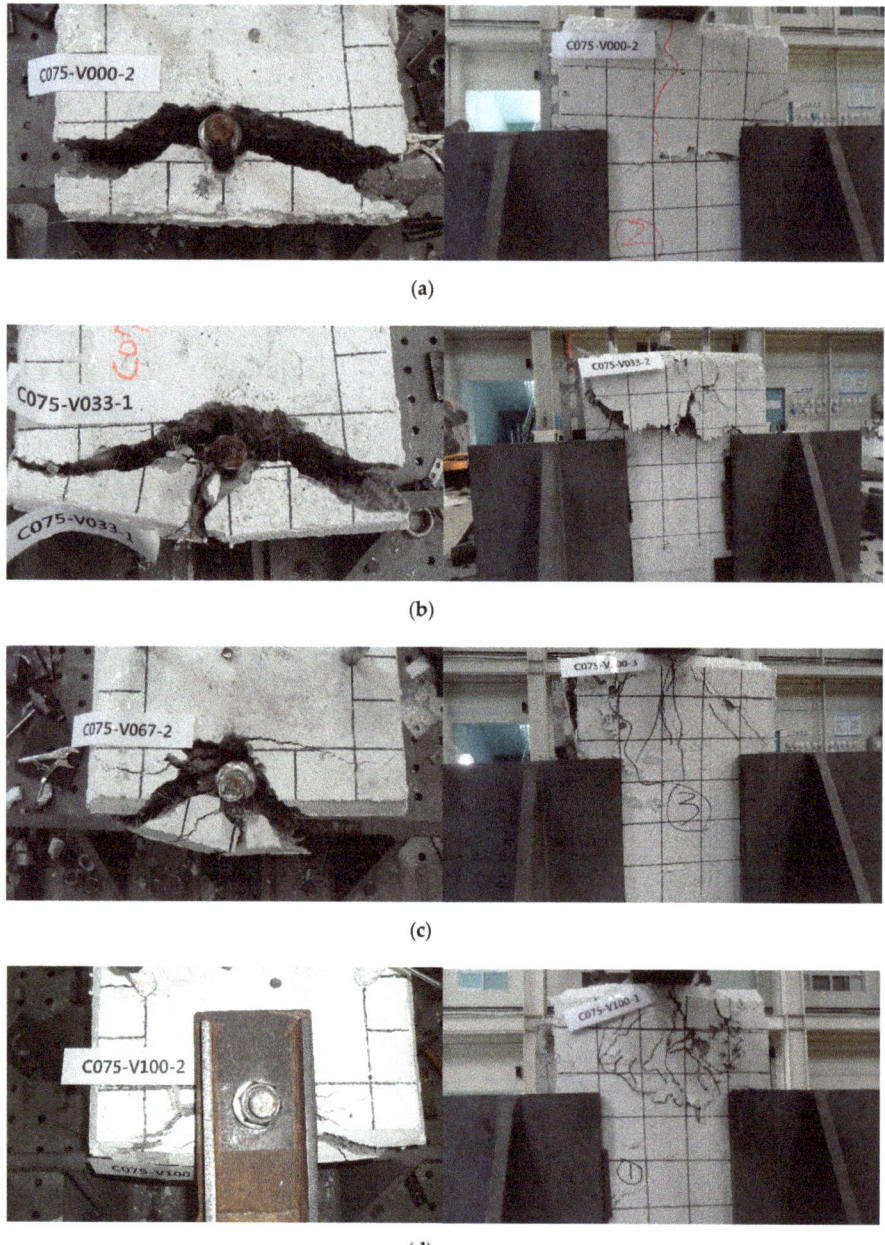

Figure 8. Crack distributions of the concrete anchor block upon concrete breakout failure: (**a**) plain concrete; (**b**) SFRC with $v_f = 0.33\%$; (**c**) SFRC with $v_f = 0.67\%$; (**d**) SFRC with $v_f = 1.00\%$.

Figure 8b–d show the crack initiation and propagation of the anchor installed in concrete reinforced with steel fibers. On the top surface of the block, additional inclined cracks occurred near the original inclined crack. The final fracture angle on the top surface tended to be lower than the 35° defined in the CCD method. On the front of the concrete block, another vertical crack was also generated along the original vertical crack. The most distinctive feature in the anchors installed in the SFRC was the generation of a new type of radially straight crack originating from the center of the edge on the front side of the concrete block, which was independent of the previous inclined, vertical, and curved cracks. In particular, the anchor in concrete with higher steel fiber volume fractions showed more cracks and more complicated crack shapes.

4.2. Load and Displacement Resistance

The effect of steel fibers on the cracking and ultimate shear loads of the anchor was investigated (V_{cr} and V_u, respectively). The linear regression analysis was performed with respect to the steel fiber volume percentages from zero to 1.00%. As shown in Figure 9, V_{cr} is almost constant in the range of approximately 31.0 to 33.1 MPa, regardless of the increasing fiber volume fractions. However, V_u increases almost linearly from approximately 36.3 kN in the non-reinforced concrete to 52.2 kN in the SFRC with a fiber volume percentage of 1.00%, which is approximately 1.44 times that of the non-reinforced concrete anchor. The slope of the line obtained from the linear regression analysis with fiber volume percentages of zero to 1.00% indicates an increase of approximately 17.6 kN in V_u per 1% increase in the fiber volume percentage.

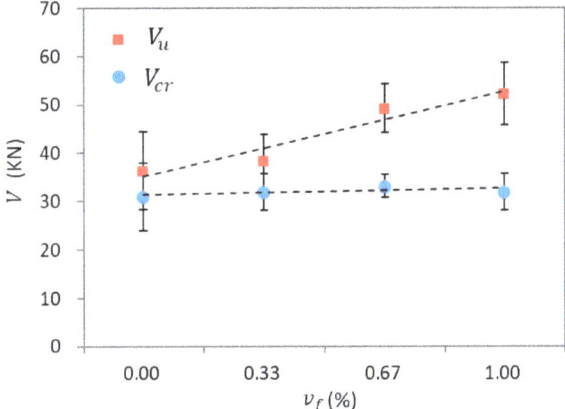

Figure 9. Cracking and ultimate loads measured from the shear tests of anchors with increasing fiber volume percentages from zero to 1.00%.

Figure 10 shows the displacement at the maximum shear load of the anchor, $\delta_{c,u}$, with respect to the steel fiber volume fractions. The linear regression line shows that the displacement of the anchor at the maximum load increases in proportion to the steel fiber content. The displacement of the anchor in the SFRC with a fiber volume percentage of 1.00% was approximately 11.4 mm at the maximum load, which is 5.74 times that of the non-reinforced concrete anchor. The slope of the linear line in the range of the fiber volume percentages from zero to 1.00% is approximately 9.92.

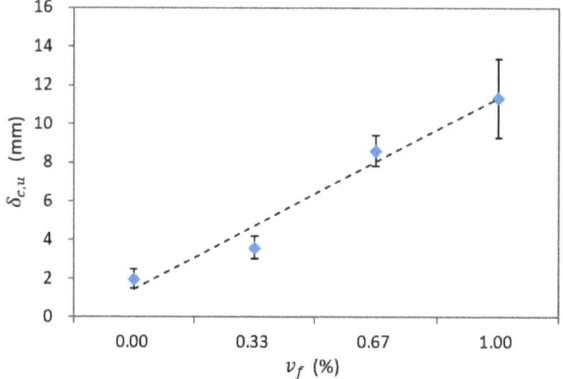

Figure 10. Displacement at the maximum shear load with increasing fiber volume percentages from zero to 1.00%.

4.3. Energy Absorption Capacity

This study also evaluated the energy absorption capacity of the plain and SFRC anchors, which can be used as an index for evaluating the fracture resistance of a material or structural member. Figure 11 shows the energy absorption and displacement curves of the plain concrete and SFRC anchors, which were calculated using the area of the load and displacement curve. The energy absorption increases almost linearly with increasing displacement, and the steel fibers significantly improve the energy absorption capacity. Since no criteria have been established for the energy absorption capacity of the concrete anchors, the evaluation methods of the energy absorption capacity in fiber-reinforced concrete beams have been reviewed.

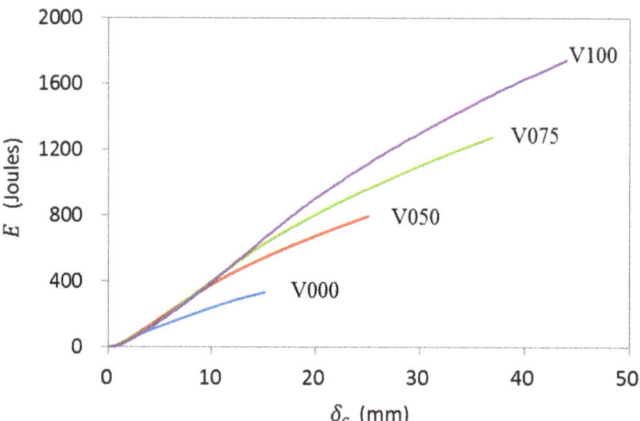

Figure 11. Typical energy absorption and displacement curves of the anchors in the plain concrete and SFRC.

Gopalaratman and Gettu [18] introduced the following three methods to evaluate the energy absorption capacity of SFRC beams: (1) the absolute energy absorption capacity until a specific deflection, (2) dimensionless indices related to energy absorption capacity, and (3) the energy absorption capacity by an equivalent flexural strength at a specified deflection in the post-cracking region. Among the three methods, Gopalaratman and Gettu [18] recommended the absolute energy absorption capacity at a certain deflection and the energy absorption capacity using the equivalent flexural strength. JSCE

SF-4 [38] adopted the absolute energy absorption method calculated at a deflection, and ASTM C 1550 [39], ASTM C 1609 [35], and BS EN 14,651 [37] adopted the evaluation method using the equivalent flexural strength at a specific deflection in the post-cracking region. The energy absorption evaluation method using the non-dimensional index was adopted in the ASTM C 1018 [40], but it is currently excluded from the ASTM recommendations.

Therefore, based on the absolute energy absorption method at a specific deflection, this study assessed the energy absorption capacity of anchors in plain concrete and SFRC. Two energy absorption values are defined: (1) the energy absorption until a displacement corresponding to the maximum shear load, E_u, and (2) the energy absorption from the displacement at the maximum shear load to a displacement when the post-cracking residual shear load reaches the level of the cracking shear load, E_r. Figure 12 shows E_u and E_r with increasing fiber volume percentages from zero to 1.00%. The linear regression lines show a linear increase of E_u and E_r with increasing fiber volume fractions. The E_u of the anchors in the SFRC with steel fiber volume percentages of 0.33 and 1.00% are 120.3 and 518.5 J, which are approximately 2.64 and 11.4 times that of the anchor in the non-reinforced plain concrete, respectively. E_r also increased from approximately 62.5 J in the plain concrete to 1387.2 J in the SFRC with a steel fiber volume percentage of 1.00%, which is 22.2 times that of the anchor in the non-reinforced concrete. The slope of the linear trend line obtained in the range of fiber volume percentages from zero to 1.00% is approximately 501.7 J per percent for E_u and 1324.9 J per percent for E_r.

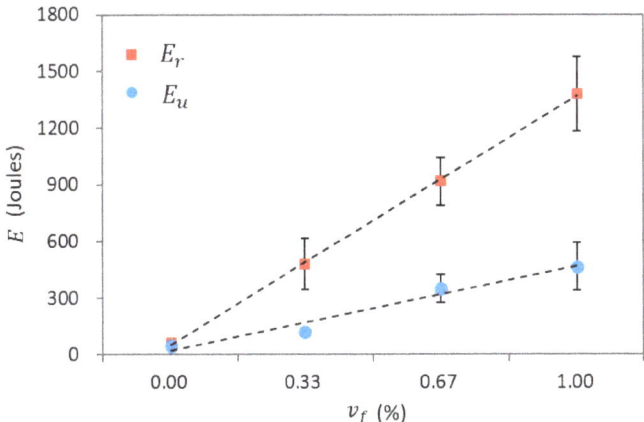

Figure 12. Energy absorption capacities E_u and E_r of the anchors in the plain concrete and SFRC with increasing fiber volume percentages from zero to 1.00%.

5. Relationship between Shear Behavior of Anchors and Tensile Performance of SFRC

The load and deflection curves of both the SFRC beams and anchors exhibited linear and elastic behavior before concrete cracking. This means that the tensile stress generated from the external loads is smaller than the cracking strength of the concrete and, thus, all of the materials including the concrete effectively endure the external loads. The inclusion of steel fibers has little effect on the cracking strength of the SFRC beams and anchors. As shown in Figures 3 and 9, the cracking strength of the beams, f_{cr}, and the cracking shear load of the anchors, V_{cr}, are almost constant regardless of the fiber volume fractions. On the other hand, the ultimate strength of the beams, f_u, and the maximum concrete breakout shear load of the anchors, V_u, increased almost linearly with increasing fiber volume fractions. According to the linear trend analysis, f_u and V_u increase by approximately 4.04 MPa and 19.90 kN per percent of fiber volume fraction, respectively.

The design shear resistance of anchors embedded in the non-reinforced plain concrete is dependent on the edge distance, embedded depth, and diameter of the anchor, as well as the compressive strength of the concrete [8,9]. As mentioned, the steel fibers had little effect on the compressive and cracking strengths but greatly improved the ultimate flexural strength of the concrete. Thus, the concrete breakout shear resistance of the anchors needs to be determined using the ultimate flexural strength of the concrete, f_u, which can account for the improvement of the tensile capacity of concrete by the steel fibers rather than the compressive strength of the concrete specified in the ACI 318 design standards [7,8].

Figure 13 shows the relationships between the ratios of f_u and $\sqrt{f_u}$ of the SFRC beams and V_u of the SFRC anchors with the increasing fiber volume fractions to those of the non-reinforced plain concrete. $f_{u,o}$ is the ultimate flexural strength of the plain concrete beam, which corresponds to f_{cr}. $V_{u,o}$ is the concrete breakout shear resistance of the plain concrete anchor. f_u tends to increase exponentially, while $\sqrt{f_u}$ and V_u show linear increases as the fiber volume fraction increases. Furthermore, the rate of increase of $\sqrt{f_u}$ is very similar to that of V_u, which corresponds to the design equation expressed as a function of $\sqrt{f'_c}$, in which f'_c is the compressive strength of concrete. Thus, based on the design equation specified in the ACI 318 [7,8], the concrete breakout shear resistance of an anchor in the SFRC can be expressed as

$$V_u = k \left(\frac{h_0}{d_0}\right)^2 \sqrt{d_0} \sqrt{f_u}(c_0)^{1.5} \tag{2}$$

where the influence factors of the anchor diameter, embedded depth, and edge distance are taken as those specified in the ACI 318 [7,8], and the term for the compressive strength of concrete, $\sqrt{f'_c}$, is replaced with a term for the ultimate flexural strength of concrete, $\sqrt{f_u}$, which includes the effect of the steel fibers on the tensile capacity of concrete. For the plain concrete anchor, $\sqrt{f_u}$ corresponds to the flexural tensile strength of concrete. The factor k in Equation (2) can be determined by comparing the measured maximum shear load of anchors with the shear capacity calculated by Equation (2) without the k term.

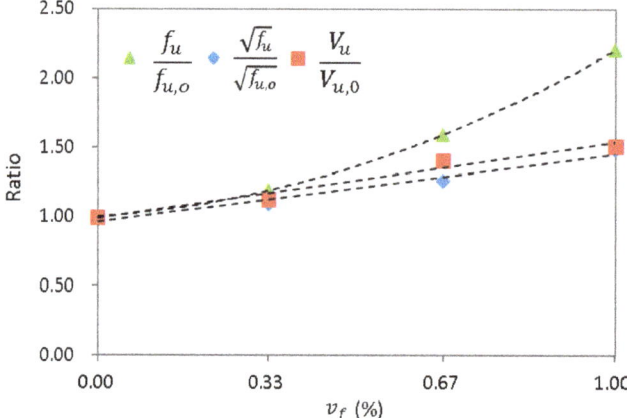

Figure 13. Ratios of the f_u and $\sqrt{f_u}$ of the SFRC beams and V_u of the SFRC anchors to those of the non-reinforced concrete with increasing fiber volume percentages from zero to 1.00%.

Figure 14 shows the variation in the factor k with the change in the fiber volume percentages from zero to 1.00%. The average and standard deviation of k are 3.83 and 0.19, respectively. Thus, k is simplified as a constant equal to 4.0. Figure 15 compares the concrete breakout shear resistance of anchors calculated using the proposed Equation (2) with the measured average maximum shear loads. The calculated shear resistance shows very good agreement with the measurements. The differences between the calculated and measured shear loads are in the range of approximately 5–6%.

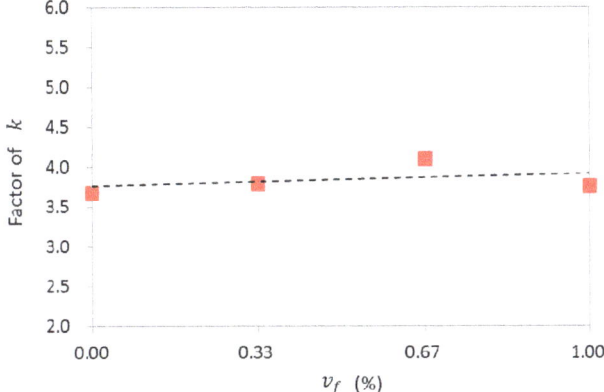

Figure 14. Variation in the factor k with the change in the fiber volume percentages from zero to 1.00%.

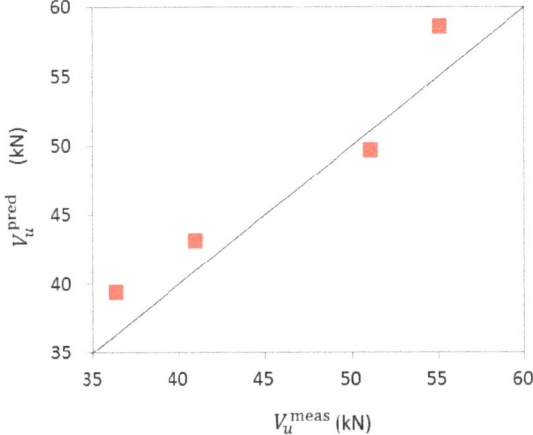

Figure 15. Comparison of the predicted and measured concrete breakout shear resistance.

In addition to the load capacity of an anchor, the energy absorption capacity can also be used to assess the fracture resistance of the anchor system. Therefore, this study evaluated the relationship between the energy absorption capacities of the anchor, E_u and E_r, with the energy absorption capacity of the beam, T_{150}, which increases proportionally to the increase in fiber volume percentages from zero to 1.00%. Figures 16 and 17 show that the E_u and E_r of the SFRC anchors is proportional to the T_{150} of the SFRC beams. Therefore, using the linear relationship with T_{150}, the energy absorption capacities E_u and E_r of an anchor can be determined as 3.63 and 9.97 times the value of T_{150}, respectively.

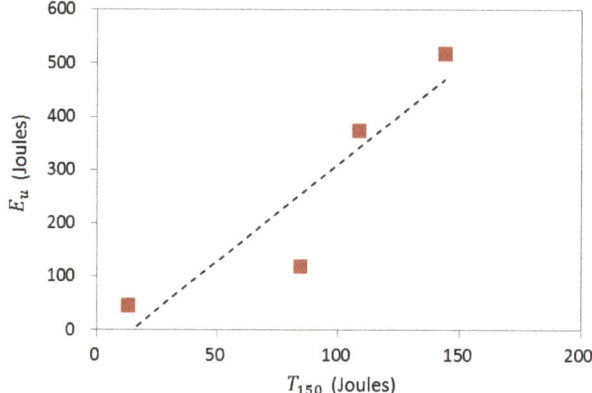

Figure 16. Relationship between the energy absorption capacity E_u of the anchor and the energy absorption capacity T_{150} of the beam.

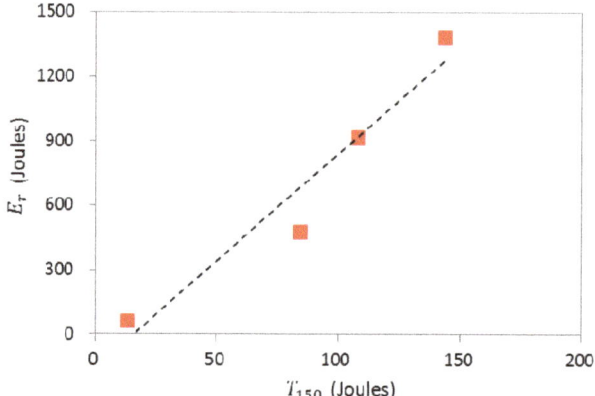

Figure 17. Relationship between the energy absorption capacity E_r of the anchor and the energy absorption capacity T_{150} of the beam.

6. Conclusions

The concrete shear resistance of an anchor is strongly dependent on the strength of concrete. The current design specifications are based on the compressive strength of concrete to determine the concrete breakout resistance of an anchor. The addition of steel fibers to concrete can greatly improve the tensile resistance of the concrete. Thus, this study investigated the effect of steel fibers on the shear failure mode and breakout resistance of anchors embedded in SFRC. Beam specimens were also prepared to assess the relationship between the tensile performance of SFRC beams and the shear resistance of SFRC anchors.

In the bending tests, the non-reinforced plain concrete beam showed a sudden decrease in the strength after cracking, which led to brittle failure. However, the beams reinforced with steel fibers showed deflection-softening or deflection-hardening behavior depending on the amount of steel fibers. The beams with fiber volume percentages of 0.67 and 1.00% continuously increased to reach the ultimate strength without decreasing the strength due to the cracking of concrete. The steel fibers had little effect on the cracking strength, but the ultimate flexural strength, post-cracking residual strength, and energy absorption capacity showed linear increases with increasing fiber volume fractions. The ultimate flexural strength f_u and the residual strength f_{150}^{eq} in the beam with a fiber volume percentage of

1.00% were approximately 2.21 and 1.70 times the cracking strength of the non-reinforced plain beam, respectively.

The shear tests of anchors also showed that the shear load and displacement capacities of the SFRC anchors increased almost linearly with the increase in the fiber volume fraction of steel fibers. The anchors in the plain concrete failed immediately after concrete cracking. The shear resistance V_u in the SFRC with a fiber volume percentage of 1.00% was approximately 55.2 kN, which is 1.52 times that in the non-reinforced plain concrete. The displacement at the maximum shear load also increased by approximately 6.03 times in the SFRC with a fiber volume percentage of 1.00%. Thus, the energy absorption capacity at the maximum shear load for the SFRC anchors at fiber volume percentages of 0.33 and 1.00% was approximately 2.64 and 11.4 times those of the plain concrete anchor, respectively.

In the bending and shear tests, the ultimate strength of the beam and the maximum concrete breakout shear resistance of the anchor increased almost linearly with increasing fiber volume fractions. The design shear resistance of the concrete anchor is based on the compressive strength of concrete, which was rarely affected by the addition of steel fibers to the concrete with the fiber volume percentage less than 1.00%. Thus, based on the ACI 318 design equation, this study utilized a term for the ultimate flexural strength of concrete instead of the compressive strength to determine the concrete breakout shear resistance of an anchor in the SFRC. The calculated shear resistance of anchors in both the plain concrete and SFRC agreed well with the measured shear loads. In addition, the energy absorption capacity of the SFRC anchor with a fiber volume percentage of 1.00% showed a linear relationship with the energy absorption capacity of the SFRC beam, which increased proportionally to the increase in fiber volume percentages from zero to 1.00%.

Author Contributions: The authors J.-H.L., E.C. and B.-S.C. conceived and designed the study; In particular, methodology and formal analysis, J.-H.L. and B.-S.C.; experimental programming and investigation, J.-H.L. and B.-S.C.; writing—original draft preparation, J.-H.L.; writing—review and editing, E.C. and B.-S.C.; funding acquisition, E.C. All authors have read and agreed to the published version of the manuscript.

Funding: This research was funded by the National Research Foundation of Korea (NRF) grant funded by the Korea government (MSIT) (Project No. NRF 2020R1A4A1018826).

Conflicts of Interest: The authors declare no conflict of interest.

References

1. Wang, D.; Wu, D.; He, S.; Zhou, J.; Ouyang, C. Behavior of post-installed large-diameter anchors in concrete foundations. *Constr. Build. Mater.* **2015**, *95*, 124–132. [CrossRef]
2. Park, Y.M.; Kim, T.H.; Kim, D.H.; Kang, C.H.; Lee, J.H. Breakout shear strength of cast-in-place anchors using shaking table tests. *Proc. Inst. Civ. Eng. Struct. Build.* **2017**, *170*, 939–950. [CrossRef]
3. Fuchs, W.; Eligehausen, R.; Breen, J.E. Concrete capacity design (CCD) approach for fastenings to concrete. *ACI Struct. J.* **1995**, *92*, 794–802.
4. Eligehausen, R.; Balogh, T. Behavior of fasteners loaded in tension in cracked reinforced concrete. *ACI Struct. J.* **1995**, *92*, 365–379.
5. Muratli, H.; Klingner, R.E.; Graves, H.L., III. Breakout capacity of anchors in concrete-Part 2: Shear. *ACI Struct. J.* **2004**, *101*, 821–829.
6. ACI 349-01. *Code Requirements for Nuclear Safety Related Concrete Structures, Appendix B: Anchoring to Concrete*; ACI (American Concrete Institute): Farmington Hills, MI, USA, 2001.
7. ACI 318M-02. *Building Code Requirements for Structural Concrete and Commentary, Appendix D Anchoring to Concrete*; 3rd Amendment; ACI (American Concrete Institute): Farmington Hills, MI, USA, 2002.
8. ACI 318M-08. *Building Code Requirements for Structural Concrete and Commentary, Appendix D: Anchoring to Concrete*; 3rd Amendment; ACI (American Concrete Institute): Farmington Hills, MI, USA, 2008.
9. ACI 318M-11. *Building Code Requirements for Structural Concrete and Commentary, Appendix D: Anchoring to Concrete*; 3rd Amendment; ACI (American Concrete Institute): Farmington Hills, MI, USA, 2011.
10. Olalusi, O.B.; Spyridis, P. Uncertainty modelling and analysis of the concrete edge breakout resistance of single anchors in shear. *Eng. Struct.* **2020**, *222*, 111112. [CrossRef]

11. Bokor, B.; Sharma, A.; Hofmann, J. Experimental investigations on concrete cone failure of rectangular and non-rectangular anchor groups. *Eng. Struct.* **2019**, *188*, 200–217. [CrossRef]
12. Pürgstaller, A.; Gallo, P.Q.; Pampanin, S.; Bergmeister, K. Seismic demands on nonstructural components anchored to concrete accounting for structure-fastener-nonstructural interaction (SFNI). *Earthq. Eng. Struct. Dyn.* **2020**, *49*, 589–606. [CrossRef]
13. Dengg, F.; Zeman, O.; Voit, K.; Bergmeister, K. Fastening application in concrete using recycled tunnel excavation material. *Struct. Concr.* **2018**, *19*, 374–386. [CrossRef]
14. Lee, J.H.; Cho, B.; Choi, E.; Kim, Y.H. Experimental study of the reinforcement effect of macro-type high strength polypropylene on the flexural capacity of concrete. *Constr. Build. Mater.* **2016**, *126*, 967–975. [CrossRef]
15. Lee, J.H.; Kighuta, K. Twin-twist effect of fibers on the pullout resistance in cementitious materials. *Constr. Build. Mater.* **2017**, *146*, 555–562. [CrossRef]
16. Naaman, A.E.; Najm, H. Bond-slip mechanisms of steel fibres in concrete. *ACI Mater. J.* **1991**, *88*, 135–145.
17. Alwan, J.M.; Naaman, A.E.; Hansen, W. Analytical investigation on pullout work of steel fibers from cementitious composites. *Cem. Concr. Compos.* **1991**, *13*, 247–255. [CrossRef]
18. Gopalaratnam, V.S.; Gettu, R. On the characterization of flexural toughness in fiber reinforced concrete. *Cem. Concr. Compos.* **1995**, *17*, 239–254. [CrossRef]
19. Lee, J.H.; Cho, B.; Choi, E. Flexural capacity of fiber reinforced concrete with a consideration of concrete strength and fiber content. *Constr. Build. Mater.* **2017**, *138*, 222–231. [CrossRef]
20. Narayanan, R.; Darwish, I.Y.S. Use of steel fibers as shear reinforcement. *ACI Struct. J.* **1987**, *84*, 216–227.
21. Sharma, A.K. Shear strength of steel fiber reinforced concrete beams. *ACI J. Proc.* **1986**, *83*, 624–628.
22. Amin, A.; Foster, S.J. Shear strength of steel fibre reinforced concrete beams with stirrups. *Eng. Struct.* **2016**, *111*, 323–332. [CrossRef]
23. Khuntia, M.; Stojadinovic, B.; Goel, S.C. Shear strength of normal and high strength fiber reinforced concrete beams without stirrups. *ACI Struct. J.* **1999**, *96*, 282–289.
24. Mansur, M.A.; Chin, M.S.; Wee, T.H. Stress-strain relationship of high-strength fiber concrete in compression. *J. Mater. Civ. Eng.* **1999**, *11*, 21–29. [CrossRef]
25. Holschemacher, K.; Mueller, T.; Ribakov, Y. Effect of steel fibers on mechanical properties of high strength concrete. *Mater. Des.* **2010**, *31*, 2604–2615. [CrossRef]
26. Lee, J.H. Influence of concrete strength combined with fiber content in the residual flexural strengths of fiber reinforced concrete. *Compos. Struct.* **2017**, *168*, 216–225. [CrossRef]
27. Jeng, F.; Lin, M.L.; Yuan, S.C. Performance of toughness indices for steel fiber reinforced shotcrete. *Tunn. Undergr. Space Technol.* **2002**, *17*, 69–82. [CrossRef]
28. Sorelli, L.G.; Meda, A.; Plizzari, G.A. Steel fiber concrete slabs on ground: A structural matter. *ACI Struct. J.* **2006**, *103*, 551–558.
29. Altoubat, S.A.; Roesler, J.R.; Lange, D.A.; Rieder, K.A. Simplified method for concrete pavement design with discrete structural fibers. *Constr. Build. Mater.* **2008**, *22*, 384–393. [CrossRef]
30. Roesler, J.R.; Altoubat, S.A.; Lange, D.A.; Rieder, K.A.; Ulreich, G.R. Effect of synthetic fibers on structural behavior of concrete slabs-on-ground. *ACI Mater. J.* **2016**, *103*, 3–10.
31. Nilforouch, R.; Nilsson, M.; Elfgren, L. Experimental evaluation of tensile behaviour of single cast-in-place anchor bolts in plain and steel fibre-reinforced normal- and high-strength concrete. *Eng. Struct.* **2017**, *147*, 195–206. [CrossRef]
32. Tóth, M.; Bokor, B.; Sharma, A. Anchorage in steel fiber reinforced concrete—concept, experimental evidence and design recommendations for concrete cone and concrete edge breakout failure modes. *Eng. Struct.* **2019**, *181*, 60–75. [CrossRef]
33. Mahrenholtz, C.; Ayoubi, M.; Müller, S.; Bachschmid, S. Tension and shear performance of anchor channels with channel bolts cast in Fibre Reinforced Concrete (FRC). *IOP Conf. Ser. Mater. Sci. Eng.* **2019**, *615*, 012089. [CrossRef]
34. Lee, J.H.; Cho, B.; Kim, J.B.; Lee, K.J.; Jung, C.Y. Shear capacity of cast-in headed anchors in steel fiber-reinforced concrete. *Eng. Struct.* **2018**, *171*, 421–432. [CrossRef]
35. ASTM C 1609. *Standard Test Method for Flexural Performance of Fiber Reinforced Concrete*; ASTM (American Society of Testing Material): West Conshonocken, PA, USA, 2007.

36. Gopalaratnam, V.S.; Shah, S.P.; Batson, G.B.; Criswell, M.E.; Ramakrishnam, V.; Wicharatana, M. Fracture toughness of fiber reinforced concrete. *ACI Mater. J.* **1991**, *88*, 339–353.
37. BS EN 14651. *Test Method for Metallic Fibered Concrete–Measuring the Flexural Tensile Strength (Limit of Proportionality (LOP), Residual)*; European Committee for Standardization: Brussels, Belgium, 2005.
38. JSCE SF-4. *Method of Tests for Flexural Strength and Flexural Toughness of SFRC.*; The Japanese Society of Civil Engineers (JSCE): Tokyo, Japan, 1985; pp. 45–74.
39. ASTM C 1550. *Standard Test Method for Flexural Toughness of Fiber Reinforced Concrete (Using centrally Loaded Round Panel)*; ASTM (American Society of Testing Material): West Conshonocken, PA, USA, 2010.
40. ASTM C 1608. *Standard Test Method for Flexural Toughness and First-Crack Strength of Fiber Reinforced Concrete*; ASTM (American Society of Testing Material): West Conshonocken, PA, USA, 1987.

© 2020 by the authors. Licensee MDPI, Basel, Switzerland. This article is an open access article distributed under the terms and conditions of the Creative Commons Attribution (CC BY) license (http://creativecommons.org/licenses/by/4.0/).

Article

Longitudinal Compressive Property of Three-Dimensional Four-Step Braided Composites after Cyclic Hygrothermal Aging under High Strain Rates

Kailong Xu, Wei Chen, Lulu Liu *, Gang Luo and Zhenhua Zhao

Jiangsu Province Key Laboratory of Aerospace Power System, College of Energy and Power Engineering, Nanjing University of Aeronautics and Astronautics, Nanjing 210016, China; xukailong@nuaa.edu.cn (K.X.); chenwei@nuaa.edu.cn (W.C.); mevislab@nuaa.edu.cn (G.L.); zhaozhenhua@nuaa.edu.cn (Z.Z.)
* Correspondence: liululu@nuaa.edu.cn

Received: 12 February 2020; Accepted: 11 March 2020; Published: 18 March 2020

Abstract: The longitudinal compressive behavior of the three-dimensional four-step braided composites after cyclic hygrothermal aging was investigated using a split Hopkinson pressure bar (SHPB) apparatus under high strain rates (1100~1250 s^{-1}, 1400~1600 s^{-1}, 1700~1850 s^{-1}, respectively). The SEM micrographs were examined to the damage evolution of the composites after cyclic hygrothermal aging. A high-speed camera was employed to capture the progressive damage process for the composites. The results indicate that the saturated moisture absorption of the composites was not reached during the whole 210 cyclic hygrothermal aging days. The composites mainly underwent epoxy hydrolysis and interfaces debonding during continuous cyclic hygrothermal aging time. The peak stress of the composites still behaved as a strain rate effect after different cyclic hygrothermal aging days, but the dynamic stiffness modulus clearly had no specific regularity. In addition, the peak stress and the dynamic stiffness modulus of the composites after 210 cyclic hygrothermal aging days almost decreased by half when subjected to longitudinal compression.

Keywords: cyclic hygrothermal aging; high strain rates; braided composites; compressive property

1. Introduction

Fiber reinforced composites (FRC) are widely applied in many engineering applications, such as aerospace, wind power and energy, due to their excellent mechanical properties in recent decades [1–3]. In particular, the laminated composites are more commonly used in aerospace. However, the traditional laminated composites have the disadvantages of relatively low fracture toughness in plane and delamination through thickness. In the recent decade, three-dimensional (3D) textile composites, e.g., weaving, knitting and braiding, have received extensive attention thanks to their excellent anti-delamination characteristics. Specifically, the 3D braided composites were generally recognized as the most promising textile composites, due to their excellent anti-delamination properties [4–6]. Composite structure in aircraft applications may be subjected to a large number of dynamic loads, e.g., bird strike, blade out and unexpected fragment impact [7,8]. Therefore, it is necessary and meaningful to understand their dynamic mechanical properties at high strain rates, to guide the design of composite structure that will be used under dynamic loadings in engineering applications.

Many efforts have been committed on the dynamic mechanical behavior of the 3D textile composites in recent years. Gu and Xu [9] conducted the ballistic perforation tests of the 3D four-step braided Twaron/epoxy composites, and adopted a "Fiber inclination model" to decompose 3D braided composites to geometrical modeling in FEM, proving that the analysis scheme in microstructures can work. Sun et al. [10–12] studied in-plane, out-plane compressive and uniaxial tensile mechanical behavior of the 3D four-step braided composites through a series of experiments at different strain rates and discussed its failure modes. The results revealed that the 3D braided composite is sensitive to strain rate and the higher strain rate, the higher failure stress and compressive stiffness. Walter et al. [13] presented dynamic indentation and small caliber ballistic impact tests for the 3D glass fiber composites; they concluded that the delamination damage at high strain rates was still a dominant failure mode in the 3D woven composites. Zhang et al. [14] reported the compressive mechanical performance and damage mode at quasi-static and high strain rates, based on a meso-structure model of the 3D four-step braided composites; the results proved the validity of meso-structure FEM model. Recently, Tan [15] experimentally studied dynamic compressive response of the 3D four-step braided composites in three directions (longitudinal, transverse and thickness, respectively), and the experiments of dynamic compressive tests revealed that with increasing the strain rate of composites, the compressive stiffness and failure stress both increased.

Most research on the above papers was focused on investigating the mechanical properties on compressive behavior in room temperature, not considering the impact of environment. However, composites structure, usually subject to long-term hygrothermal exposures during their service life, such as the aero-engine fan blade with carbon fiber reinforced polymer material, operating in the temperature and humidity environment, is continuously exposed to a hygrothermal environment. Epoxy in composites, as a typical thermoset resin, will absorb the moisture, inducing swelling and plasticization [16–18]. Composites' mechanical advantages such as stiffness and strength will significantly be affected because of hygrothermal aging [19–22]. Yilmaz and Sinmazcelik [23] investigated the effects of moisture on thermal and mechanical properties and the moisture absorption of glass-fiber/polyetherimide (PEI) laminated composites, and the laminated composites were exposed to hydrothermal aging environment under two different temperatures and high moisture rates. The hydrothermal aging laminates comprised a large number of moisture, causing a decrease in the glass transition temperature and degradation in mechanical properties. Sun et al. [24] studied the hygrothermal aging properties of carbon fiber/bismaleimide (BMI) composites, and measured the water diffusivity through the three wet-dry cycles of BMI resin, reinforced with unidirectional carbon fiber CCF300/QY9511 composites. The results showed that the re-absorption of CCF300/QY9511 composites exhibit a higher amplitude of diffusivity but a lower saturated moisture content. Li et al. [25] extensively investigated the low velocity impact tests of laminated composites in an ambient hygrothermal environment, and the unidirectional and crossply glass fiber reinforced plastic (GFRP) laminates were placed in a conditioning chamber for a maximum of eight moisture cycles from 50 to 100 °C. The impact loading decreases and the deflection increases with cycling moisture levels. Liotier et al. [26] analyzed the micro-cracking of polymer matrix composites through using multiaxial multi-ply stitched carbon preforms, subjected to purely cyclic hygrothermal loadings. The specific cracks occurred within the resin transverse channels sur-rounding the stitches, showing that hygrothermal damage is different to that caused by the mechanical loadings [27]. Patel [28] designed a 24 h cyclic hygrothermal aging spectrum, containing cyclical variations between temperature and moisture, which aimed to simulate a mission environment for an advanced subsonic aircraft. Durability research on the hygrothermal aging material system was carried out. Experiments showed that the initial and residual tensile properties after hygrothermal aging were almost unaffected by hygrothermal aging. However, residual strength and dynamic stiffness were significantly decreased at high temperature. Recently, Liu et al. [29] experimentally investigated the high velocity impact performance of the T700/TDE85 carbon/epoxy composites after cyclic hygrothermal aging in an environmental chamber at a constant temperature and humidity (70 °C, 95%RH) for the first time. They found that the ballistic limit of a carbon/epoxy composite decreases after long-time hygrothermal aging. Zhang et al. [30] reported the thermal aging on the compressive behavior of pure epoxy and 3D braided composites, the epoxy, and the 3D braided composites were aged in air for 1, 2, 4, 8, and 16 days at 180 °C. The experiments and the micrograph of samples by using the scanning electron microscope (SEM) revealed that epoxy initiated degradation in regions close to the surface layer, and the compressive behavior degradation was mainly from interface degradation and crack propagation after high temperature aging. Song [31] carried out tensile tests of 3D four-step braided composite after heat accelerated aging under 150 and 180 °C for 60, 120 and 180 h. After aging, they found that the structure of braided composites still kept integrity, and the tensile performance degraded less. To our knowledge, much research has been focused on the mechanical properties considering the hygrothermal aging effects of resin-based laminated composites except the above two papers, with respect to 3D braided composites. There are even fewer articles on the 3D braided composite after cyclic hygrothermal aging. Therefore, it is of a certain significance to research the 3D braided composite after cyclic hygrothermal aging.

The objective of this paper is to investigate the dynamic compressive mechanical properties of the 3D four-step braided composites along the longitudinal direction after cyclic hygrothermal aging. All of the research in this paper can be divided into three stages: first, identify the moisture absorption of 3D four-step braided composites after cyclic hygrothermal aging, observe morphology evolution of the cross section for the 3D four-step braided composite by using SEM; then, quantify the dynamic mechanical property degradation caused by the cyclic hygrothermal aging effect; finally, evaluate the cyclic hygrothermal aging effects on the dynamic compressive mechanical performance of 3D four-step braided composites.

In this work, the 3D four-step carbon/epoxy (T700-12K/TDE-86) braided composites were exposed to cyclic hygrothermal environment. An accelerated hygrothermal aging spectrum for a military aircraft was applied. SEM was used to observe composites' morphology changes at different cyclic hygrothermal aging days. A split Hopkinson pressure bar (SHPB) was employed to evaluate the dynamic compressive mechanical property along the longitudinal direction of the 3D four-step braided composites at various cyclic hygrothermal aging days.

2. Experimentation

2.1. Material and Specimens

The 3D performs of composites in this paper were manufactured by four-step 1×1 method, with 12K T700 carbon fibers (made by Toray Industries) as braiding yarns. The schematic of the four-step 1×1 braiding process can be found at Figure 1. During the braiding process, a perform for braiding was hung above the braided machine bed, on which yarn carriers were arranged in a preset style, in order to satisfy the cross-section shape of the braiding perform. More specifically, the first step involved the motion of yarn carriers in alternate rows, and the second step involved the motion of yarn carriers in alternate columns. The motion of yarn carriers in the third and fourth steps were similar to the first and second steps, respectively, except for the motion compared with the first and second steps in the opposite direction. The epoxy resin (TDE-86, from Tianjin Jingdong chemical composites Co., LTD of China, the epoxy value $\geq 0.90 \pm 0.02$) was injected into the performs through vacuum assisted resin transform modeling (VARTM); the fiber volume fraction in the 3D four-step braided plate is $60 \pm 2\%$. The specimens for measuring moisture absorption were cut from the nominal dimension of 380 mm (longitudinal direction) × 180 mm (transverse direction) × 4 mm (thickness direction) plate, with a braiding angle of 20°. Finally, the specimens were cut in the shape of 50 mm × 50 mm × 4 mm, as shown in Figure 2, according to the ASTM DD5229/D5299M standard [32]. As for the dimensions of specimens for SHPB tests is 10 mm × 10 mm × 10 mm, as shown Figure 3, further details about manufacturing or other related information can be found in our previous work [15].

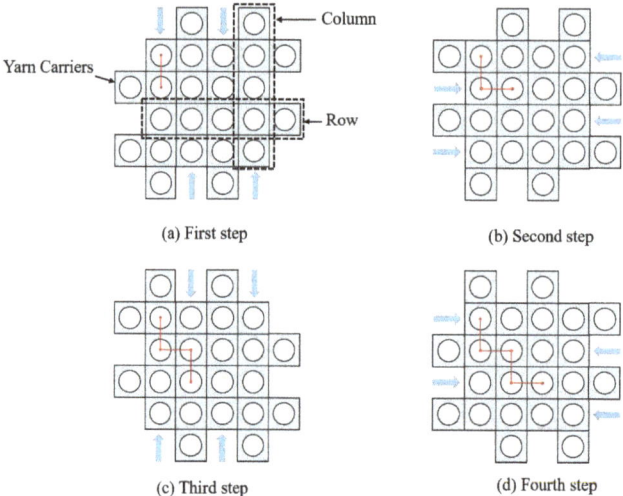

Figure 1. Schematic of four-step 1 × 1 braiding process, (**a**) First step, (**b**) Second step, (**c**)Third step, (**d**) Fourth step.

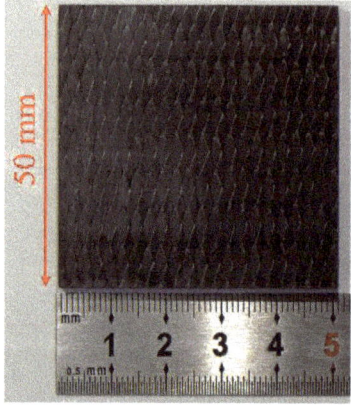

Figure 2. Specimen of 3D four-step braided composite for measuring moisture absorption.

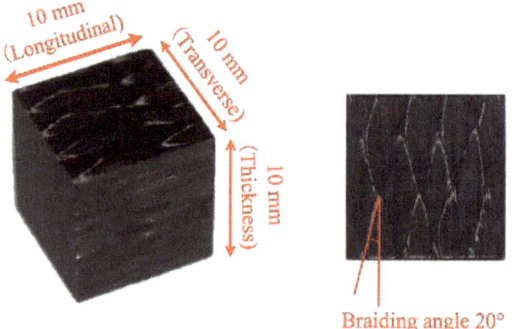

Figure 3. Specimen of 3D four-step braided composite for SHPB tests.

2.2. Hygrothermal Cycling

Military aircrafts experience, during service, the following; pre-flight time of the aircraft on the ground, the takeoff, the combat air patrols, and finally, the landing [33]. During the flight cycle, the composites on aircrafts experience temperature and humidity circulations alternately, therefore, it is absolutely necessary to investigate the mechanical performance of composites in-service, under the specific hygrothermal aging condition.

The hygrothermal cycling tests took place in the LRHS-101D-LJS high and low temperature chamber, alternately. Before placing the specimens into the chamber, the specimens were all dried at 70 °C, until the variation of weight was just under 0.02% [32]. An accelerated hygrothermal aging spectrum first proposed by Li [34] was employed. Likewise, we investigated the inter-laminar shear property and high-velocity impact resistance of carbon fiber-reinforced epoxy polymer (CFRP) composites after cyclic hygrothermal aging for the first time, by the same aging spectrum as in our previous work [35]. Figure 4 illustrates the set points prescribed temperature and humidity profile of one hygrothermal aging cycling process, in which the lowest temperature is 70 °C, and the highest temperature is 110 °C. A hygrothermal aging cycling process continues for 1440 min (one day), and the highest temperature is 110 °C lower than the glass transition temperature of the epoxy resin; more details about description of this hygrothermal aging spectrum can be found in our previous works [35], which are not repeated here. The 14 cycles (same as 14 days) of accelerated hygrothermal aging represent the effects of natural aging for one year at the same environmental state.

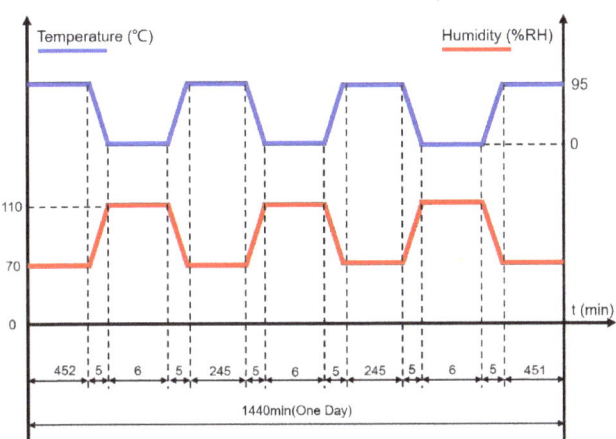

Figure 4. Accelerated hygrothermal aging spectrum for the 3D braided composites.

Five specimens for moisture absorption were employed for the moisture absorption tests. The weight of these specimens were measured at intervals of 24 h (one hygrothermal aging cycle) by using an analytical balance (accuracy 0.1 mg). The weights obtained from the five specimens were averaged. All specimens for SHPB tests were divided into four groups, aged for 14, 70, 140, 210 days in the hygrothermal chamber, respectively. When the target aging time reached, the specimens were cooled down to room temperature and put into sealed plastic bags to prevent moisture absorption again.

2.3. Microscopy Observation

A scanning electron microscope (SEM, JapanElectron Optics Laboratory Co. Ltd.,Tokyo, Japan) was used to observe the morphologies of specimens after cyclic hygrothermal aging.

Appl. Sci. **2020**, *10*, 2061

2.4. SHPB Tests

The SHPB apparatus [36] has been widely used to investigate the dynamic behaviors of materials at high strain rate (10^2–10^4 s^{-1}). The SHPB apparatus consists of a strike bar, an incident bar, a transmission bar, an absorbing bar and a shock absorber. Figure 5 shows a schematic of SHPB apparatus for dynamic compressive tests. In this study, all bars have diameters of 14.5 mm, and the incident bar and transmission bar both have a length of 1.5 m. The striker bar is fired from the gas gun system, and then the striker bar impacts the free end of the incident bar. Upon impact, a longitudinal elastic compressive stress wave is created, which propagates along the incident bar toward the incident bar/specimen interface; once the compressive stress wave has reached the incident bar/specimen interface, and the compressive has been divided into two parts: one part is reflected back into the incident bar as a tensile stress wave, and the other part of the stress wave which is not reflected is transmitted into the specimen, and the wave that is transmitted into specimen propagates reaches the specimen/transmission bar interface along the loading direction of the specimen. When the wave reaches the interface between the transmission bar and the specimen, one part of the stress wave is reflected into the specimen, and the other part of the stress wave is transmitted into the transmission bar. Finally, the transmission bar along its axis displaces until the other end of it reaches the absorbing bar, and the stress wave is absorbed by the shock absorber. The stress wave within the specimen undergoes a complex reverberation, until the dynamic compressive strength of the specimen is reached. It should be noted that the dynamic deformation of 3D braided composites is in a longitudinal direction; more specific details are shown in Figure 13.

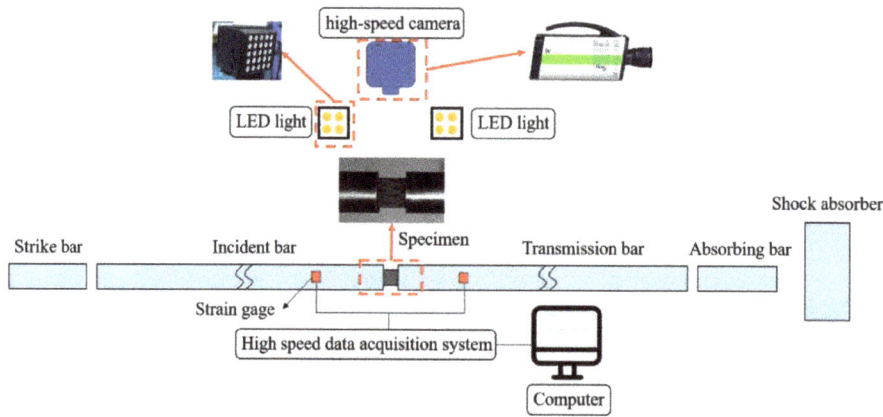

Figure 5. Schematic of SHPB(split Hopkinson pressure bar)apparatus.

During dynamic compressive tests, two strain gauges, which were mounted diametrically opposite to each other on the incident bar and transmission bar to cancel any bending strains, were employed to measure the incident, reflect and transmit signals, respectively. The data were captured by using a high-speed data acquisition card, employing a sampling rate of 1 MHz. A high-speed camera was employed to capture the deformation and damage evolution process during impact. In this paper, we chose a resolution of 1792 × 448, and the sampling frequency was 20000 fps. Besides, two LED lights were placed on both sides of the high-speed camera to enhance the light on the specimen, in order to obtain clearer photos.

In this study, the strain signals from strain gauges were analyzed based on one-dimensional wave propagation assumption. The stress (σ_s), strain (ε_s) and strain rate ($\dot{\varepsilon}_s$) could be obtained through the following equations, respectively [37]:

$$\sigma_s(t) = \frac{A_0}{A_s} E_0 \varepsilon_T(t), \tag{1}$$

$$\varepsilon_s(t) = -2\frac{C_0}{l_s} \int_0^t (\varepsilon_I(t) - \varepsilon_T(t))\,dt, \tag{2}$$

$$\dot{\varepsilon}_s(t) = -2\frac{C_0}{l_s}(\varepsilon_I(t) - \varepsilon_T(t)), \tag{3}$$

where $C_0 = \sqrt{E_0/\rho_0}$ is the longitudinal stress wave velocity in the bar, E_0 is Young's modulus of the bar, $E_0 = 206$ GPa, ρ_0 is density of the bar, $\rho_0 = 7850$ kg/m^3, l_s is the length of the specimen, A_0 and A_s are the cross-sectional area of the bar and the specimen, respectively. Moreover, $\varepsilon_I(t)$ and $\varepsilon_T(t)$ are strain gauge signals of the incident and transmitted pules, respectively.

2.5. Experimental Procedure for SHPB

The different strain rates were obtained through adjusting the air pressure from 0.6 to 1.3 MPa. In order to reduce the friction between the bars and specimen, some Vaseline lubrication oil was adopted on the ends of bars. Then, the 3D four-step braided composites were imbedded between the incident bar and transmitted bar, ensuring the specimen was centered between the bars to enhance the reliability. The typical strain waves were measured by the stain gauges for the specimen after 210 cyclic hygrothermal aging days, at the strain rate 1845 s^{-1}, which is shown in Figure 6. It can be seen that the incident wave is very close to a square wave, and the amplitude of the incident wave is a function of the impact velocity. The reflected wave represents the strain rate vs. time curve. The reflected wave is oscillating at a certain constant value when the maximum value is reached, which implies that the specimen satisfies the constant strain rate assumption. The stress vs. time curve is presented through the transmitted wave. In order to guarantee the repeatability of the tests, we carried out the repeatable tests. Figure 7 shows the contrast of the stress-strain curves for the specimens of 14 cyclic hygrothermal aging days under the same pressure, 0.6 MPa. It can be found that the stress-strain curves are almost identical, and the strain rates were similar, which is done to enhance the repeatability of the tests and to further ensure the reliability of the test data.

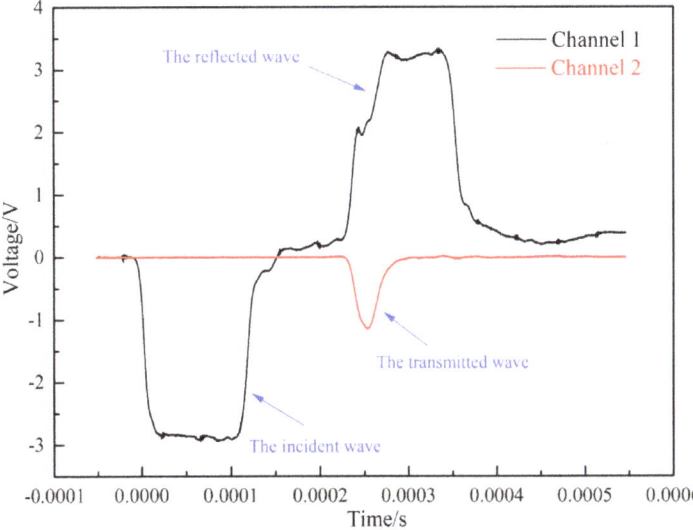

Figure 6. Typical signals on the incident and transmission bars for 210 cyclic hygrothermal aging days under 1845 s^{-1}.

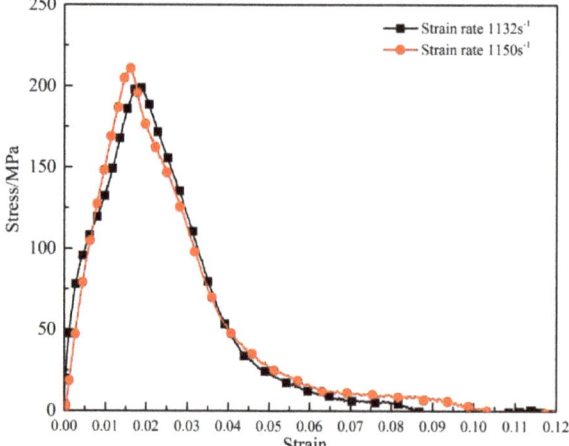

Figure 7. Contrast of the stress-strain curves for the specimens of 14 cyclic hygrothermal aging days under the same pressure 0.6 MPa.

Figure 8 shows the strain rate history curves under dynamic loading for specimens after 14, 70, 140, 210 cyclic hygrothermal aging days (abbreviation, HA14, HA70, HA140, HA210, respectively.) under typical strain rate 1700–1800 s^{-1}. It can be found that the curves of the strain rate vs. time under different hygrothermal aging days were similar at the air pressure 1.3 MPa (strain rate 1700–1800 s^{-1}.). In this study, the average values of the plateau region highlighted by dot lines were adopted as strain rates, whose values were shown in corresponding boxes in Figure 8.

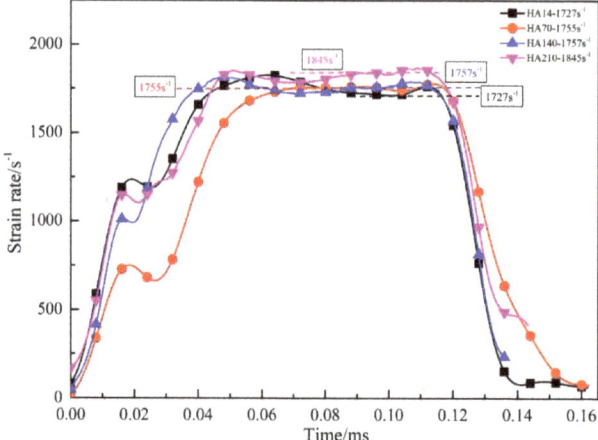

Figure 8. The strain rate history curves under dynamic loading for specimens after 14, 70, 140, 210 hygrothermal aging days under typical strain rate 1700–1800 s^{-1}.

3. Results and Discussion

3.1. Moisture Absorption Behavior

The water content M defined as:

$$M = \frac{W - W_d}{W_d} \times 100\%, \tag{4}$$

where W_d is the weight of the dry specimen and W is the weight of the specimen after hygrothermal aging. The weights and moisture absorption rates of the specimens can be found in Table 1.

Table 1. Moisture absorption parameters of 3D four-step braided composites after different cyclic hygrothermal aging days.

Hygrothermal Aging Days	Average Weight W (g)	Standard Deviation	Water Content (%)
0	15.6710	0.0320	0
14	15.7069	0.0323	0.2288
70	15.7267	0.0348	0.3550
140	15.7564	0.0344	0.5448
210	15.7799	0.0343	0.6944

The 210 days of moisture absorption of 3D four-step braided composite, as a function of square root of time (hour 0.5), is shown in Figure 9, and the moisture absorption of the T700/TDE85 laminated composite after cyclic hygrothermal aging by our previous work [35] was also added into Figure 9, in order to analyze contrastively moisture absorption behavior of 3D four-step braided composites more comprehensively. It can be found that the 3D four-step braided composite did not achieve the saturated moisture absorption, and the relationship between moisture absorption and aging time was almost linear; the maximum moisture absorption was 0.6944%. However, the T700/TDE85 laminated composite not only reached the saturated moisture absorption (1.261%, at 180 cyclic hygrothermal aging days), but also the maximum moisture absorption (1.261%) [35], which was almost twice that of the 3D four-step braided composite at the same cyclic hygrothermal aging days. In this study, the magnitude of the diffusion coefficient of the 3D four-step braided composite could not be computed, for the reason that the saturated moisture absorption was not reached under 210 cyclic hygrothermal aging days, according to the Fick law. The reason that the saturation moisture absorption of 3D four-step braided composites was not reached during this study, is that its spatial structure is more sophisticated, which may cause more porosity of the matrices that may act as secondary absorption locations and lead to deviations from the Fick diffusion behavior [38,39] compared with the laminated composites.

It is well recognized that the diffusion mechanism of water molecules with the polymer composites is obviously different from that in a pure polymer, because the interfaces between fibers and resin play a decisive role in the whole moisture absorption. The moisture absorption process of the composites mainly contains three phases: epoxy plasticization, epoxy hydrolysis and interfaces debonding (causing cracks). At the beginning, the water entered into the specimen from the specimen surface, then the water molecules diffused into the resin through microflaws (e.g., void or microcracks). With the absorption process of water molecules continuing, the epoxy had produced plasticization, this is because the weak interactions between water molecules and polymer chains inside epoxy ranged from Van der Waals bonds into single hydrogen bonds. In addition, the water molecules which entered into epoxy acted as a corrosion agent and promoted epoxy hydrolysis. The volume of epoxy increased with the increased polymer chain mobility and disruption of interchain Van der Waals, which produced differential swelling stresses (since the carbon fibers do not absorb water), that might drive the interfaces debonding between fiber and epoxy [18,40,41].

Figure 10 presents the damage evolution on the cross section of the 3D four-step composites after different hygrothermal aging days. Some broken fibers were observed on the surface of the composite, due to defects caused during the manufacturing process, as shown (Figure 10a). Before hygrothermal aging, the surface of epoxy was relatively smooth, except that there were some impurities. After 14 cyclic hygrothermal aging days, some potholes appeared on the surface of epoxy, with a length below 100 μm, which was the phenomenon after epoxy hydrolysis. During water absorption, both free and bound water molecules existed in the epoxy network, based on the formation of either one or two hydrogen bonds with polymer chains [42]. Two types between water molecules and polymeric chains are shown in Figure 11. The black lines stand for the backbone chains of epoxy, and the water molecules are represented in red color. Figure 11a,b explain the case of one hydrogen bond for water molecules, and two hydrogen bonds, respectively. Continuing hygrothermal aging days, some fibers were exposed when the area and depth of epoxy hydrolysis increased, as shown in Figure 10c. Later on in the cyclic hygrothermal aging days, the cracks in both Figure 10d,e can be found, and the cracks were on the junction of the braiding yarns and epoxy; differential swelling stresses can be employed to explain the phenomenon. The combined effects of epoxy plasticization, epoxy hydrolysis and cracks caused by interfaces debonding, contribute to a change in dynamic mechanical performance after cyclic hygrothermal aging.

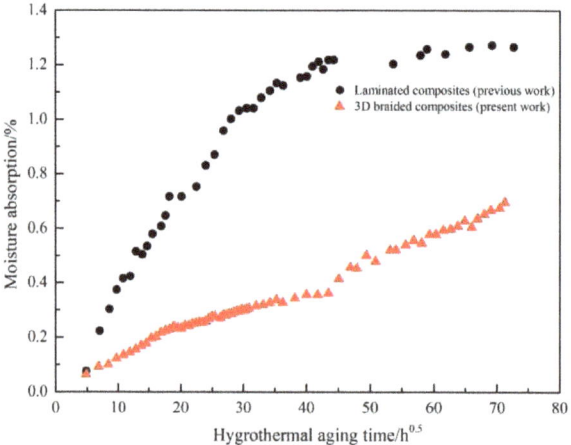

Figure 9. Contrast of moisture absorption between the 3D four-step braided composites and the T700/TDE85 laminated composites during 210 cyclic hygrothermal aging days.

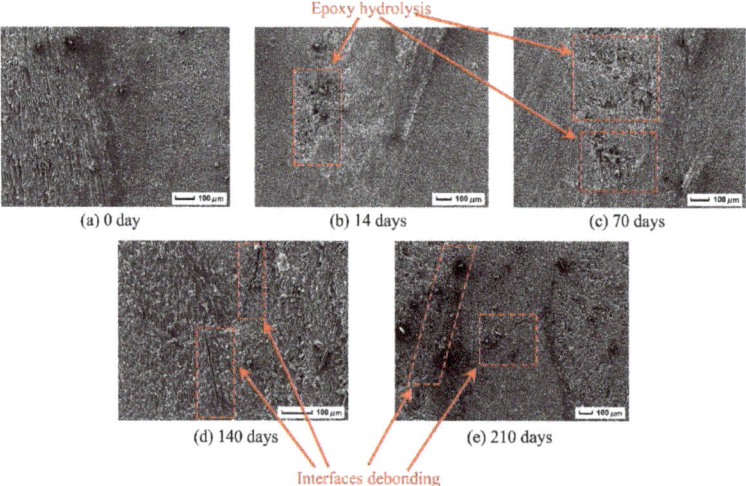

Figure 10. Damage evolution on the cross section of the 3D four-step braided composites after different cyclic hygrothermal aging days, (**a**) 0 day, (**b**) 14 days, (**c**) 70 days, (**d**) 140 days, (**e**) 210 days.

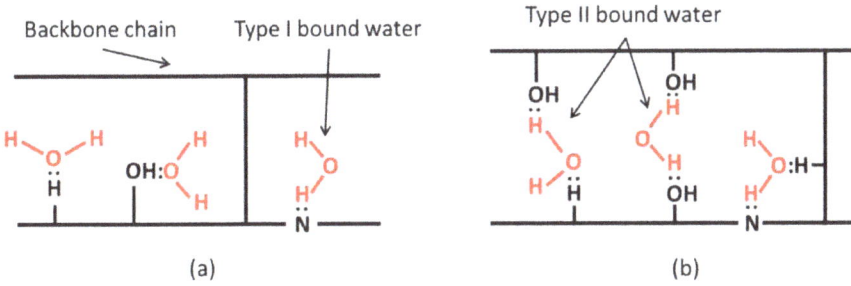

Figure 11. Two types between water molecules and polymeric chains, (**a**) Type I bound water forming one hydrogen bond, (**b**) Type II bound water forming two hydrogen bonds [42].

3.2. Typical Dynamic Compressive Process for Braided Composites Captured by High-Speed Camera

Figure 12 shows the typical dynamical compressive process under strain rate 1845 s^{-1}, after 210 cyclic hygrothermal aging days. From this figure, it can be seen that the 3D four-step braided composite mainly experienced matrix failure, braiding yarns kinking and finally disassembling when subjected to compression along the longitudinal direction. When the incident bar impacted the specimen, the surface of the specimen no longer remained intact and some surface braiding yarns [43] kinked, due to matrix failure under compressive stress wave propagation. At 0.1 ms, all surface braiding yarns kinked, the matrix could no longer maintain structural integrity of the 3D four-step braided composite, and the composite entered the phase of composite disassembly. From 0.125 ms, the braiding yarns struggled to get rid of the constraints of the matrix, and during this process, matrix failure was the most damaged mode, and the fibers in braiding yarns had almost no damage.

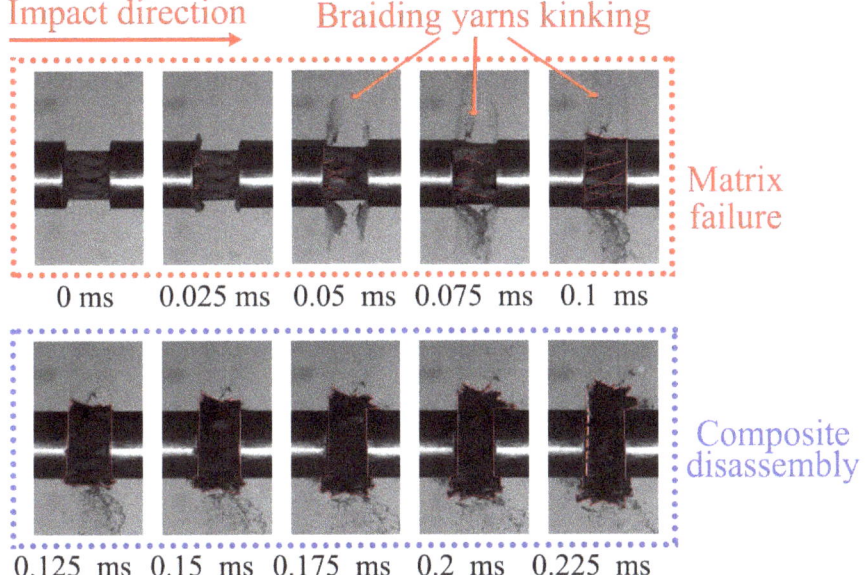

Figure 12. Typical dynamical compressive process for the 3D four-step braided composite under strain rate 1845 s^{-1}, after 210 cyclic hygrothermal aging days.

Figure 13 gives the typical fracture morphology for the 3D four-step braided composite under strain rate 1845 s^{-1} after 210 cyclic hygrothermal aging days. It can be found that the matrix failure was the main damage mode when the specimens were subjected to impact in the longitudinal direction. The yarns were scarcely damaged at the macro level. It is important to emphasize that the specimens for the SHPB tests were cut from the whole braided composite plate, which could decrease the strength and stiffness of the 3D braided composites due to the cut-edge effect. This may be also the reason that the specimens broken up when subjected to the longitudinal compression under high strain rates. There is no difference in the fracture morphology for the 3D braided composites between un-hygrothermal aging and hygrothermal aging, compared to our previous work [15].

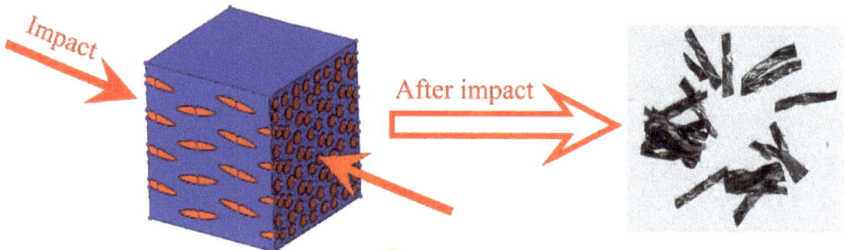

Figure 13. The typical fracture morphology for the 3D four-step braided composite under strain rate 1845 s^{-1} after 210 cyclic hygrothermal aging days.

3.3. Dynamic Mechanical Properties for Braided Composites with Different Hygrothermal Aging Days

Figure 14 shows the dynamic compressive stress vs. strain curves for different hygrothermal aging days. All stress strain curves, whether hygrothermaling aging or not, reveal that the 3D four-step braided composites almost behave in a linear response in the initial stage and have no significant yield in the impact process. However, the composites had non-linear behavior at the location close to the peak stress, and then the stress declined gradually beyond the peak stress. This is because the degradation of the matrix and debonding between yarns and the epoxy increase gradually, which exhibits a progressive loss of stiffness. It is worth noting that a plateau appears near the peak stress at the HA70 specimen under 1755 s^{-1}, which meant that the stiffness loss of the HA70 specimen was progressive before the peak stress. The reason that this occurred is that the epoxy in the specimen endured plasticization, that enhanced the stiffness to a certain extent. Combined with Figure 15a, it can be found that with increasing strain rate, the peak stress shows a significant strain rate effect at different hygrothermal aging days. However, there is no obvious trend in the dynamic stiffness modulus, and no remarkable strain rate effect after hygrothermal aging in Figure 15b. From the above, we conclude that the peak stress of the 3D four-step braided composite is still sensitive after cyclic hygrothermal aging, when compressed in the longitudinal direction at high strain rates.

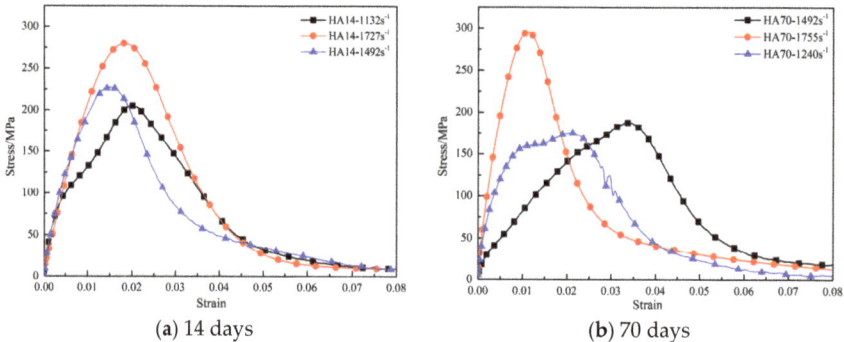

(a) 14 days (b) 70 days

Figure 14. *Cont.*

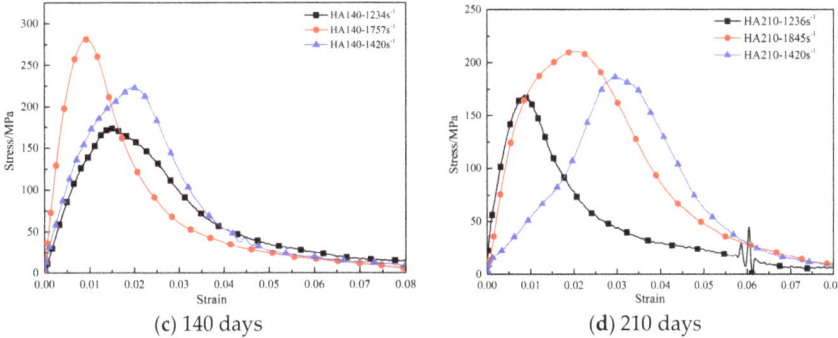

(c) 140 days   (d) 210 days

Figure 14. Dynamic compressive stress vs. strain curves for the 3D four-step braided composites after different hygrothermal aging days, (**a**) 14 days, (**b**) 70 days, (**c**) 140 days, (**d**) 210 days.

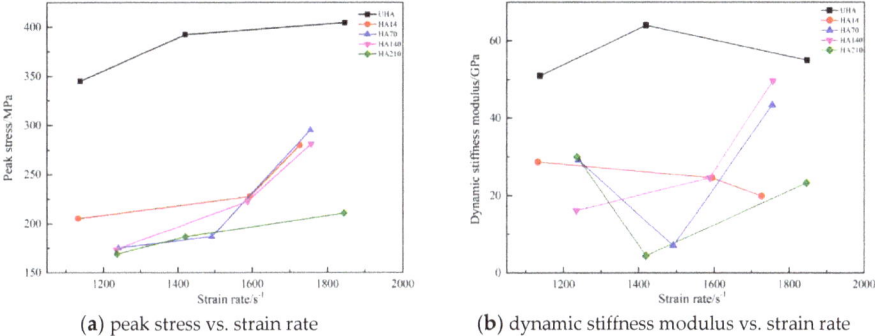

(**a**) peak stress vs. strain rate   (**b**) dynamic stiffness modulus vs. strain rate

Figure 15. Dynamic mechanical property vs. strain rate for the 3D four-step braided composites under different hygrothermal aging days, (**a**) peak stress vs. strain rate, (**b**) dynamic stiffness modulus vs. strain rate.

3.4. Hygrothermal Aging Effects for Braided Composites under High Strain Rates

In order to evaluate the hygrothermal aging effects of 3D four-step braided composites when compressed on the longitudinal direction under high strain rates, in this study, the difference of dynamic mechanical properties of the composites between hygrothermal aging and un-hygrothermal aging (abbreviation, UHA) at three high strain rates that correspond to 1100~1250 s^{-1}, 1400~1600 s^{-1}, 1700~1850 s^{-1}, respectively. Due to the limit on the number of the specimens, the dynamic experimental results of the un-hygrothermal aging specimens are from our previous work [15]. However, the previous work lacked the experimental results under the strain rate 1400~1600 s^{-1}, therefore, we did the SHPB tests of un-hygrothermal aging specimens under strain rate 1400~1600 s^{-1}. It is important to emphasize that all specimens, whether they are the specimens in our previous work [15] or the specimens in this study, came from the same batch of 3D four-step braided composites. The validity and objectivity of the experimental tests can be guaranteed this way. Figures 16a, 17a and 18a present the stress vs. strain curves of the composites between hygrothermal aging and un-hygrothermal aging under three high strain rates. The stress vs. strain curves of un-hygrothermal aging specimens also present in a linear manner up to failure and have no clear yield at three high strain rates. The peak stresses are 345.1 MPa, 392.3 MPa, and 404.4 MPa, corresponding to the strain rates 1138 s^{-1}, 1420 s^{-1}, and 1847 s^{-1}, respectively. The peak stress of 3D four-step braided composites without hygrothermal aging also behaves like the strain rate effect. It reveals that the experimental result of the un-hygrothermal aging specimen under the strain rate 1420 s^{-1} in this study is in accordance with the conclusion that peak stress has a strain rate effect in our previous work [15]. At the same time, it can obviously be seen that the peak stress and the dynamic stiffness modulus of un-hygrothermal aging specimens are both greater than that hygrothermal aging specimen. It can be concluded that the dynamic compressive strength in a longitudinal direction was weakened when subjected to cyclic hygrothermal aging. Figure 16b,c, Figure 17b,c and Figure 18b,c shows the effects of hygrothermal aging on the peak stress and dynamic stiffness modulus. It can be seen that no matter the peak stress or the dynamic stiffness, the moduli have both decreased significantly after cyclic hygrothermal aging. However, the peak stress and dynamic stiffness modulus have not shown any specific regularity from 14 to 210 cyclic hygrothermal aging days at three high strain rates. Although the peak stress of the specimens under strain rate 1100~1250 s^{-1} decreases with respect to the increase of cyclic hygrothermal aging days, the same value shows no obvious up or down trend with the increase of cyclic hygrothermal aging days under strain rate 1400~1600 s^{-1} and 1700~1850 s^{-1}. In particular, the peak stress and the dynamic stiffness modulus reached a minimum value at 210 cyclic hygrothermal aging days, under three high strain rates. The peak stress decreased by 51.06%, 52.43%, 47.90%, compared with that without hygrothermal aging under the strain rate 1100~1250 s^{-1}, 1400~1600 s^{-1}, 1700~1850 s^{-1}, respectively. The dynamic stiffness modulus decreased by 41.18%, 93.13%, 58.02% compared with that without hygrothermal aging under the strain rates 1100~1250 s^{-1}, 1400~1600 s^{-1}, and 1700~1850 s^{-1}, respectively. All of these results demonstrate that the peak stress and the dynamic stiffness modulus of 3D four-step braided composites after 210 cyclic hygrothermal aging days almost decrease by half when subjected to longitudinal compression.

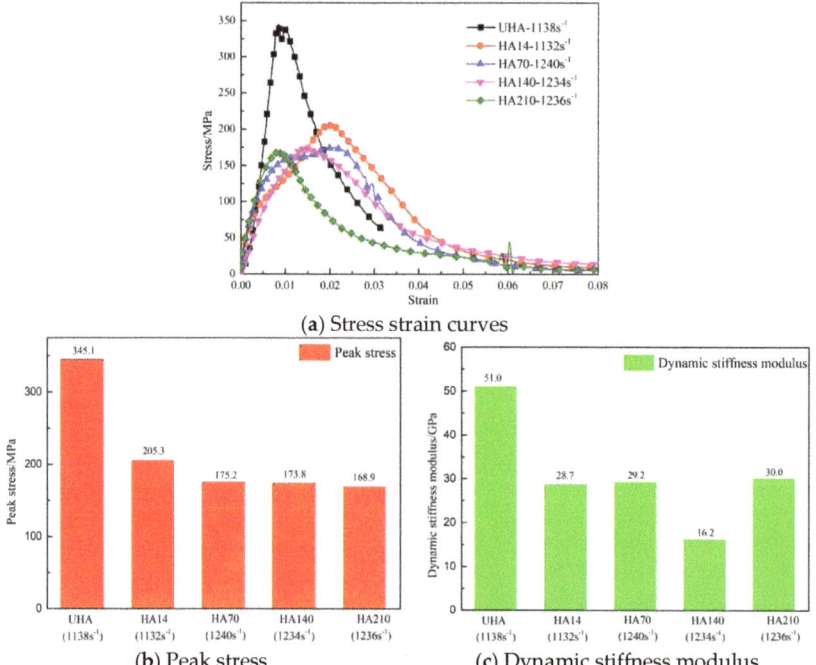

Figure 16. The effect of cyclic hygrothermal aging on the dynamic compressive properties under 1100~1250 s^{-1}, (**a**) Stress strain curves, (**b**) Peak stress, (**c**) Dynamic stiffness modulus.

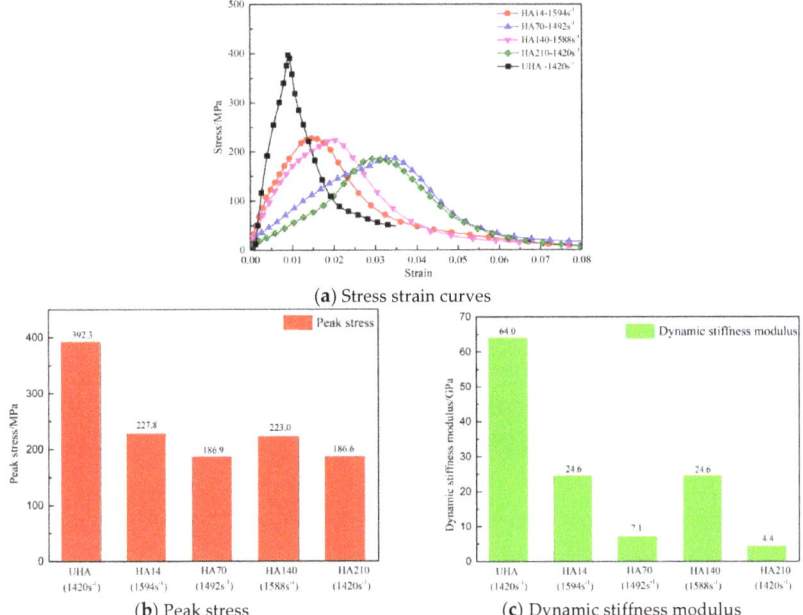

Figure 17. The effect of cyclic hygrothermal aging on the dynamic compressive properties under 1400~1600 s^{-1}, (**a**) Stress strain curves, (**b**) Peak stress, (**c**) Dynamic stiffness modulus.

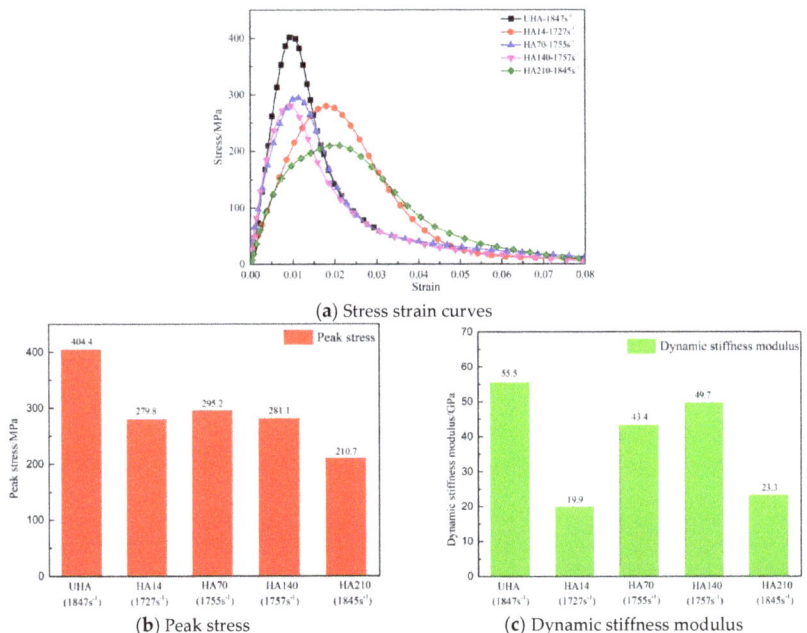

Figure 18. The effect of cyclic hygrothermal aging on the dynamic compressive properties under 1700~1850 s^{-1}, (**a**) Stress strain curves, (**b**) Peak stress, (**c**) Dynamic stiffness modulus.

4. Conclusions

In this study, the 3D four-step braided composites were aged in a cyclic hygrothermal environment, and the microscopic damage morphologies of the composites were examined by SEM, to determine the damage evolution with cyclic hygrothermal aging days. A split Hopkinson pressure bar (SHPB) apparatus was employed, to determine the dynamic mechanical properties of the composites under three high strain rates, when subjected to the longitudinal compression. The main findings are summarized in the following:

(1) The saturated moisture absorption of 3D four-step braided composites was not reached over the full duration of 210 cyclic hygrothermal aging days, the maximum moisture absorption was 0.6944%, and the composites did not show very much non-linear moisture behavior, compared with the T700/TDE85 laminated composites in our previous work.

(2) The damage evolution of 3D four-step composites during the whole cyclic hygrothermal aging days mainly included epoxy hydrolysis and interfaces debonding.

(3) The dynamic compressive stress-strain curves for the 3D four-step braided composites both behaved in a similar way, whether cyclic hygrothermal aging or not, and a linear response in the initial stage had no significant yield before reaching the peak stress.

(4) The peak stress of the 3D four-step braided composites still behaved as a strain rate effect after cyclic hygrothermal aging, however, the dynamic stiffness modulus clearly had no specific regularity.

(5) The peak stress and the dynamic stiffness modulus of 3D four-step braided composites after 210 cyclic hygrothermal aging days almost decreased by half when subjected to longitudinal compression.

Author Contributions: Conceptualization, L.L.; methodology, K.X. and W.C.; software, K.X.; validation, K.X., G.L. and Z.Z.; formal analysis, K.X.; investigation, K.X. and L.L.; resources, L.L.; data curation, K.X., G.L. and Z.Z.; writing—original draft preparation, K.X.; writing—review and editing, W.C. and L.L.; visualization, K.X.; supervision, W.C.; project administration, L.L.; funding acquisition, L.L. and W.C. All authors have read and agreed to the published version of the manuscript.

Funding: This research was funded by the National Natural Science Foundation of China (Grant No. 51975279, and 51605218) and Equipment Pre-research field Foundation of Equipment Development Department of China (61409220203).

Acknowledgments: The authors would like to thank Chunbo Wu for his suggestions about image processing.

Conflicts of Interest: The authors declare no conflict of interest.

References

1. Schultheisz, C.R.; Waas, A.M. Compressive failure of composites, Part I: Testing and micromechanical theories. *Prog. Aerosp. Sci.* **1996**, *32*, 1–42. [CrossRef]
2. Wang, H.; Hazell, P.J.; Shankar, K.; Morozov, E.V.; Escobedo, J.P. Impact behaviour of Dyneema® fabric-reinforced composites with different resin matrices. *Polym. Test.* **2017**, *61*, 17–26. [CrossRef]
3. Li, Z.; Khennane, A.; Hazell, P.J.; Remennikov, A.M. Performance of a hybrid GFRP-concrete beam subject to low-velocity impacts. *Compos. Struct.* **2018**, *206*, 425–438. [CrossRef]
4. Huang, H.J.; Waas, A.M. Modeling and predicting the compression strength limiting mechanisms in Z-pinned textile composites. *Compos. Part B Eng.* **2009**, *40*, 530–539. [CrossRef]
5. Pankow, M.; Waas, A.M.; Yen, C.F.; Ghiorse, S. Shock loading of 3D woven composites: A validated finite element investigation. *Compos. Struct.* **2011**, *93*, 1347–1362. [CrossRef]
6. Zhang, D.; Waas, A.M.; Yen, C.F. Progressive damage and failure response of hybrid 3D textile composites subjected to flexural loading, part I: Experimental studies. *Int. J. Solids Struct.* **2015**, *75–76*, 309–320. [CrossRef]
7. Xuan, H.J.; Wu, R.R. Aeroengine turbine blade containment tests using high-speed rotor spin testing facility. *Aerosp. Sci. Technol.* **2006**, *10*, 501–508. [CrossRef]
8. Guan, Y.P.; Zhao, Z.H.; Chen, W.; Gao, D.P. Foreign object damage to fan rotor blades of aeroengine part I: Experimental study of bird impact. *Chin. J. Aeronaut.* **2008**, *21*, 328–334.

9. Gu, B.H.; Xu, J.Y. Finite element calculation of 4-step 3-dimensional braided composite under ballistic perforation. *Compos. Part B Eng.* **2004**, *35*, 291–297. [CrossRef]
10. Sun, B.Z.; Liu, F.; Gu, B.H. Influence of the strain rate on the uniaxial tensile behavior of 4-step 3D braided composites. *Compos. Part A Appl. Sci. Manuf.* **2005**, *36*, 1485. [CrossRef]
11. Sun, B.Z.; Gu, B.H. High strain rate behavior of 4-step 3D braided composites under compressive failure. *J. Mater. Sci.* **2007**, *42*, 2463–2470. [CrossRef]
12. Sun, B.Z.; Gu, B.H. In-plane compressive behaviors of 3-D textile composites at various strain rates. *Appl. Compos. Mater.* **2007**, *14*, 193–207. [CrossRef]
13. Walter, T.R.; Subhash, G.; Sankar, B.V.; Yen, C.F. Damage modes in 3D glass fiber epoxy woven composites under high rate of impact loading. *Compos. Part B Eng.* **2009**, *40*, 584–589. [CrossRef]
14. Zhang, F.; Wan, Y.; Gu, B.H.; Sun, B.Z. Impact compressive behavior and failure modes of four-step three-dimensional braided composites-based meso-structure model. *Int. J. Damage Mech.* **2015**, *24*, 805–827. [CrossRef]
15. Tan, H.C.; Huang, X.; Liu, L.L.; Guan, Y.P.; Chen, W. Dynamic compressive behavior of four-step three-dimensional braided composites along three directions. *Int. J. Impact Eng.* **2019**, *134*, 103366. [CrossRef]
16. Diamant, Y.; Marom, G.; Broutman, L.J. The effect of network structure on moisture absorption of epoxy resins. *J. Appl. Polym. Sci.* **1981**, *26*, 3015–3025. [CrossRef]
17. Davies, P.; Mazéas, F.; Casari, P. Sea water aging of glass reinforced composites: Shear behaviour and damage modelling. *J. Compos. Mater.* **2001**, *35*, 1343–1372. [CrossRef]
18. Joliff, Y.; Rekik, W.; Belec, L.; Chailan, J.F. Study of the moisture/stress effects on glass fibre/epoxy composite and the impact of the interphase area. *Compos. Struct.* **2014**, *108*, 876–885. [CrossRef]
19. Abdel, M.B.; Ziaee, S.; Gass, K.; Schneider, M. The combined effects of load, moisture and temperature on the properties of E-glass/epoxy composites. *Compos. Struct.* **2005**, *71*, 320–326. [CrossRef]
20. Tsenoglou, C.J.; Pavlidou, S.; Papaspyrides, C.D. Evaluation of interfacial relaxation due to water absorption in fiber-polymer composites. *Compos. Sci. Technol.* **2006**, *66*, 2855–2864. [CrossRef]
21. Ogi, K. Influence of thermal history on transverse cracking in a carbon fiber reinforced epoxy composite. *Adv. Compos. Mater.* **2003**, *11*, 265–275. [CrossRef]
22. Zhang, A.; Lu, H.; Zhang, D. Effects of voids on residual tensile strength after impact of hygrothermal conditioned CFRP laminates. *Compos. Struct.* **2013**, *95*, 322–327. [CrossRef]
23. Yilmaz, T.; Sinmazcelik, T. Effects of hydrothermal aging on glass-fiber/polyetherimide (PEI) composites. *J. Mater. Sci.* **2010**, *45*, 399–404. [CrossRef]
24. Sun, P.; Zhao, Y.; Luo, Y.; Sun, L. Effect of temperature and cyclic hygrothermal aging on the interlaminar shear strength of carbon fiber/bismaleimide (BMI) composite. *Mater. Des.* **2011**, *32*, 4341–4347. [CrossRef]
25. Li, G.; Pang, S.S.; Helms, J.E.; Ibekwe, S.I. Low velocity impact response of GFRP laminates subjected to cycling moistures. *Polym. Compos.* **2000**, *21*, 686–695. [CrossRef]
26. Liotier, P.J.; Vautrin, A.; Beraud, J.M. Microcracking of composites reinforced by stitched multiaxials subjected to cyclical hygrothermal loadings. *Compos. Part A Appl. Sci. Manuf.* **2011**, *42*, 425–437. [CrossRef]
27. Mikhaluk, D.S.; Truong, T.C.; Borovkov, A.I.; Lomov, S.V.; Verpoest, I. Experimental observations and finite element modelling of damage initiation and evolution in carbon/epoxy non-crimp fabric composites. *Eng. Fract. Mech.* **2008**, *75*, 2751–2766. [CrossRef]
28. Patel, S.R.; Case, S.W. Durability of hygrothermally aged graphite/epoxy woven composite under combined hygrothermal conditions. *Int. J. Fatigue* **2002**, *24*, 1295–1301. [CrossRef]
29. Liu, L.L.; Zhao, Z.Z.; Chen, W.; Shuang, C.; Luo, G. An experimental investigation on high velocity impact behavior of hygrothermal aged CFRP composites. *Compos. Struct.* **2018**, *204*, 645–657. [CrossRef]
30. Zhang, M.; Zuo, C.; Sun, B.Z.; Gu, B.H. Thermal ageing degradation mechanisms on compressive behavior of 3-D braided composites in experimental and numerical study. *Compos. Struct.* **2016**, *140*, 180–191. [CrossRef]
31. Song, L.; Li, J. Effects of heat accelerated aging on tensile strength of three dimensional braided/epoxy resin composites. *Polym. Compos.* **2012**, *3*, 1635–1643.
32. ASTM Standard. *ASTM D 5229–29M—Standard Test Method for Moisture Absorption Properties and Equilibrium Conditioning of Polymer Matrix Composite Materials*; Annual Book of ASTM Standards; ASTM International: West Conshohocken, PA, USA, 2010. [CrossRef]

33. De Sarmot, G.; Salvia, M.; Vautrin, A.; Colombaro, A.M. Influence of long term maintenance on the performances of composites for Super Sonic Transportation. In Proceedings of the First National Colloquium of Aeronautic Researches on the Supersonic Aircraft, Paris, France, 6–7 February 2002.
34. Li, Y.; Chen, Y.; Sheng, G.; Li, G. Hygrothermal aging effects for aircraft composite structures. In Proceedings of the 11th National Conference on Composite Materials, Hefei, China, 21–25 October 2000. (In Chinese).
35. Liu, L.L.; Zhao, Z.Z.; Chen, W.; Xue, M.F.; Shuang, C. Interlaminar shear property and high-velocity impact resistance of CFRP laminates after cyclic hygrothermal aging. *Int. J. Crashworthiness* **2019**, 1–14. [CrossRef]
36. Kolsky, H. An investigation of the mechanical properties of materials at very high rates of loading. *Proc. Phys. Soc. Sect. B* **1949**, *62*, 676–700.
37. Gama, B.A.; Lopatnikov, S.L.; Gillespie, J.W. Hopkinson bar experimental technique: A critical review. *Appl. Mech. Rev.* **2004**, *57*, 223–250. [CrossRef]
38. Kumosa, L.; Benedikt, B.; Armentrout, D.; Kumosa, M. Moisture absorption properties of unidirectional glass/polymer composites used in composite (non-ceramic) insulators. *Compos. Part A Appl. Sci. Manuf.* **2004**, *35*, 1049–1063. [CrossRef]
39. Gautier, L.; Mortaigne, B.; Bellenger, V. Interface damage study of hydrothermally aged glass-fibre-reinforced polyester composites. *Compos. Sci. Technol.* **1999**, *59*, 2329–2337.
40. Gagani, A.I.; Echtermeyer, A.T. Fluid diffusion in cracked composite laminates—Analytical, numerical and experimental study. *Compos. Sci. Technol.* **2018**, *160*, 86–96. [CrossRef]
41. Morii, T.; Ikuta, N.; Kiyosumi, K.; Hamada, H. Weight-change analysis of the interphase in hygrothermally aged FRP: Consideration of debonding. *Compos. Sci. Technol.* **1997**, *57*, 985–990. [CrossRef]
42. Zhou, J.; Lucas, J.P. Hygrothermal effects of epoxy resin. Part I: The nature of water in epoxy. *Polymer* **1999**, *40*, 5505–5512. [CrossRef]
43. Tan, H.C.; Liu, L.L.; Guan, Y.P.; Chen, W.; Zhao, Z.Z. Investigation of three-dimensional braided composites subjected to steel projectile impact: Automatically modelling mesoscale finite element model. *Compos. Struct.* **2019**, *209*, 317–327.

© 2020 by the authors. Licensee MDPI, Basel, Switzerland. This article is an open access article distributed under the terms and conditions of the Creative Commons Attribution (CC BY) license (http://creativecommons.org/licenses/by/4.0/).

Article

Compression Shear Properties of Bonded–Bolted Hybrid Single-Lap Joints of C/C Composites at High Temperature

Yanfeng Zhang [1], Zhengong Zhou [1,*] and Zhiyong Tan [2]

[1] National Key Laboratory of Science and Technology on Advanced Composites in Special Environments, Harbin Institute of Technology, Harbin 150000, China; zhangyanfenghit@126.com
[2] Beijing Institute of Near-space Vehicle's System Engineering, Beijing 100076, China; doctortanzhiyong@126.com
* Correspondence: zyfztc@hit.edu.cn

Received: 13 December 2019; Accepted: 27 January 2020; Published: 5 February 2020

Abstract: Based on previous research, in this paper, the compressive shear failure behavior and mechanical properties of bonded–bolted hybrid single-lap joints of C/C composites at high temperature were studied. The compression shear test was performed on the joints at 800 °C to obtain the load–displacement curve and failure morphology. The failure modes of joints were observed by digital microscopy and scanning electron microscopy. A numerical analysis model was implemented in finite element code Abaqus/Explicit embedded with the user material subroutine (VUMAT). The numerical results were compared with the test results to verify the correctness of the model. The interrelationship of the compression shear loading mechanism and the variations in stress distribution between bonded joints and bonded–bolted hybrid joints at high temperature were explored. The progressive damage of hybrid joints and the variations in the ratio of the bolt load to the total load with displacement were obtained.

Keywords: compression shear properties; bonded–bolted hybrid; C/C composites; high temperature

1. Introduction

The design of a composite connection structure is one of the difficulties in the research of composite application, and the bearing capacity of the joint directly affects the reliability of composite components. A great deal of research has been conducted on the mechanical properties of the conventional bonded [1–6] and bolted [7–10] joints of composite materials. Many adhesives undergo instantaneous brittle fracture when they fail, resulting in sudden damage to the structure. The bonded–bolted hybrid [11] connection has better security and integrity than a bonded and bolted connection. Although the hybrid connection has been used in aviation, shipbuilding, and many other industrial sectors [12,13], studies on composite hybrid joint structure design are lacking due to its complex force transmission mechanism. Most research [14–18] on hybrid joints has been limited to the tensile failure process of joints with a resin-based fiber-reinforced composite plate as an adherend and metal as a fastener at normal temperature, and the failure modes are usually adhesive fracture and hole extrusion failure. Failure analysis of joints in which both the adherend and fasteners [19,20] are high-temperature-resistant composites (C/C composites, high-temperature ceramics) under compression shear is rare. In addition to the complexity and high price of the preparation process [21] of highly temperature-resistant composites, these studies are rare due to the high brittleness, which makes the cutting and fine processing of specific shapes difficult, and damage to composites, which reduces their strength. Carbon-fiber-reinforced carbon (CFRC), or carbon–carbon [22,23], is a unique composite material consisting of carbon fibers embedded in a carbonaceous matrix. Originally developed for aerospace applications, its low density, high thermal conductivity, and excellent mechanical properties at elevated temperatures make it an ideal material for the automobile, ship, and aerospace industries [24–28]. Moreover, to improve the mechanical properties of the structure, it is essential to reduce the bending effect of single-lap joints due to eccentric loading and improve the shear bearing capacity of the adhesive, and there are few reports [29] on this aspect. Based on the research of [30], in this study, the failure modes and force transfer mechanism of C/C composites single-lap joints at 800 °C were obtained by means of tests and numerical simulation. The experimental results were compared with the calculated results to verify the correctness of the numerical model. The relationship of compression shear mechanical properties between the bonded joint and bonded–bolted hybrid joint at high temperature was determined.

2. Materials and Methods

2.1. Experiment

The material of the adherend was C/C piercing woven composites used in [30]. The material of the bolt is C/C orthogonal three-directional woven composites. The x, y, and z correspond to three principal directions of the material, and the three directions are perpendicular T300-1k (Toray, Tokyo) carbon fibers. The preform weaving mode and fiber direction are shown in Figure 1a, and the microstructure of the fastener material observed by the digital microscopic system (VHX-7000, KEYENCE, Japan, Osaka) is shown in Figure 1b. The two kinds of preform were densified using the chemical vapor infiltration (CVI) and then graphitized to form C/C woven composites. To prevent the test piece from being damaged, chemical vapor deposition (CVD) was used to coat the surface of the adherend and fastener with a silicon carbide coating to obtain the antioxidant C/C composites required for the high-temperature test. The material performance of the adherend and fastener is shown in Table 1. The process parameters of CVI and CVD are listed in Table 2. The overlapped plate with a hole was shaped into the letter L. Two C/C plates and a bolt were assembled into one single-lap test piece. The prepared test piece and geometry dimensions are shown in Figure 2. The total length of the adherend (L_d) was 60 mm, the overlapping length (L_o) was 40 mm, the thickness of the adherend (T_d) was 10 mm, the thickness of the adhesive (T_e) was 0.2 mm, the width of the adherend (W) was 40 mm, and the bolt diameter D was 12 mm. Due to the high cost of C/C composites, three sets of test pieces were provided. In order to compare the bonded–bolted hybrid joint test results with the bonded joint test results at 800 °C in [30], in this study, the adhesive type, the adhesive curing method, the surface treatment method of test pieces, the testing equipment, and the fixture and loading method (Figure 3) were all completely consistent with that study. This study differed in that holes needed to be made in the center of the upper and lower cover plates to connect the adherend with the fasteners to control the adhesive thickness of the hybrid joint. In order to make the cured adhesive fill the bolt hole clearance, and to make the adhesive layer and bolt bear the load simultaneously during the test, holes in the adherend were drilled in advance, and the bolt was mounted before curing. The operation method is shown in Figure 4.

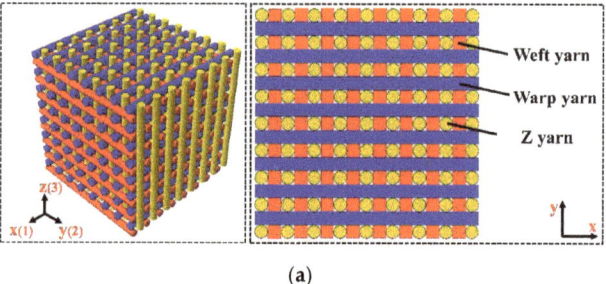

(a)

Figure 1. Cont.

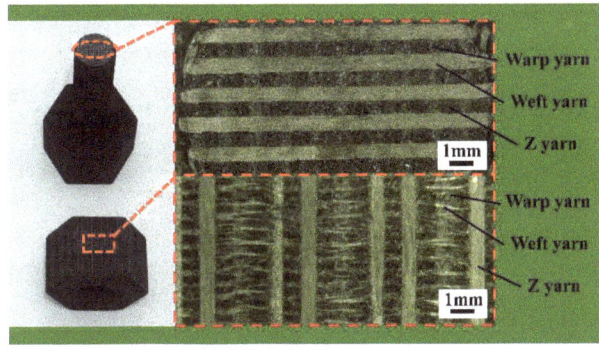

(b)

Figure 1. (**a**) The preform weaving mode and fiber direction; (**b**) microstructure of C/C fasteners.

Table 1. Material performance of the adherend and fastener at 800 °C.

Variable	Adherend	Fastener	Variable	Adherend	Fastener
Density (g/cm^3)	1.65	1.78	Tensile strength X_T (MPa)	271.3	254.6
Coefficient of thermal expansion (10^{-6} °C)	0.19	0.17	Compression strength X_C (MPa)	224	256
Elastic modulus E_{11} (GPa)	95	98.2	Tensile strength Y_T (MPa)	271.3	254.6
Elastic modulus E_{22} (GPa)	95	98.2	Compression strength Y_C (MPa)	224	256
Elastic modulus E_{33} (GPa)	10.3	61.8	Tensile strength Z_T (MPa)	84.9	80.7
Shear modulus G_{12} (GPa)	27.6	13.2	Compression strength Z_C (MPa)	358	308
Shear modulus $G_{13} = G_{23}$ (GPa)	6.8	7.1	Shear strength S_{12} (MPa)	54.5	48.7
Poisson ratio ν_{12}	0.035	0.036	Shear strength $S_{13} = S_{23}$ (MPa)	18.2	65.9
Poisson ratio $\nu_{12} = \nu_{23}$	0.032	0.035			

Table 2. Process parameters of chemical vapor infiltration (CVI) and chemical vapor deposition (CVD).

		Equipment	Temperature(°C)	Source Gas	Diluting Gas
Adherend	CVI	ZRHC-1500 CVD system (SANTE VACUUM TECHNOLOGY)	1000~1300	C_2H_2	Ar
	CVD		673~1173	CH_3SiCl_3	Ar
Fastener	CVI		900~1100	CH_4	Ar
	CVD		673~1173	CH_3SiCl_3	Ar

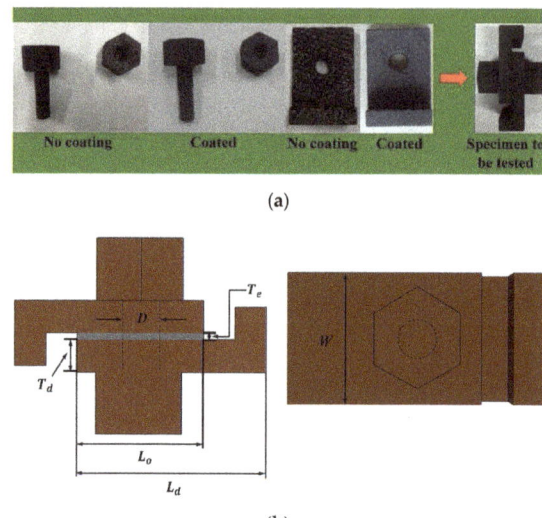

Figure 2. (**a**) C/C plate and fastener before and after anti-oxidation treatment; (**b**) geometric dimensions of the specimen.

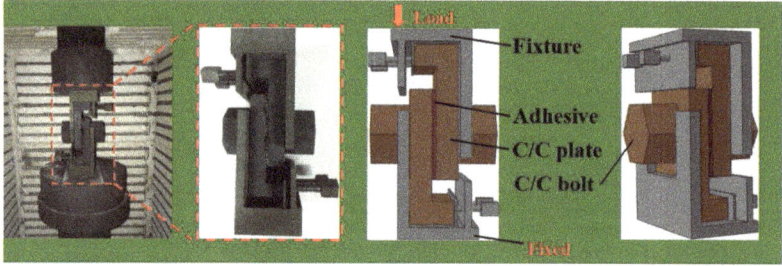

Figure 3. Fixture and loading method.

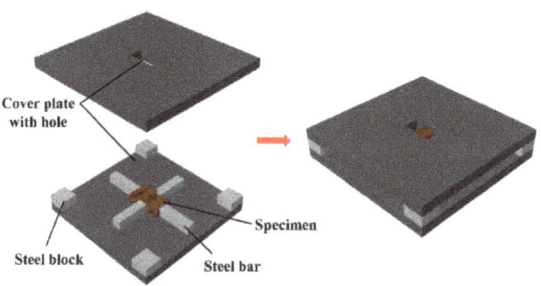

Figure 4. Adhesive thickness control method.

2.2. Finite Element Analysis

2.2.1. Model Establishment

For the above test process, the modeling was performed using ABAQUS 6.14 (Dassault, Paris, France,) finite element software. The bonded–bolted hybrid joint model consisted of four parts, including two lapped plates, an adhesive layer, and a bolt. A tie constraint existed between the lapped plates and the adhesive layer as shown in Figure 5a. Since the thread has little effect on the failure, the nut and screw were set as one part to speed up the calculation. To model the composites, an eight-node linear brick reduced integration element (C3D8R) was used. This element is suitable for simulating the three-dimensional loading process, and the calculation time is less than that of the quadratic element. In addition, it can reduce the hourglass problem and is not sensitive to element distortion [31]. For the adhesive layer, the same cohesive element as in [30] was adopted for modeling, and the element type was COH3D8R. As shown in Figure 5b, meshes of the adhesive layer, hole, and bolt rod were refined, respectively. The fixture slightly deformed during the loading and was set as a rigid body. The contacts in the model were achieved by defining contacting surfaces. The setting of the master and slave surfaces of the bolt and lapped plates is shown in Figure 5c. A finite slide was set between surfaces with a friction coefficient of 0.2 [32]. The hybrid joint contains two kinds of C/C composites and an adhesive layer, and the divergences of mechanical properties make the structure present rather complex mechanical behavior under load, coupled with the nonlinear contact between the bolt and the perforated plate and the adhesive layer. If an implicit analysis was adopted, the degradation and failure of the materials would easily lead to the non-convergence of the analysis, such that the calculation would stop before the maximum bearing capacity of the structure is reached. In this study, the ABAQUS/explicit module was used, the embedded user material subroutine VUMAT, combined with the Hashin [33] failure criterion and the stiffness degradation law of materials in [34]. Two analysis steps were set up throughout the loading process. Firstly, a temperature load was applied to the model using a predefined field. The initial temperature was room temperature, which was raised to 800 °C, and the model was then loaded.

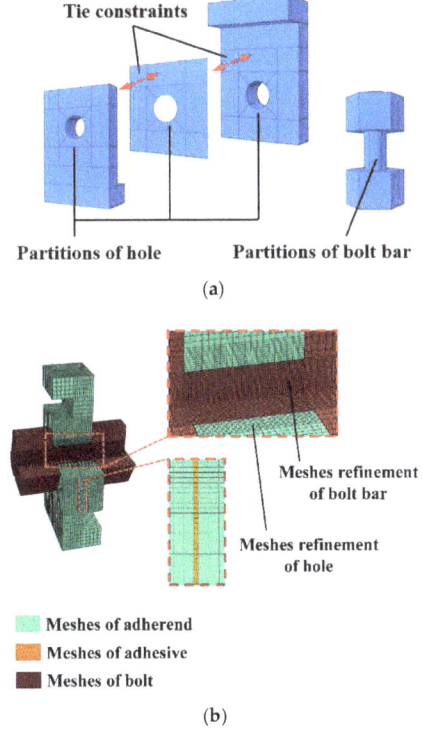

Figure 5. *Cont.*

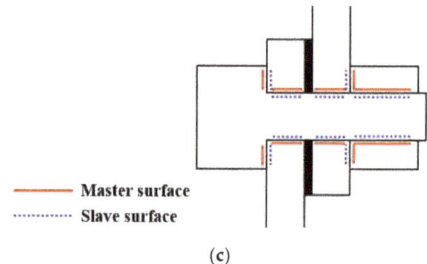

(c)

Figure 5. (**a**) Parts and constraints; (**b**) mesh generation and boundary conditions; (**c**) master and slave surfaces in each contact pair.

2.2.2. Failure Criteria

In order to facilitate the comparison of the calculation results of the hybrid joint and the adhesive joint, the same bilinear constitutive relation as in [30] is used to simulate the failure process of the adhesive layer. For damage to the bolt, the three-dimensional Hashin failure criterion is used, and it can be expressed as follows:

(1) Fiber tensile failure ($\sigma_{11} > 0$)

$$\left(\frac{\sigma_{11}}{X_T}\right)^2 + \left(\frac{\tau_{11}}{X_T}\right)^2 + \left(\frac{\tau_{13}}{S_{13}}\right)^2 \geq 1 \qquad (1)$$

(3) Fiber compressive failure ($\sigma_{11} < 0$)

$$\left(\frac{\sigma_{11}}{X_C}\right)^2 \geq 1 \qquad (2)$$

(5) Matrix tensile failure ($\sigma_{22} > 0$)

$$\left(\frac{\sigma_{22}}{Y_C}\right)^2 + \left(\frac{\tau_{12}}{S_{12}}\right)^2 + \left(\frac{\tau_{23}}{S_{23}}\right)^2 \geq 1 \qquad (3)$$

(7) Matrix compressive failure ($\sigma_{22} < 0$)

$$\left(\frac{\sigma_{22}}{Y_C}\right)^2 + \left(\frac{\tau_{12}}{S_{12}}\right)^2 + \left(\frac{\tau_{23}}{S_{23}}\right)^2 \geq 1 \qquad (4)$$

(8) Matrix/fiber shear failure ($\sigma_{11} < 0$)

$$\left(\frac{\sigma_{11}}{X_C}\right)^2 + \left(\frac{\tau_{12}}{S_{12}}\right)^2 + \left(\frac{\tau_{13}}{S_{13}}\right)^2 \geq 1 \qquad (5)$$

(9) Tensile delamination failure ($\sigma_{33} > 0$)

$$\left(\frac{\sigma_{33}}{Z_T}\right)^2 + \left(\frac{\tau_{13}}{S_{13}}\right)^2 + \left(\frac{\tau_{23}}{S_{23}}\right)^2 \geq 1 \qquad (6)$$

(10) Compressive delamination failure ($\sigma_{33} < 0$)

$$\left(\frac{\sigma_{33}}{Z_C}\right)^2 + \left(\frac{\tau_{13}}{S_{13}}\right)^2 + \left(\frac{\tau_{23}}{S_{23}}\right)^2 \geq 1 \qquad (7)$$

where σ_{11}, σ_{22}, σ_{33}, τ_{12}, τ_{13}, and τ_{23} are the respective normal stress and shear stress in the three directions of the element. X_T, X_C, Y_T, Y_C, Z_T, and Z_C are the strength parameters of the element, as shown in Table 1. The element stiffness degradation model of Camanho [34] was adopted for the damage to the bolt element, as shown in Table 3.

Table 3. Element stiffness degradation mode.

Failure Mode	Stiffness Degradation
Fiber tension	$E_{11} = 0.07 E_{11}$
Fiber compression	$E_{11} = 0.07 E_{11}$
Fiber tension	$E_{22} = 0.2 E_{22}, G_{22} = 0.2 G_{22}, G_{23} = 0.2 G_{23}$,
Fiber compression	$E_{22} = 0.4 E_{22}, G_{12} = 0.4 G_{12}, G_{23} = 0.4 G_{23}$,
Matrix/fiber shear	$G_{12} = 0, \nu_{12} = 0$
Delamination	$E_{22} = 0, G_{13} = 0, G_{23} = 0, \nu_{13} = 0$

The calculation process of progressive damage of composites is demonstrated in Figure 6.

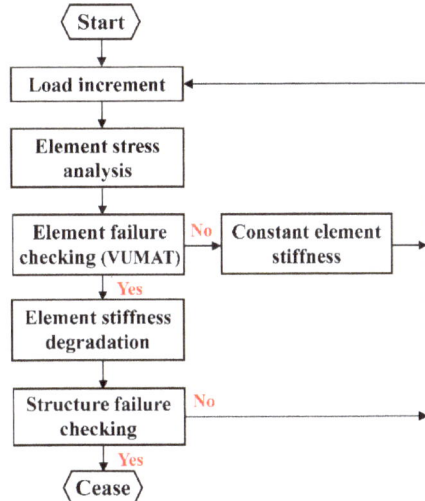

Figure 6. Calculation process of the bolt progressive damage.

3. Results and Discussion

Due to the interrelation of the load-carrying capacity between the hybrid joint and the bonded joint, the failure mechanism and mechanical properties of the hybrid joint are revealed by comparing the experimental and numerical results of the two kinds of joint.

3.1. Failure Mode and Mechanical Response

The compression shear test at 800 °C was conducted on the C/C composite bonded–bolted hybrid single-lap joints. The failure morphology of the test piece after cooling was observed using a digital microscopic system and a scanning electron microscope (Helios Nanolab 600i, FEI, Hillsboro, Oregon, USA), as shown in Figure 7. It can be observed that shear failure occurred in both the adhesive layer and the bolt, and the shear plane of the bolt is flat and coincides with the middle plane of the adhesive layer. The bolt rod had no obvious deformation, and the bolt hole had no obvious extrusion failure, indicating that the bolt rod had a brittle fracture. In contrast to [30], for hybrid joints, the whole adhesive layer exhibited cohesive failure (adhesives were observed on both overlapped plates after the failure), and no interface failure occurs (adhesives were only found on one overlapped plate after the failure). This is due to the secondary bending caused by the eccentricity of the load, which causes the bolt to compress the adhesive layer perpendicular to the lap surface, reducing the peeling behavior of the adhesive layer as shown in Figure 8.

(a)

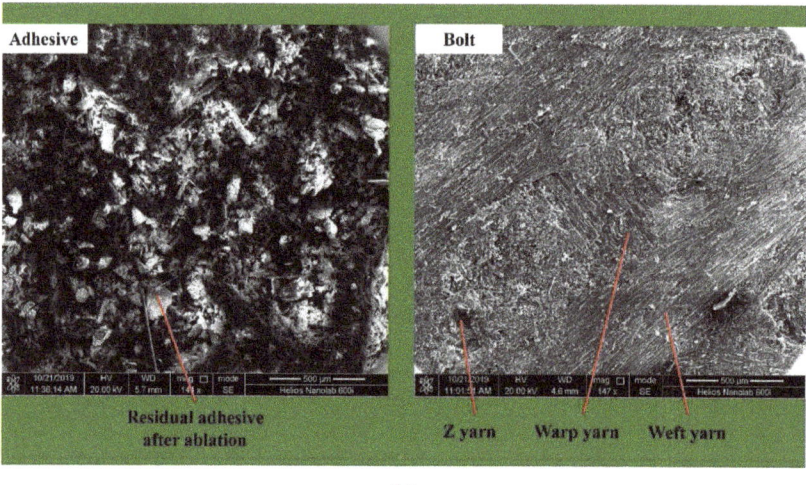

(b)

Figure 7. (a) Failure morphology observed by the digital microscopic system; (b) failure morphology observed by scanning electron microscopy.

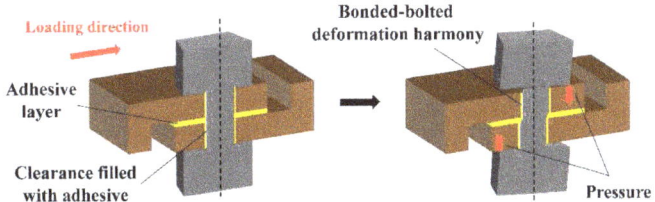

Figure 8. Bolt loading mechanism.

The test and simulation load–displacement curves of C/C composite bonded joints [30] and bonded–bolted hybrid joints at 800 °C are shown in Figure 9, respectively. It can be seen that the load on the hybrid joints had two peaks with the increase of displacement. The load shared by the adhesive layer and the bolt resulted from the hole clearance filled with the cured adhesive. After reaching the first peak, a brittle fracture occurred in the adhesive layer, the load-bearing capacity dropped rapidly, the curve presented as a falling straight line, and the load was then entirely taken by the bolt. Subsequently, the load rose again, the hole wall was compressed, and the bolt was sheared. The eccentricity of the load caused the bolt to bend slightly, resulting in stress concentration on the middle plane of the bolt. Finally, the bolt failed in the shear, and the shear plane was the middle plane of the bolt rod. Similar to bonded joints, the load on the hybrid joints increased nonlinearly until the first peak. In contrast to the bonded joint, the hybrid joint can be seen as replacing the original adhesive layer with a higher modulus adhesive at the center of the adhesive layer. The stiffness of the bonded joints was less than that of the hybrid joints before reaching the first peak, which is because the existence of the bolt increased the overall stiffness of the hybrid joint. The displacement corresponding to the failure point of the adhesive layer in the hybrid joint is greater than the bonded joint, because the existence of the bolt slowed down the crack propagation in the adhesive layer and improved the bearing performance of the joint. For hybrid joints, the failure load of the adhesive layer is defined as P_{max}, the load at the point where the curve descended after the failure of the adhesive layer as P_a, and the difference between P_{max} and P_a as ΔP. Therefore, ΔP was the load shared by the adhesive layer in the hybrid joints. The experimental limit load and the simulated limit load of the bonded joints and the hybrid joints, as well as ΔP, are given in Figure 10, respectively. The bearing capacity of the adhesive layer in the hybrid joints was slightly higher than the bonded joints. This was also due to the presence of the bolt improving the bearing capacity of the adhesive layer. It can be simultaneously seen that the calculated ultimate load was slightly larger than the data measured in the test due to bubbles, the uneven distribution of the adhesive thickness, and the initial defects of C/C composites, and these factors had not been fully considered in the simulation. Despite these factors, the simulation results maintained good consistency with the test results, reflecting the bearing capacity of the test pieces.

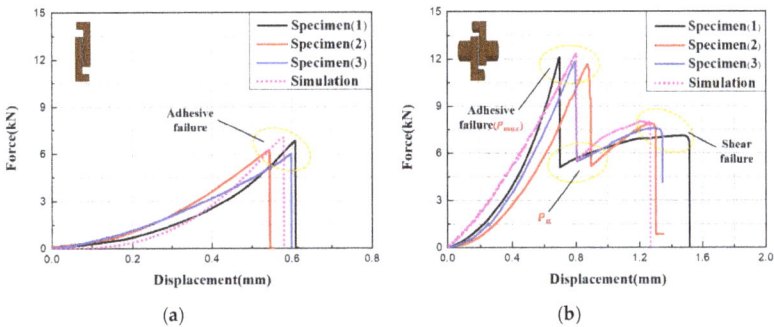

Figure 9. Test and simulation load–displacement curve: (**a**) bonded joints; (**b**) hybrid joints.

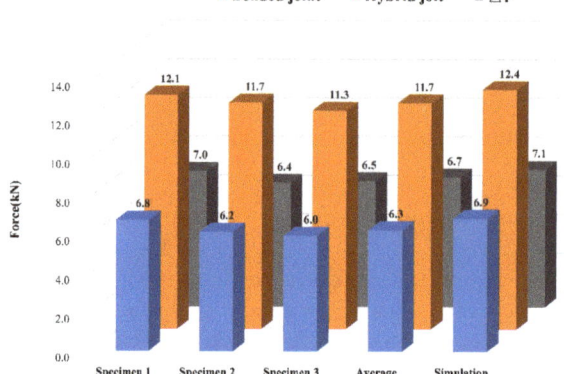

Figure 10. The comparison of the bearing capacity of the two kinds of joints.

3.2. Stress Distribution and Progressive Damage

Taking the center point of the adhesive layer as the coordinate origin, the loading direction as the x-axis, and the width direction of the lap plate as the y-axis, the coordinate system was established, as shown in Figure 11. The normalized shear stress peeling stress of the adhesive layer in the bonded joint and the variations in normalized shear stress and in peel stress of the adhesive layer in the hybrid with y are also given in Figure 11. When $y = 0$ mm and $y = 3$ mm, the existence of the bolt reduced the area of the adhesive layer, which increased the shear stress gradient. The adhesive layer on the edge of the hole ($y = 6$ mm) indicated an obvious shear stress concentration. The shear stress distribution of the adhesive layer far from the center was similar to that of the bonded joint. The peel stress gradient is lower than that of the bonded joint.

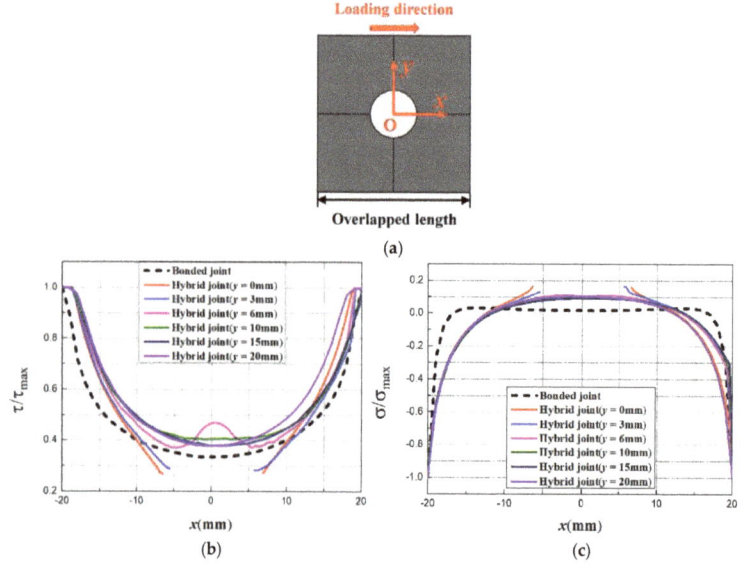

Figure 11. The schematic of overlapped length (**a**); the normalized shear stress (**b**) and peeling stress (**c**) of the adhesive layer in the two kinds of joints.

The simulation load–displacement curve of the hybrid joint is shown in Figure 12. Four points A, B, C, and D were taken on the curve at Stage 1 in order to present progressive damage to the adhesive layer. The shear stress of the adhesive layer reached shear strength at Point A, which is the initial point of the damage, and Point D is the adhesive layer failure load point. Similar to [30], the damage evolution process of the adhesive layer was also distributed symmetrically, and the failure region expanded from the edge to the center of the adhesive layer. However, the failure of the adhesive layer near the centerline was slowed down by the bolt. Points E, F, G, and H were taken on the curve at Stage 2. It can be observed that the bolt was slightly damaged when the adhesive layer failed completely, and the bolt damage was aggravated with the further increase of the load. Damage to the element occurred in the middle plane of the bolt and extended from the middle to both sides in the circular section. Due to the eccentric load resulting in the slight deflection of the bolt, the damage is not completely symmetrical. At Point H, shear failure of the bolt rod occurred.

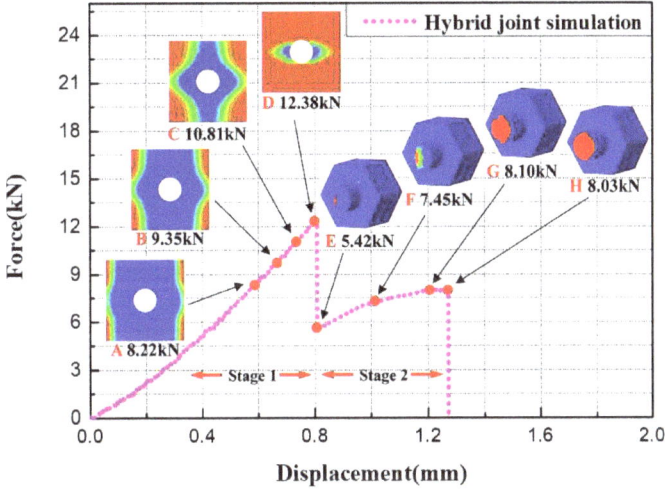

Figure 12. Progressive damage of the hybrid joint.

3.3. Bolt Load Transfer

The load-bearing parts of the bonded–bolted hybrid joint include the adhesive layer and the bolt. The load is transferred to the bolt through the adhesive layer when the structure is loaded. Compared with bonded joints, hybrid joints replaced the adhesive layer at the hole in the center of the lap area with a bolt. The adhesive layer in the central region had little effect on the adhesive bearing capacity, such that the load shared by the bolt directly affected the bearing performance of the hybrid joint. The variations in the ratio (P_b/P_t) of the load shared by the bolt to the total load with displacement are shown in Figure 13. The load shared by the bolt is the sum of nodal forces on the middle plane of the bolt [31]. It can be seen that the value of (P_b/P_t) increased nonlinearly with the displacement. This is because the stiffness of the bolt was greater than that of the adhesive, so the bolt shared more load with the same deformation. The rate at which the value increased decreased. The probable cause is the eccentricity of the load resulting in the bolt deflecting slightly, which caused the bolt head to produce the pressure perpendicular to the adhesive layer. The pressure slowed down the crack propagation in the adhesive layer and increased the bearing capacity of the adhesive layer.

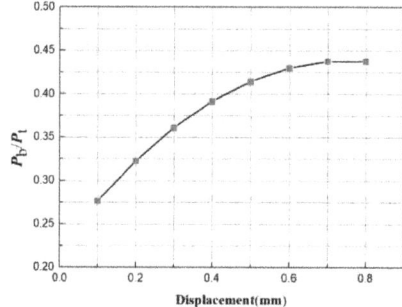

Figure 13. The variations in the ratio of the load shared by the bolt to the total load with displacement.

4. Conclusions

The compression shear test of C/C composite bonded–bolted hybrid single-lap joints was carried out at 800 °C, the failure process of the joint was obtained through finite element analysis, and the numerical results were compared with the test results to verify the correctness of the model. The failure mode, adhesive layer stress distribution, and progressive damage of the joint were determined. The variations in the ratio of the load shared by the bolt to the total load with displacement were obtained. Some conclusions derived from the experimental and numerical studies can be summarized as follows:

1. The shear plane first appeared on the bonding surface with cohesive failures occurring in the adhesive layer. Shear failure then occurred on the bolt, and the shear plane appeared on the middle plane of the bolt rod.
2. Compared with the bonded joint, the existence of the bolt in the hybrid joint increased the shear stress gradient of the adhesive layer when the y value was smaller, and the distribution of shear stress was close to that of the bonded joint. The peel stress gradient was lower than that of the bonded joint. The failure region expanded from the edge to the center in the adhesive layer, while the failure of the adhesive layer at the centerline was slowed down. The failure region on the middle plane of the bolt rod expanded from the middle to both sides.
3. The ratio of the load shared by the bolt to the total load increased nonlinearly with displacement.

Author Contributions: Conceptualization, Y.Z., Z.Z., and Z.T.; data curation, Y.Z. and Z.Z.; formal analysis, Y.Z.; funding acquisition, Z.Z.; investigation, Y.Z.; methodology, Y.Z. and Z.Z.; project administration, Z.Z. and Z.T.; resources, Z.Z. and Z.T.; software, Y.Z.; validation, Y.Z.; writing—original draft, Y.Z.; writing—review & editing, Y.Z. and Z.Z. All authors have read and agreed to the published version of the manuscript.

Funding: This work was supported by the National Natural Science Foundation of China under Grant No. 11572101.

Acknowledgments: In this study, the supplier of all materials and parameters was the China Academy of Launch Vehicle Technology.

Conflicts of Interest: The authors declare that there is no conflict of interest.

References

1. Ozel, A.; Yazici, B.; Akpinar, S.; Aydin, M.D. A study on the strength of adhesively bonded joints with different adherends. *Compos. Part B Eng.* **2014**, *62*, 167–174. [CrossRef]
2. Kim, K.S.; Yoo, J.S.; Yi, Y.M.; Kim, C.G. Failure mode and strength of uni-directional composite single lap bonded joints with different bonding methods. *Compos. Struct.* **2006**, *72*, 477–485. [CrossRef]
3. Turan, K.; Pekbey, Y. Progressive Failure Analysis of Reinforced-Adhesively Single-Lap Joint. *J. Adhes.* **2015**, *91*, 962–977. [CrossRef]
4. Da Silva, L.F.M.; Adams, R.D. Adhesive joints at high and low temperatures using similar and dissimilar adherends and dual adhesives. *Int. J. Adhes. Adhes.* **2007**, *27*, 216–226. [CrossRef]

5. Guilpin, A.; Franciere, G.; Barton, L.; Blacklock, M.; Birkett, M. A Numerical and Experimental Study of Adhesively-Bonded Polyethylene Pipelines. *Polymers* **2019**, *11*, 1531. [CrossRef]
6. Anasiewicz, K.; Kuczmaszewski, J. Adhesive Joint Stiffness in the Aspect of FEM Modelling. *Materials* **2019**, *12*, 3911. [CrossRef]
7. Egan, B.; McCarthy, C.T.; McCarthy, C.T.; Gray, P.J.; Frizzell, R.M. Modelling a single-bolt countersunk composite joint using implicit and explicit finite element analysis. *Comput. Mater. Sci.* **2012**, *64*, 203–208. [CrossRef]
8. McCarthy, C.T.; McCarthy, M.A.; Lawlor, V.P. Progressive damage analysis of multi-bolt composite joints with variable bolt–hole clearances. *Compos. Part B Eng.* **2005**, *36*, 290–305. [CrossRef]
9. Egan, B.; McCarthy, C.T.; McCarthy, M.A.; Frizzell, R.M. Stress analysis of single-bolt, single-lap, countersunk composite joints with variable bolt-hole clearance. *Compos. Struct.* **2012**, *94*, 1038–1051. [CrossRef]
10. Nezhad, H.Y.; Egan, B.; Merwick, F.; McCarthy, C.T. Bearing damage characteristics of fibre-reinforced countersunk composite bolted joints subjected to quasi-static shear loading. *Compos. Struct.* **2017**, *166*, 184–192. [CrossRef]
11. Hart-Smith, L.J. Bonded-Bolted Composite Joints. *J. Aircraft.* **1985**, *22*, 993–1004. [CrossRef]
12. Gomez, S.; Onoro, J.; Pecharroman, J. A simple mechanical model of a structural hybrid adhesive/riveted single lap joint. *Int. J. Adhes. Adhes.* **2007**, *27*, 263–267. [CrossRef]
13. Chowdhury, N.; Chiu, W.K.; Wang, J.; Chang, P. Static and fatigue testing thin riveted, bonded and hybrid carbon fiber double lap joints used in aircraft structures. *Compos. Struct.* **2015**, *121*, 315–323. [CrossRef]
14. Vallee, T.; Tannert, T.; Meena, R.; Hehl, S. Dimensioning method for bolted, adhesively bonded, and hybrid joints involving Fibre-Reinforced-Polymers. *Compos. Part B Eng.* **2013**, *46*, 179–187. [CrossRef]
15. Chowdhury, N.M.; Chiu, W.K.; Wang, J.; Chang, P. Experimental and finite element studies of bolted, bonded and hybrid step lap joints of thick carbon fibre/epoxy panels used in aircraft structures. *Compos. Part B Eng.* **2016**, *100*, 68–77. [CrossRef]
16. Ucsnik, S.; Scheerer, M.; Zaremba, S.; Pahr, D.H. Experimental investigation of a novel hybrid metal–composite joining technology. *Compos. Part A Appl. Sci. Manuf.* **2010**, *41*, 369–374. [CrossRef]
17. Marannano, G.; Zuccarello, B. Numerical experimental analysis of hybrid double lap aluminum-CFRP joints. *Compos. Part B Eng.* **2015**, *71*, 28–39. [CrossRef]
18. Kelly, G. Quasi-static strength and fatigue life of hybrid (bonded/bolted) composite single-lap joints. *Compos. Struct.* **2006**, *72*, 119–129. [CrossRef]
19. Mei, H.; Cheng, L.; Ke, Q.; Zhang, L. High-temperature tensile properties and oxidation behavior of carbon fiber reinforced silicon carbide bolts in a simulated re-entry environment. *Carbon* **2010**, *48*, 3007–3013. [CrossRef]
20. Li, S.; Chen, X.; Chen, Z. The effect of high temperature heat-treatment on the strength of C/C to C/C–SiC joints. *Carbon* **2010**, *48*, 3042–3049. [CrossRef]
21. Bruneton, E.; Narcy, B.; Oberlin, A. Carbon-carbon composites prepared by a rapid densification process I: Synthesis and physico-chemical data. *Carbon* **1997**, *35*, 1593–1598. [CrossRef]
22. Virgil'ev, Y.S.; Kalyagina, I.P. Carbon–Carbon Composite Materials. *Inorg. Mater.* **2004**, *40*, S33–S49. [CrossRef]
23. Buckley, J.D.; Edie, D.D. *Carbon–Carbon Materials and Composites*; Noyes Publications: New Jersey, NJ, USA, 1993.
24. Lim, D.W.; Kim, T.H.; Choi, J.H.; Kweon, J.H.; Park, H.S. A study of the strength of carbon–carbon brake disks for automotive applications. *Compos. Struct.* **2008**, *86*, 101–106. [CrossRef]
25. Li, K.Z.; Shen, X.T.; Li, H.J.; Zhang, S.Y.; Feng, T.; Zhang, L.L. Ablation of the carbon/carbon composite nozzle-throats in a small solid rocket motor. *Carbon* **2011**, *49*, 1208–1215. [CrossRef]
26. Christin, F. Design, fabrication, and application of thermostructural composites (TSC) like C/C, C/SiC, and SiC/SiC composites. *Adv. Eng. Mater.* **2002**, *4*, 903–912. [CrossRef]
27. Windhorst, T.; Blount, G. Carbon-carbon composites: A summary of recent developments and applications. *Mater. Des.* **1997**, *18*, 11–15. [CrossRef]
28. Kumar, S.; Kushwaha, J.; Mondal, S.; Kumar, A.; Jain, R.K.; Devi, J.G. Fabrication and ablation testing of 4D C/C composite at 10MW/m^2 heat flux under aplasma arc heater. *Mater. Sci. Eng. A* **2013**, *566*, 102–111. [CrossRef]
29. Sun, C.T.; Kumar, B. Development of Improved Hybrid Joints for Composite Structures. *Compos. Struct.* **2005**, *35*, 1–20.

30. Zhang, Y.; Zhou, Z.; Tan, Z. Compression Shear Properties of Adhesively Bonded Single-Lap Joints of C/C Composite Materials at High Temperatures. *Symmetry* **2019**, *11*, 1437. [CrossRef]
31. Dassault Systemes Simulia Corp. *Abaqus Analysis User's Manual 6.14*; Dassault Systemes Simulia Corp: Providence, RI, USA, 2014.
32. Kelly, G. Load transfer in hybrid (bonded/bolted) composite single-lap joints. *Compos. Struct.* **2005**, *69*, 35–43. [CrossRef]
33. Hashin, Z. Failure Criteria for Unidirectional Fiber Composites. *J. Appl. Mech.* **1980**, *47*, 329–334. [CrossRef]
34. Camanho, P.P.; Matthews, F.L. A progressive damage model for mechanically fastened joints in composite laminates. *J. Compos. Mater.* **1999**, *33*, 2248–2280. [CrossRef]

© 2020 by the authors. Licensee MDPI, Basel, Switzerland. This article is an open access article distributed under the terms and conditions of the Creative Commons Attribution (CC BY) license (http://creativecommons.org/licenses/by/4.0/).

Article

Synthesis and Processing of Melt Spun Materials from Esterified Lignin with Lactic Acid

Panagiotis Goulis, Ioannis A. Kartsonakis, George Konstantopoulos and Costas A. Charitidis *

Research Unit of Advanced, Composite, Nano-Materials and Nanotechnology, School of Chemical Engineering, National Technical University of Athens, 9 Heroon Polytechniou St., Zographos, GR-15773 Athens, Greece; pgoulis@chemeng.ntua.gr (P.G.); ikartso@chemeng.ntua.gr (I.A.K.); gkonstanto@chemeng.ntua.gr (G.K.)
* Correspondence: charitidis@chemeng.ntua.gr; Tel.: +30-210-772-4046

Received: 14 October 2019; Accepted: 5 December 2019; Published: 8 December 2019

Abstract: In this study, the carbon fiber manufacturing process is investigated, using high-density polyethylene (HDPE) and esterified lignin either with lactic acid (LA) or with poly(lactic acid) (PLA) as precursors. More specifically, lignin was modified using either LA or PLA in order to increase its chemical affinity with HDPE. The modified compounds were continuously melt spun to fibrous materials by blending with HDPE in order to fabricate a carbon fiber precursor. The obtained products were characterized with respect to their morphology, as well as their structure and chemical composition. Moreover, an assessment of both physical and structural transformations after modification of lignin with LA and PLA was performed in order to evaluate the spinning ability of the composite fibers, as well as the thermal processing to carbon fibers. This bottom–up approach seems to be able to provide a viable route considering large scale production in order to transform lignin in value-added product. Tensile tests revealed that the chemical lignin modification allowed an enhancement in its spinning ability due to its compatibility improvement with the commercial low-cost and thermoplastic HDPE polymer. Finally, stabilization and carbonization thermal processing was performed in order to obtain carbon fibers.

Keywords: carbon fibers; lignin; melt spinning; carbonization; Raman; micro-CT

1. Introduction

Carbon fibers are constantly being investigated considering their potential of high prospect technological application such as in the airspace or car industry, which require high performance and long-serving advanced materials for construction. Carbon fibers of exceptional mechanical properties have already been produced; however, new low-cost precursors are under investigation. At present, the need to reduce the cost of the carbon fibers for technological applications is of paramount importance and provides the opportunity to broaden their market. As a consequence, eco-friendly, renewable, and low-cost precursors of lignin have attracted major interest in order to be incorporated in the synthesis of the carbon fibers, considering both high carbon content and dense aromatic structure [1–3].

Synthesis of carbon fibers consists of several processing paths, such as chemical synthesis of the precursor polymer, fiber spinning (dry, wet, melt), stabilization in oxidative atmosphere to induce crosslinking and thermal resistance, and carbonization in inert atmosphere to induce graphitic planes in fiber structure. Regarding the chemical synthesis step, it should be mentioned that it requires careful selection and design of the reaction mechanism of the reagents in order to improve the linearity of molecular chains and reduce the polar character induced by lignin hydroxyls [4].

A potential low-cost precursor that could be used for carbon fiber production is Softwood Kraft Lignin (SKL). However, this precursor has several particularities due to the fact that it consists of a complex three-dimensional network, while it becomes brittle upon thermal processing. Therefore, in order for lignin to be used for structural carbon–fiber production, it has to be modified so that the high-extent reactive C–O bonds in interunit lignin linkages, such as b-O-aryl ether, can be reduced [4]. On the other hand, lignin provides a low-cost and widely available alternative that could be utilized in industries, assuming that a viable modification route is established to enhance process ability and compounding to form fibers with enhanced mechanical properties and behavior overall. In general, lignin composites with epoxy resins, derivatives, and blends have already been investigated [5–8], but there is no sufficient reference to their spinning ability, while Zhou et al. overviewed the research done on plant fiber composites [9]. Moreover, Megiatto et al. investigated chemical modification of Sisal (a Mexican agave) with lignin and the effect of phenolic composite properties on fibers. It was concluded that phenolic functional groups enhance fiber properties.

In other works considering biopolymers similar to lignin, cellulose was studied as a binder in green composites [10,11], whereas application of plant fibers, composites, and their life cycle assessment was also reported [12–18]. Additionally, the properties of elastomeric composites coming from widely used commercial polymers are already known, as well as the effect of lignin enrichment in foams, composites, and fibers [19–21]. Furthermore, high-tech grafting of carbon nanotubes on lignin structures to manufacture fibers with advanced electrical or mechanical properties has been performed [22–24].

The motivation of the present work is the synthesis of carbon fibers using low-cost and eco-friendly precursors. Therefore, the aim in the present research is to follow a relatively simple and low-cost procedure to reach a carbon fiber production. Lignin was esterified with lactic acid (LA) and poly(lactic acid) (PLA) in order to improve its chemical affinity with high-density polyethylene (HDPE) and pave the way for the preparation of a blend that could be melt spun and then produce carbon fibers. The aforementioned organic acids were implemented in the precursor lignin-enriched fiber production via chlorination with thionyl chloride ($SOCl_2$) and esterification with lignin. Evaluations of esterification reaction, spinning process, as well as stabilization and carbonization procedures were conducted. The originality and novelty of the present study lies in the fact that a melt-spun lignin conjugation with high molecular weight commercial spinnable polymers was accomplished, using esterification after a chlorination reaction, so as to enhance the intermolecular bonds between the compounds. Finally, it could be mentioned that the added value of our work resides in the fact that the chemical synthesis that was conducted in this study provided thermal process ability to the precursor fiber without the use of plasticization.

2. Materials and Methods

2.1. Materials

Analytical reagent grade chemicals were used. Lactic acid (LA, Fisher Chemicals, Waltham, MA, USA), high-density polyethylene (HDPE, Sigma Aldrich, St. Louis, MO, USA, Mw~125,000), poly(lactic acid) (PLA, Sigma Aldrich, St. Louis, MO, USA, Mw~125,000), hydrochloric acid (HCl 37%, Fisher Chemicals, Waltham, MA, USA), thionyl chloride ($SOCl_2$, Acros Organics, Geel, Belgium), chloroform ($CHCl_3$, Fisher Chemicals, Waltham, MA, USA), and pyridine (Fisher Chemicals, Waltham, MA, USA) were used as received without further purification. Softwood kraft lignin (SKL, Westvaco Corp., Indulin AT, MWV, Norcross, GA, USA) was washed with HCl 0.1 M before being used in the blends. All blends intended for melt spinning were placed in a vacuum oven before the extrusion took place and were left there overnight.

2.2. Synthesis of Fibers

In a 1000 mL three-neck round bottom flask, 100 mL of LA was dissolved in 230 mL of $CHCl_3$, under vigorous stirring at room temperature for 1 h. Then, 100 mL of $SOCl_2$ was added to the mixture, dripping slowly and in small amounts in order to perform the chlorination reaction. As the $SOCl_2$ was added to the flask, vapors of SO_2 were released from the chemical system according to the reaction in Figure 1 [25]. According to this reaction, a nucleophilic substitution pathway occurs, in which the carboxylic acid is first converted into a chlorosulfite intermediate. Then, the chlorosulfite reacts with a nucleophilic chloride ion in order lactic acid chloride (LCl) to be synthesized.

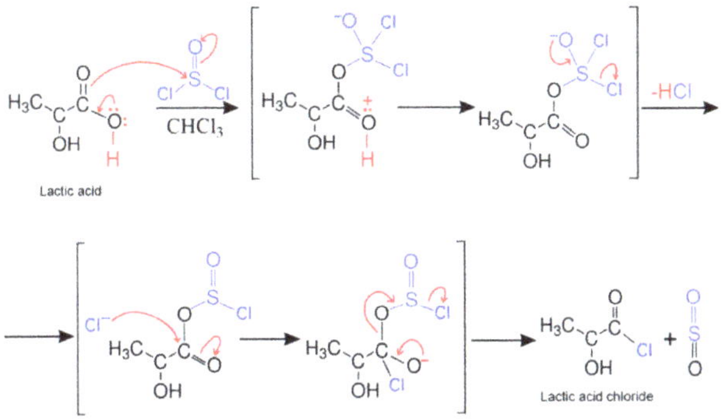

Figure 1. Chemical reaction of the chlorination process.

The released vapors indicate that the chlorination reaction has progressed. After partial evaporation of the solvent ($CHCl_3$), the solution was added to 250 mL of pyridine. Then, 100 g of washed SKL was added to the mixture, and the blend was left under vigorous stirring at room temperature for 1 h. The temperature was then elevated at 120 °C, and a reflux condenser was adapted onto the flask. The esterification reaction duration was 5 h at the aforementioned conditions; the mechanism of the corresponding esterification reaction (alcoholysis) between the hydroxyl groups of SKL and lactic chloride is depicted in Figure 2 [26]. According to the esterification process, an addition of oxygen (nucleophile) to the carbonyl group occurs, yielding a tetrahedral intermediate. Then, an electron pair from oxygen displaces the chloride (leaving group), generating a new carbonyl compound as a product. After the completion of the reaction, the solvent (pyridine) was evaporated. The solid product (SKL esterified with lactic chloride, SKL–LA) was oven-dried overnight at 70 °C.

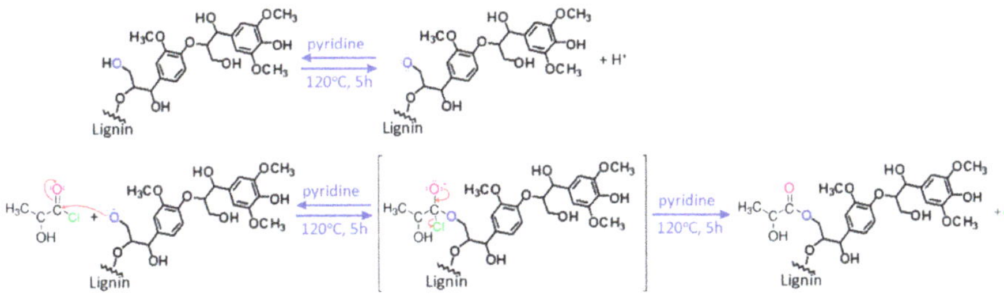

Figure 2. Esterification reaction between Softwood Kraft Lignin (SKL) and lactic acid chloride (LCl).

In a 1000 mL glass beaker, 45 g of poly(lactic acid) (PLA) was dissolved in 600 mL of chloroform under vigorous mechanical stirring. The blend was left under stirring overnight in order to ensure complete dissolution. Afterwards, 100 mL of $SOCl_2$ was added dripping in the solution under mechanical stirring. The $SOCl_2$ reacted with the hydroxyl groups of PLA and produced poly(lactic acid) chloride (PLCl) according to the reaction in Figure 3 [25]. During the $SOCl_2$ addition, there was a release of SO_2 vapors that indicated the reaction progress. After completion of the reaction, the obtained product (PLCl) was dried and pulverized. Then, 25 g of the synthesized PLCl was dissolved in 270 mL of pyridine in a 500 mL three-neck round bottom spherical flask and was left under vigorous stirring overnight at room temperature. In the next step, 40 g of SKL was added, and the mixture was left to react at 120 °C for 5 h under vigorous stirring. After that, the solvent was evaporated, and the solid collected product of SKL–PLA (SKL esterified with PLA) was dried at 80 °C. The chemical reaction of the esterification process (alcoholysis) is illustrated in Figure 4 [26].

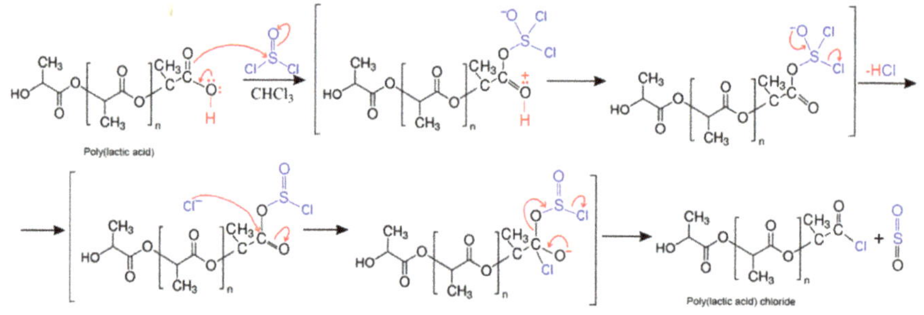

Figure 3. Chemical reaction of the poly(lactic acid) (PLA) chlorination process.

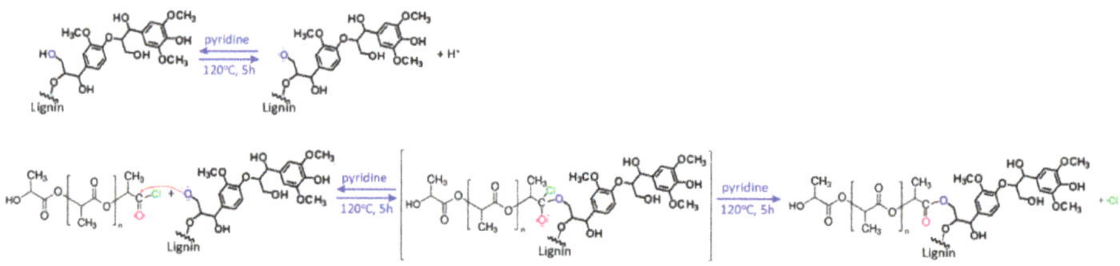

Figure 4. Esterification reaction between SKL and PLCl.

The abovementioned materials were mixed with HDPE, and the resulting blends were loaded in an extruder line, optimizing spinning parameters of extruder temperature, roller distance from the extruder, and extrusion speed for each precursor blend. The parameters and conditions for the optimization process for the HDPE spinning for the production of HDPE/SKL–LA (80:20) and HDPE/SKL–PLA (80:20) fibers are tabulated in Tables A1 and A2. The best results concerning the spinning ability of the blends were obtained for the conditions that are presented in Table 1. The blend mass ratio of the HDPE/esterified lignin was 80/20 by weight (HDPE/SKL–LA (80:20), HDPE/SKL–PLA (80:20)).

Table 1. Tabulated parameters and conditions of melt spinning process.

Blend	Stage 1 Temperature (°C)	Stage 2 Temperature (°C)	Extrusion Speed (% rpm)	Roller Distance (cm)
HDPE/SKL–LA	226	230	100	23
HDPE/SKL–PLA	222	225	85	25

The extruded precursor HDPE/SKL–PLA fibers were fixed in bundles and were thermally treated following four different thermal stages. Firstly, they were subjected to a 1 h long pre-stabilization step at 150 °C, followed by a second stage at 220 °C for 3 h and finally at 300 °C for 10 min to maximize crosslinking. This thermal procedure enhanced crosslinking through oxidation and prepared the structure to withstand carbonization. Carbonization was performed at 800 °C, under nitrogen atmosphere with a residence time of 5 min. The thermal treatment effect is presented in Table 2.

Table 2. Tabulated thermal treatment results.

Sample	150 °C, 1 h	220 °C, 3 h	300 °C, 10 min	Proceeded to Carbonization	Successful Carbonization
HDPE/SKL–LA	sustained	sustained	sustained	Yes	need for optimization
HDPE/SKL–PLA	sustained (fiber shrinkage)	sustained	sustained	Yes	Yes

The carbonization of the fibers was conducted via a high-temperature furnace (Protherm—PTF Series, Single-Zone tube furnace, Maximum Temperature: 1200 °C) under nitrogen atmosphere at 800 °C for 5 min.

The melt spinning experiments were carried out on a Dynisco extruder. The Dynisco LME (Laboratory Mixing Extruder) is a laboratory tool developed to produce fibers continuously derived from thermoplastic polymers. The specimen is conveyed in a cooled hopper right after exiting the hot surface of the cylindrical rotor, against the inclined surface of the stationary scroll, and moves toward the outlet die.

2.3. Characterization

The fibers composition and their morphology were determined with a Hitachi Electron Microscope TM3030 coupled with an Ultra-High Resolution Scanning Electron Microscope (UHR-SEM) using NOVA NANOSEM 230 (FEI Company, Hillsboro, United States, 2009) and with an Energy Dispersive X-ray Spectrophotometer (EDS) (QUANTAX 70, BRUKER, Billerica, Massachusettes, United States, 2016). The Attenuated Total Reflectance Fourier Transform Infrared Spectroscopy (FTIR) instrument Cary 630 spectrometer (Agilent, Santa Clara, California, United States, 2016) was used to analyze the FTIR spectra of the fibers. This instrument has a resolution of 4 cm^{-1} and an operating wavelength range of 4000–400 cm^{-1}.

The Thermogravimetric/Differential Scanning Calorimetry (TGA/DSC) instrument STA 449 F5 Jupiter (NETZSCH-Gerätebau GmbH, Selb, Germany, 2017) was used for the thermal analysis of the fibers. The system consists of a SiC furnace with an operation temperature varying between 25 and 1550 °C. The maximum heating rate is 50 K/min. The gas used was nitrogen at a volumetric rate of 50 mL/min and at a heating rate of 20 °C/min.

A Renishaw inVia spectrometer was used to perform micro-Raman measurements coupled with a green diode laser with emission at 532 nm (Renishaw plc, Gloucestershire, United Kingdom, 2013). The backscattering configuration was used. A beam spot size of 1 micrometer for a 1 mW beam at 1% intensity was used. The MATLAB 8.0 and Statistics Toolbox 8.1 (The MathWorks, Inc., Natick, Massachusetts, United States, 2012) was used for fitting the Raman signal to calculate the integrals of D and G peaks and quantify ID/IG ratio.

The structure of the fibers was observed and collected by the instrument SkyScan 1272, a 3D X-ray scan system (BRUKER, Billerica, Massachusettes, United States, 2017). The system disposes an X-ray source which operates at 20–100 kV and 10 W (<5 µm spot size @ 4W), an X-ray detector with a maximum resolution of 11 Mp (4032 × 2688 pixels), and a CCD fiber of 14 bits, cooled and optically joined to a scintillator. During the observation, the specimens were fixed on a rotational stage, the distance between the rotation axis and the radiation source is set from 6 to 12 mm, and the pixel resolution is set to 2016 × 1344 in order to obtain reliable images. Several stacked images were used for rendering, calculations, and image analysis for each sample.

The tensile strength at the breaking point of the fiber was determined according to the scale of the dynamometer of the MPM-10 tensile test machine, based on the standards ISO 11566:1996 and ASTM D 3379-89.

3. Results and Discussion

3.1. Scanning Electron Microscopy and Energy Dispersive X-ray Spectroscopy

Taking into account the SEM images of the fibers HDPE/SKL–LA, HDPE/SKL–PLA (precursor) and HDPE/SKL–PLA (carbonized) (Figures 5–7, respectively), it can be observed that the diameter of the fibrous materials ranges from 158 to 170 µm (Table 3). It should be stated that the SEM characterization of the fibers in respect of their diameter (Table 3) includes the measurement of 10 fibers of each sample. Moreover, two measurements were conducted in each fiber.

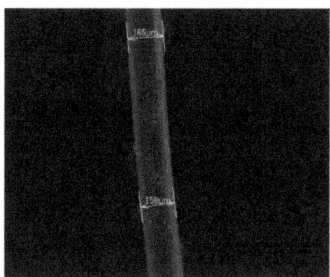

Figure 5. SEM image of high-density polyethylene (HDPE)/SKL–LA fiber.

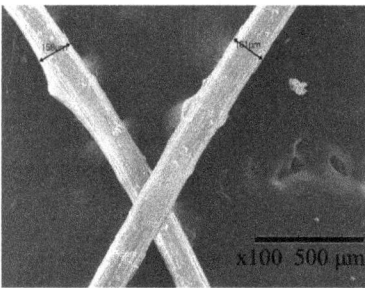

Figure 6. SEM image of HDPE/SKL–PLA (precursor) fiber.

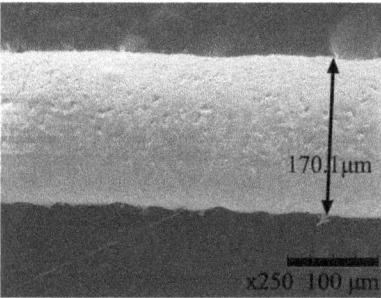

Figure 7. SEM image of HDPE/SKL–PLA carbonized fiber.

Table 3. Tabulated SEM analysis results of the produced fibers.

Sample	Min. Diameter (µm)	Max. Diameter (µm)	Average Diameter (µm)
HDPE/SKL–LA	158	165	161.5
HDPE/SKL–PLA (precursor)	155	164	159.5
HDPE/SKL–PLA (carbon fiber)	168.5	170.1	169.3

Although the typical diameter of a PAN-based precursor fiber is only 15 µm, materials with a bigger diameter than that were produced because of the low elasticity of the HDPE and the use of a 1.8 mm spinneret. Additionally, the addition of lignin reduces the tension capacity of the material [27]. The thermal behavior of HDPE/SKL–LA fibers was not proper to withstand carbonization, due to the formation of agglomerates and heterogeneity on the extruded fiber. On the other hand, HDPE/SKL–PLA fibers were carbonized due to the fact that SKL–PLA presents enhanced mixing ability with HDPE blends. Furthermore, it should be mentioned that HDPE/SKL–PLA carbon fiber shrunk in length about 6% compared to the precursor, owing to densification and the increase of sturdiness after carbonization treatment [28]. Due to the same fact, the diameter of the fiber is increased. Furthermore, combined SEM-EDS analysis was conducted on the produced samples in order to provide information of the chemical composition, with the percentage of each chemical element in the fibers, as well as to support the chlorination mechanism of LA and PLA reaction and to confirm the successful synthesis of modified lignin by the corroboration of the absence of Cl element in the observed specimens. Table 4 presents the tabulated values of wt.% elemental composition of the produced HDPE/SKL–PLA fibers prior to and after carbonization process. Regarding the obtained results, it should be remarked that the percentage of chlorine in the samples was practically zero for both the HDPE/SKL-PLA precursor and carbon fiber specimens.

Table 4. Tabulated values of wt.% elemental composition of the produced HDPE/SKL–PLA fibers.

Sample	Element	wt.%
HDPE/SKL–PLA (precursor)	C	57.60
	O	41.53
	Cl	0.87
HDPE/SKL–PLA (carbon fiber)	C	87.89
	O	11.60
	Cl	0.51

3.2. Thermogravimetric Analysis and Differential Scanning Calorimetry

The TGA and DSC diagrams of lignin compounds are illustrated in Figure 8. The gas used was nitrogen and the temperature ramp was 10 °C/min. Taking into consideration the TGA and DSC diagrams of the compounds SKL–LA (Figure 8b) and SKL–PLA (Figure 8c), it may be remarked that both compounds present an initial peak of thermal degradation at 280 °C, whereas their thermal degradation end temperature is located at 800 °C leaving a residue of approximately 22% and 35% mass, respectively (Table 5). The residual mass is attributed to the autocatalytic stabilization effect of lignin derived by high elemental composition in oxygen [29–32]. The PLCl is degraded almost quantitatively at 400 °C (Figure 8a). The DSC curve of PLCl presents two characteristic peaks, an endothermic at 340 °C and an exothermic at 480 °C, which correspond to its thermal degradation peak and crosslinking, respectively [33] (Figure 8a). As far as SKL–LA is concerned, the DSC graph presents a wide exothermic curve, which implies a steady rate and slow evolution of gases as crosslinking byproducts of the material, considering the mass loss as provided by TGA (Figure 8b). Finally the DSC graph for the SKL–PLA sample demonstrates a more intense exothermic peak ascribed to oxygen-derived autocatalytic crosslinking, generated at multiple starting points due to its size, but there is only a small endothermic peak at 700 °C, relating to its degradation and state transition from crystal-like to amorphous [34,35] (Figure 8c). Concerning the TGA/DSC graphs of the HDPE/SKL–PLA carbon fiber, it can be easily observed that the carbon residue after thermal treatment is about 85%, meaning that the whole procedure successfully produced a carbon fiber with desired thermal behavior, resistance, and adequate carbon yield. The wide exothermic DSC curve is ascribed to intramolecular bond forming reactions as well as to polycondensation and denitrification reactions (Figure 8d) [36].

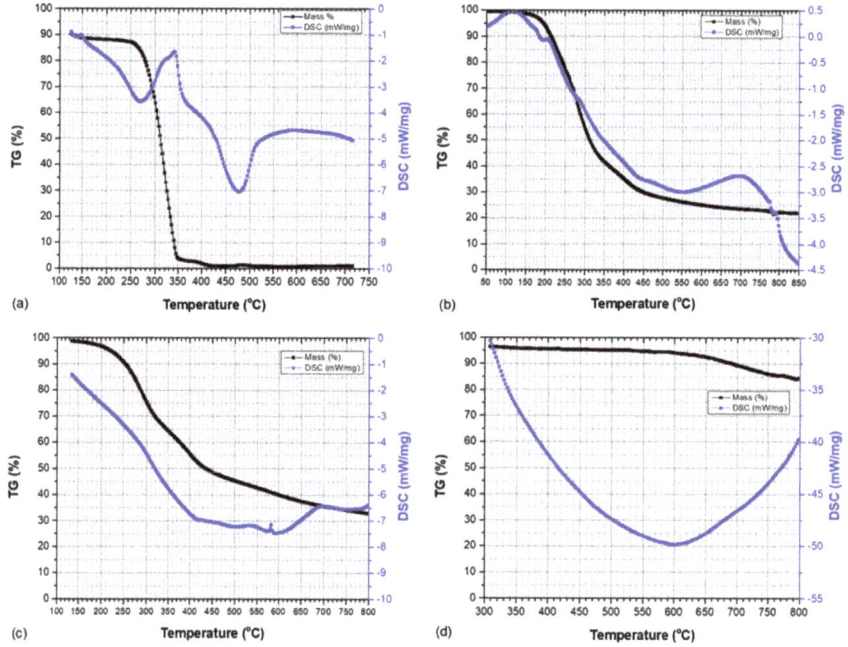

Figure 8. TGA/DSC graphs of: (a) PLCl compound, (b) SKL–LA compound, (c) SKL–PLA compound, and (d) HDPE/SKL–PLA carbon fiber.

Table 5. Thermal analysis of the samples.

Sample	Onset Temperature of Thermal Degradation (°C)	Peak Temperature of Thermal Degradation (°C)	End Temperature of Thermal Degradation (°C)	Mass Change (wt.%)
PLCl	284.9	320.8	399.3	95
SKL–LA	205.1	277.9	740.0	75
SKL–PLA	236.1	285.7	781.8	69
HDPE/SKL–PLA (carbon fiber)	625.0	725	800.0	15

3.3. Attenuated Total Reflectance Fourier Transform Infrared Spectroscopy

As evidenced in the aforementioned SEM/EDS micrographs as well as in the FTIR spectrogram in Figure 9, the chlorination reaction of PLA to PLCl was successful. The major differences in FTIR spectra after chlorination of PLA and LA are pointed out in the peaks at 1558 and 1640 cm^{-1} [37] that are attributed to asymmetric CO_2^- stretching vibration, which is mainly reacted with $SOCl_2$ [38–41]. Further peaks are similar between PLA and PLCl, slightly more intense in case of PLCl. This is due to the –Cl group electronegativity and the use of $SOCl_2$ in the chlorination reaction [25,42]. The FTIR analysis results are presented in Table 6.

The FTIR spectra of SKL–LA and SKL–PLA in Figure 10 demonstrate similar peaks. The small band at about 3500 cm^{-1} is attributed to the hydroxyl group of phenolic and aliphatic structures of lignin. This peak is more intense for SKL–LA due to the fact that the LA monomers which are esterified to the lignin hydroxyl groups have one hydroxyl group at a single end. The PLCl chains esterified to lignin (SKL–PLA) own only one hydroxyl group at the end of each chain. Therefore, the only signal that SKL–PLA presents at 3500 cm^{-1} is ascribed almost completely on the lignin hydroxyls, because the contribution from the PLCl hydroxyls to this signal is negligible. Moreover, SKL–LA demonstrates a peak of medium intensity at 2900 cm^{-1}, which is attributed to C–H stretches. The peaks at 2900 cm^{-1} result from various C–H bonds due to the aromatic methoxy groups and the methyl/methylene side groups. The characteristic absorption bands for aromatic rings of lignin are at 1600, 1500, and 1410 cm^{-1}. The peak at 1600 cm^{-1} also implies the presence of a conjugated structure. There is also a shoulder at 1750 cm^{-1}, corresponding to lignin's carbonyl groups [43–45]. The FTIR most important peaks are tabulated in Table 7.

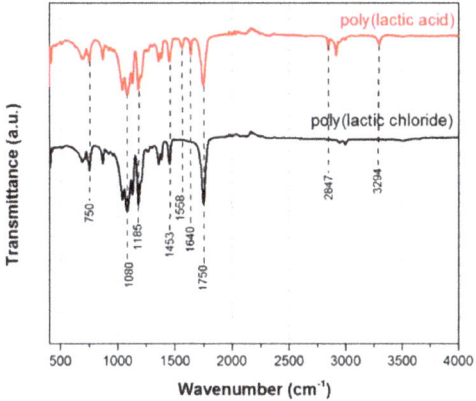

Figure 9. FTIR spectra of PLA and PLCl.

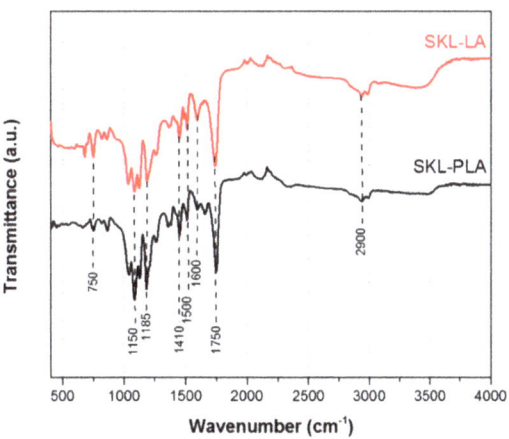

Figure 10. FTIR spectra of SKL–LA and SKL–PLA materials.

Table 6. Identified FTIR peaks ascribed to the peaks of PLA and PLCl materials.

Wavenumber (cm^{-1})	Type of Bond Vibration
3294	–COOH (PLA)
2847	–COOH (PLA)
1750	C=O (both PLA and PLCl)
1640	asymmetric CO_2^- stretching vibration (PLA)
1558	asymmetric CO_2^- stretching vibration (PLA)
1453	C–H (both PLA and PLCl)
1185	C–O (both PLA and PLCl)
1080	C–O (both PLA and PLCl)
750	C–H (both PLA and PLCl)

Table 7. Identified FTIR peaks ascribed to the SKL–LA and SKL–PLA materials.

Wavenumber (cm^{-1})	Type of Bond
3500	O–H (both SKL–LA and SKL–PLA)
2900	C–H stretch
1750	C=O (both SKL–LA and SKL–PLA)
1600	aromatic lignin rings (conjugated structure)
1500	aromatic lignin rings
1410	aromatic lignin rings
1150–1185	C–O (both SKL–LA and SKL–PLA)
750	C–H bend (both SKL–LA and SKL–PLA)

3.4. Thermal Processing (Stabilization—Carbonization)

Taking into consideration Table 5, it may be remarked that the HDPE/SKL–PLA fibers present an enhanced thermal behavior compared to those of HDPE/SKL–LA, because they present a higher onset, peak, and end temperature of thermal degradation and considering the fact that HDPE has the same thermal behavior in both samples. Moreover, taking into consideration the two fibers, the HDPE/SKL–PLA precursor is expected to exhibit an extended linear-like polymeric chain and bigger molecular weight, in comparison with the much smaller HDPE/SKL–LA fibers, which were not successfully carbonized, due to the aforementioned reasons. The yield of the carbon fiber after carbonization of the HDPE/SKL–PLA fibers was 85%. Based on the higher end temperature and lower weight loss as given in Table 5, HDPE/SKL–PLA is more thermally stable compared to HDPE/SKL–LA. According to the literature, although the carbonization of PAN precursor fibers is carried out between 1000 and 1800 °C, in our study, the fibrous materials were carbonized at 800 °C for 5 min (in a horizontal furnace used for CVD), due to the different texture and thermal behavior of the materials.

3.5. Tensile Testing of Precursor Fibers

A tensile test was performed on the precursor fibers. Regarding the obtained results, the measured average tensile strength at the breaking point varied between the fibers. The fibers which presented the optimum properties were HDPE/SKL–PLA fibers, at 32.33 MPa demonstrating an average diameter of 159.5 μm, while the HDPE/SKL–LA demonstrated 25.12 MPa of tensile strength. The pure HDPE fibers showed a tensile strength of 19.87 MPa. All tests were conducted three times, and then the average value was calculated (3 fibers of each different type were used, 9 fibers in total, 27 measurements in total). The crosshead speed used for testing was 0.2 mm/min. According to the literature [46,47], as far as the orientation is concerned, it was not substantially modified after the tensile test, which was a promising indication for potential thermal treatment (stabilization, carbonization), because during these processes, the fibers tend to shrink and become thicker. Furthermore, these values are related and comply with the tensile properties of natural fiber composites [48], while the difference between the precursor fiber in tensile strength is ascribed to their slightly different microstructure, which is affected by extrusion speed, temperature, and residence time of the materials in the extrusion line and the size of the spinneret [49]. Further, the tensile deformation and damage of fibers are strongly affected by both defects and orientation [50], while the chemical bonding has a severe impact on stiffness and tensile strength of fiber network [51].

3.6. Micro-CT Analysis

The micro-CT analysis results are depicted in Table 8, and the volume rendering images of carbon fibers are presented in Figure 11. Concerning the X-ray tomography obtained results, micro-CT studies on inner geometry and morphology of carbon fiber tapes [52] demonstrated a similar morphology as the one obtained in the present work. Additionally, tomography performed on low-density carbon fiber materials and composites exhibits similar physical properties with home-grown carbon fibers [53,54]. Furthermore, investigation of carbon fiber failure and breakage via micro-CT modeling showed similar values [55–57] and supports effective carbonization in the present study. The surface density is the mass per unit area (mm^2). Linear density is the density that the material demonstrates in relation to its length [58]. In the presented work, it is the amount of mass per millimeter. The degree of anisotropy (DA) is an index that shows the alignment of the materials (fibers) into the total scanned volume [59], and in our study, it is a measure of the orientation of the fibers within the rectangles of the Figure 8. Fractal dimension is a measure showing the change in the detail of a specific object (fibers in this case) as the measuring scale changes [60]. The micro-CT is useful for the scope of the current research as it gives information about the porosity and the morphology of the final material. The innovation of the presented technique applied to the samples produced in this study is the demonstration of a porosity below 2%. This fact underlines the durability and the sturdiness of the synthesized materials.

Table 8. Micro-CT arithmetical values.

Parameters	HDPE/SKL–PLA Carbonized
Object Surface Density (mm^{-1})	17.19
Structure Linear Density (mm^{-1})	12.68
Degree of Anisotropy	1.599
Fractal Dimension	2.600
Total Porosity (%)	1.995

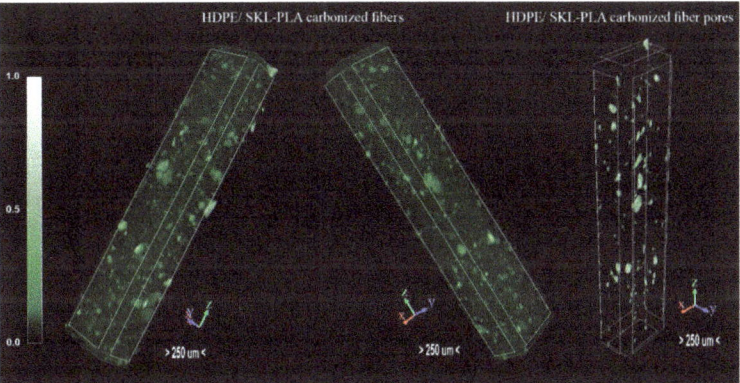

Figure 11. Micro-CT volume rendering images of the HDPE/SKL–PLA carbonized fiber.

3.7. Raman Spectroscopy

Raman spectroscopy measurements provided insight into the carbonization procedure (Figure 12). More specifically, taking into consideration that the presence of G-bands around 1650 cm^{-1} is a straightforward indication of graphite planes formation, it should be mentioned that the I_D/I_G ratio of HDPE/SKL–PLA carbon fibers was 0.847, denoting a lower intensity of D-band at 1300 cm^{-1} compared to G-band, denoting that the sp^3 bonds are more than the sp^2 bonds.

Raman investigation of the produced carbon fibers was performed in order to characterize the structural transformation of the precursor fiber to carbon fiber; the formation of a graphite-like sp^2 plane is expected to be formed as a result of carbonization. Lignin-based carbon fibers presented a promising response to electromagnetic radiation in Raman characterization, with all D, G, and 2D peaks to appear. A trend of forming a sharper and of higher intensity G to D peak was recorded in Raman spectra, which seems to provoke a proper structure reorientation [61].

Looking at the envelope of fit reported in Figure 12, it can be clearly stated that D and G regions encompass four different Gaussian components. Two of them were used to fit the D peak and two of them were used for the G peak. The D peak is described by two components centered at 1330 and 1360 cm^{-1} which present low-intensity variation changing from LA to PLA. The G peak demonstrates a more stable component around 1590–1600 cm^{-1} [62]. To obtain quantitative information by Raman spectroscopy, the ratio of the intensity of D to G peak (I_D/I_G) is measured, and I_D/I_G obtained a value of 0.847, indicating the strong presence of graphitic-like planes present in carbon fiber structure [63–65].

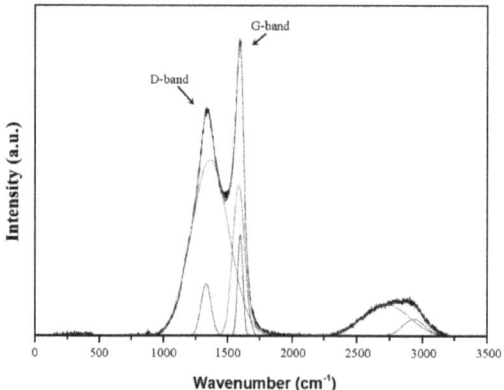

Figure 12. Raman spectroscopy of HDPE/SKL–PLA carbon fiber. The different colored lines correspond to the Gaussian curves used for fitting the density plot of Raman signal via a MATLAB script. The integrals were used to determine I_D/I_G ratio.

3.8. Outcome Analysis

In the present research, a relatively low-cost procedure was followed to reach a carbon fiber production, including a comprehensive stage of organic synthesis, three stabilization stages, and one carbonization stage, with the precursor fiber containing 20 wt.% content of lignin. Moreover, the aim of this study is the industrial upscalability along with simplicity (high profit, high spinning speed and production rate, low cost, one pot reactions), a fact which seems feasible according to this study. According to the literature, despite the fact that chemical modifications of lignin have been investigated [24], there are no references of successful melt-spun lignin conjugation with high molecular weight commercial spinnable polymers, using esterification after a chlorination reaction, so as to enhance the intermolecular bonds between the compounds. The successful chemical synthesis, which was conducted in this study without the use of plasticization [66] and confirmed by characterization, provided thermal process ability to the precursor fiber. Anthracene-oil-derived pitches as precursors have led to green carbon fibers [67], but the diameter of the fibers varied depending on the winding speed, and overall, that method was more expensive than using lignin in the blends. Additionally, carbon fibers from low molecular weight polymers by solvent extraction [68] and from polyacrylonitrile combined with CNTs [69] showed good strength but lacked the low-cost synthesis. While cellulose has been used in the fabrication of carbon fibers [70], the potential of lignin has been acknowledged already [2], but its handling and treatment remained an issue. Finally, lignin, compared to cellulose, is more suitable as a precursor for carbon fiber due to its aromatic groups, which strengthen the material structure after carbonization (higher carbon yield) [71].

4. Conclusions

In the present study, lactic acid and poly(lactic acid) were modified by chlorination, in order to produce the corresponding organic chlorides. Then, these compounds were esterified with SKL. Subsequently, the obtained product was compounded and spun with HDPE thermoplastic polymer to form composite precursors suitable for carbon fiber production. Stabilization and carbonization were conducted, and the obtained HDPE/SKL–PLA fibers demonstrated lower I_D/I_G, by Raman characterization for carbonization at 800 °C, which was adopted as optimal thermal processing. The relative low I_D/I_G ratio, compared to carbon fibers reported in literature, indicates that graphitic sp^2 planes are substantially formed and oriented, and thus, it can be feasible to enhance the electrical properties of composites. The results from this work can constitute a solid basis for scaling up similar procedures.

Author Contributions: P.G., I.A.K., and C.A.C. conceived and designed the experiments. P.G. conducted the experiments. P.G., G.K., and I.A.K. performed the characterization and evaluation of the data. P.G., I.A.K., and C.A.C. discussed the data and wrote the paper.

Funding: This research was funded by the HORIZON 2020 Collaborative project "LORCENIS" (Long Lasting Reinforced Concrete for Energy Infrastructure under Severe Operating Conditions, Grant agreement no.: 685445).

Conflicts of Interest: The authors declare no conflict of interest.

Appendix A

Table A1. Tabulated parameters and conditions for the optimization process for the HDPE spinning (HDPE/SKL–LA (80:20)).

Blend	Stage 1 Temperature (°C)	Stage 2 Temperature (°C)	Extrusion Speed (% rpm)	Roller Distance (cm)
HDPE/SKL–LA	220 226	230	100	23
	220	230 235	100	23
	220	230	80 100	23
	220	230	100	23 28
	226	230 235	100	23
	226	230	80 100	23
	226	230	100	23 28

Table A2. Tabulated parameters and conditions for the optimization process for the HDPE spinning (HDPE/SKL–PLA (80:20)).

Blend	Stage 1 Temperature (°C)	Stage 2 Temperature (°C)	Extrusion Speed (% rpm)	Roller Distance (cm)
HDPE/SKL–PLA	215 222	225	85	25
	215	225 230	85	25
	215	225	65 85	25
	215	225	85	25 30
	222	225 230	85	25
	222	225	65 85	25
	222	225	85	25 30

References

1. Thunga, M.; Chen, K.; Grewell, D.; Kessler, M.R. Bio-renewable precursor fibers from lignin/polylactide blends for conversion to carbon fibers. *Carbon* **2014**, *68*, 159–166. [CrossRef]
2. Mainka, H.; Täger, O.; Körner, E.; Hilfert, L.; Busse, S.; Edelmann, F.T.; Herrmann, A.S. Lignin—An alternative precursor for sustainable and cost-effective automotive carbon fiber. *J. Mater. Res. Technol.* **2015**, *4*, 283–296. [CrossRef]
3. Maradur, S.P.; Kim, C.H.; Kim, S.Y.; Kim, B.H.; Kim, W.C.; Yang, K.S. Preparation of carbon fibers from a lignin copolymer with polyacrylonitrile. *Synth. Met.* **2012**, *162*, 453–459. [CrossRef]
4. Ragauskas, A.J.; Beckham, G.T.; Biddy, M.J.; Chandra, R.; Chen, F.; Davis, M.F.; Davison, B.H.; Dixon, R.A.; Gilna, P.; Keller, M.; et al. Lignin valorization: Improving lignin processing in the biorefinery. *Science* **2014**, *344*, 1246843. [CrossRef]
5. Naseem, A.; Tabasum, S.; Zia, K.M.; Zuber, M.; Ali, M.; Noreen, A. Lignin-derivatives based polymers, blends and composites: A review. *Int. J. Biol. Macromol.* **2016**, *93 Pt A*, 296–313. [CrossRef]
6. Datta, J.; Parcheta, P.; Surówka, J. Softwood-lignin/natural rubber composites containing novel plasticizing agent: Preparation and characterization. *Ind. Crop. Prod.* **2017**, *95*, 675–685. [CrossRef]
7. Ferdosian, F.; Yuan, Z.; Anderson, M.; Xu, C.C. Synthesis and characterization of hydrolysis lignin-based epoxy resins. *Ind. Crop. Prod.* **2016**, *91*, 295–301. [CrossRef]
8. Simionescu, C.I.; Rusan, V.; Macoveanu, M.M.; Cazacu, G.; Lipsa, R.; Vasile, C.; Stoleriu, A.; Ioanid, A. Lignin/epoxy composites. *Compos. Sci. Technol.* **1993**, *48*, 317–323. [CrossRef]
9. Zhou, Y.; Fan, M.; Chen, L. Interface and bonding mechanisms of plant fibre composites: An overview. *Compos. B Eng.* **2016**, *101*, 31–45. [CrossRef]
10. Megiatto, J.D.; Silva, C.G.; Rosa, D.S.; Frollini, E. Sisal chemically modified with lignins: Correlation between fibers and phenolic composites properties. *Polym. Degrad. Stab.* **2008**, *93*, 1109–1121. [CrossRef]
11. Lee, K.-Y.; Ho, K.K.C.; Schlufter, K.; Bismarck, A. Hierarchical composites reinforced with robust short sisal fibre preforms utilising bacterial cellulose as binder. *Compos. Sci. Technol.* **2012**, *72*, 1479–1486. [CrossRef]
12. Migneault, S.; Koubaa, A.; Perré, P.; Riedl, B. Effects of wood fiber surface chemistry on strength of wood–plastic composites. *Appl. Surf. Sci.* **2015**, *343*, 11–18. [CrossRef]
13. Alshaaer, M.; Mallouh, S.A.; Al-Kafawein, J.A.; Al-Faiyz, Y.; Fahmy, T.; Kallel, A.; Rocha, F. Fabrication, microstructural and mechanical characterization of Luffa Cylindrical Fibre—Reinforced geopolymer composite. *Appl. Clay Sci.* **2017**, *143*, 125–133. [CrossRef]
14. Hervy, M.; Evangelisti, S.; Lettieri, P.; Lee, K.-Y. Life cycle assessment of nanocellulose-reinforced advanced fibre composites. *Compos. Sci. Technol.* **2015**, *118*, 154–162. [CrossRef]
15. Ye, X.; Wang, H.; Zheng, K.; Wu, Z.; Zhou, H.; Tian, K.; Su, Z.; Tian, X. The interface designing and reinforced features of wood fiber/polypropylene composites: Wood fiber adopting nano-zinc-oxide-coating via ion assembly. *Compos. Sci. Technol.* **2016**, *124*, 1–9. [CrossRef]
16. Soman, S.; Chacko, A.S.; Prasad, V.S. Semi-interpenetrating network composites of poly(lactic acid) with cis-9-octadecenylamine modified cellulose-nanofibers from Areca catechu husk. *Compos. Sci. Technol.* **2017**, *141*, 65–73. [CrossRef]
17. Zhang, Y.; Gan, T.; Luo, Y.; Zhao, X.; Hu, H.; Huang, Z.; Huang, A.; Qin, X. A green and efficient method for preparing acetylated cassava stillage residue and the production of all-plant fibre composites. *Compos. Sci. Technol.* **2014**, *102*, 139–144. [CrossRef]
18. Vaikhanski, L.; Lesko, J.J.; Nutt, S.R. Cellular polymer composites based on bi-component fibers. *Compos. Sci. Technol.* **2003**, *63*, 1403–1410. [CrossRef]
19. Del Saz-Orozco, B.; Oliet, M.; Alonso, M.V.; Rojo, E.; Rodríguez, F. Formulation optimization of unreinforced and lignin nanoparticle-reinforced phenolic foams using an analysis of variance approach. *Compos. Sci. Technol.* **2012**, *72*, 667–674. [CrossRef]
20. Wang, R.; Zhang, J.; Kang, H.; Zhang, L. Design, preparation and properties of bio-based elastomer composites aiming at engineering applications. *Compos. Sci. Technol.* **2016**, *133*, 136–156. [CrossRef]
21. Meek, N.; Penumadu, D.; Hosseinaei, O.; Harper, D.; Young, S.; Rials, T. Synthesis and characterization of lignin carbon fiber and composites. *Compos. Sci. Technol.* **2016**, *137*, 60–68. [CrossRef]
22. Wang, S.; Zhou, Z.; Xiang, H.; Chen, W.; Yin, E.; Chang, T.; Zhu, M. Reinforcement of lignin-based carbon fibers with functionalized carbon nanotubes. *Compos. Sci. Technol.* **2016**, *128*, 116–122. [CrossRef]

23. Asp, L.E.; Greenhalgh, E.S. Structural power composites. *Compos. Sci. Technol.* **2014**, *101*, 41–61. [CrossRef]
24. Laurichesse, S.; Avérous, L. Chemical modification of lignins: Towards biobased polymers. *Prog. Polym. Sci.* **2014**, *39*, 1266–1290. [CrossRef]
25. McMurry, J. Carboxylic Acid Derivatives: Nucleophilic Acyl Substitution Reactions of Carboxylic Acids. In *Organic Chemistry*, 7th ed.; Brooks/Cole: Belmont, CA, USA, 2008; 21.3; pp. 794–800.
26. McMurry, J. Carboxylic Acid Derivatives: Chemistry of Halide Acids. In *Organic Chemistry*, 7th ed.; Brooks/Cole: Belmont, CA, USA, 2008; 21.4; pp. 800–806.
27. Goliszek, M.; Wiącek, A.E.; Wawrzkiewicz, M.; Sevastyanova, O.; Podkościelna, B. The impact of lignin addition on the properties of hybrid microspheres based on trimethoxyvinylsilane and divinylbenzene. *Eur. Polym. J.* **2019**, *120*, 109200. [CrossRef]
28. Wang, X.; Zhai, M.; Wang, Z.; Dong, P.; Lv, W.; Liu, R. Carbonization and combustion characteristics of palm fiber. *Fuel* **2018**, *227*, 21–26. [CrossRef]
29. Szycher, M. *Szycher's Handbook of Polyurethanes*, 2nd ed.; CRC Press, Taylor & Francis Group: Boca Raton, FL, USA, 2013; pp. 87–134.
30. Chauhan, M.; Gupta, M.; Singh, B.; Singh, A.K.; Gupta, V.K. Effect of functionalized lignin on the properties of lignin–isocyanate prepolymer blends and composites. *Eur. Polym. J.* **2014**, *52*, 32–43. [CrossRef]
31. Dodda, J.M.; Bělský, P. Progress in designing poly(amide imide)s (PAI) in terms of chemical structure, preparation methods and processability. *Eur. Polym. J.* **2016**, *84*, 514–537. [CrossRef]
32. Puszka, A.; Kultys, A. New thermoplastic polyurethane elastomers based on aliphatic diisocyanate. *J. Therm. Anal. Calorim.* **2016**, *128*, 407–416. [CrossRef]
33. Ouyang, Q.; Wang, X.; Wang, X.; Huang, J.; Huang, X.; Chen, Y. Simultaneous DSC/TG analysis on the thermal behavior of PAN polymers prepared by aqueous free-radical polymerization. *Polym. Degrad. Stab.* **2016**, *130*, 320–327. [CrossRef]
34. Tripathi, A.K.; Tsavalas, J.G.; Sundberg, D.C. Quantitative measurements of the extent of phase separation during and after polymerization in polymer composites using DSC. *Thermochim. Acta* **2013**, *568*, 20–30. [CrossRef]
35. Schawe, J.E.K. Remarks regarding the determination of the initial crystallinity by temperature modulated DSC. *Thermochim. Acta* **2017**, *657*, 151–155. [CrossRef]
36. Dumas, J.P.; Gibout, S.; Cézac, P.; Franquet, E. New theoretical determination of latent heats from DSC curves. *Thermochim. Acta* **2018**, *670*, 92–106. [CrossRef]
37. Klásek, A.; Rege, D.V. Reaction of 4-hydroxy-2-quinolones with thionyl chloride—Preparation of new spiro-benzo[1,3]oxathioles and their transformations. *Tetrahedron* **2013**, *69*, 492–499. [CrossRef]
38. Merchant, J.R.; Rege, D.V. Reaction of thionyl chloride with flavone. *Tetrahedron Lett.* **1969**, *10*, 3589–3591. [CrossRef]
39. Chern, Y.-T.; Huang, C.-M. Synthesis and characterization of new polyesters derived from 1,6- or 4,9-diamantanedicarboxylic acyl chlorides with aryl ether diols. *Polymer* **1998**, *39*, 2325–2329. [CrossRef]
40. Sutter, P.; Weis, C.D. The chlorination of 1,4-dihydroxyanthraquinone with thionyl chloride. *Dye. Pigment.* **1985**, *6*, 435–443. [CrossRef]
41. Freeman, J.H.; Smith, M.L. The preparation of anhydrous inorganic chlorides by dehydration with thionyl chloride. *J. Inorg. Nucl. Chem.* **1958**, *7*, 224–227. [CrossRef]
42. Zhao, J.; Xiuwen, W.; Hu, J.; Liu, Q.; Shen, D.; Xiao, R. Thermal degradation of softwood lignin and hardwood lignin by TG-FTIR and Py-GC/MS. *Polym. Degrad. Stab.* **2014**, *108*, 133–138. [CrossRef]
43. Jin, W.; Shen, D.; Liu, Q.; Xiao, R. Evaluation of the co-pyrolysis of lignin with plastic polymers by TG-FTIR and Py-GC/MS. *Polym. Degrad. Stab.* **2016**, *133*, 65–74. [CrossRef]
44. Ravindar Reddy, M.; Subrahmanyam, A.R.; Maheshwar Reddy, M.; Siva Kumar, J.; Kamalaker, V.; Jaipal Reddy, M. X-RD, SEM, FT-IR, DSC Studies of Polymer Blend Films of PMMA and PEO. *Mater. Today Proc.* **2016**, *3*, 3713–3718. [CrossRef]
45. Soulis, S.; Anagnou, S.; Milioni, E.; Mpalias, C.; Kartsonakis, I.A.; Kanellopoulou, I.; Markakis, V.; Koumoulos, E.P.; Kontou, E.; Charitidis, C.A. Strategies towards Novel Carbon Fiber Precursors: The Research Results on the Synthesis of PAN Copolymers via AGET ATRP and on Lignin as a Precursor. *NanoWorld J.* **2015**, *1*, 88–94. [CrossRef]
46. Abrishambaf, A.; Pimentel, M.; Nunes, S. Influence of fibre orientation on the tensile behaviour of ultra-high performance fibre reinforced cementitious composites. *Cem. Concr. Res.* **2017**, *97*, 28–40. [CrossRef]

47. Jang, S.Y.; Ko, S.; Jeon, Y.P.; Choi, J.; Kang, N.; Kim, H.C.; Joh, H.-I.; Lee, S. Evaluating the stabilization of isotropic pitch fibers for optimal tensile properties of carbon fibers. *J. Ind. Eng. Chem.* **2017**, *45*, 316–322. [CrossRef]
48. Torres, J.P.; Vandi, L.J.; Veidt, M.; Heiztmann, M.T. Statistical data for the tensile properties of natural fibre composites. *Data Brief* **2017**, *12*, 222–226. [CrossRef]
49. Zhong, Y.; Bian, W. Analysis of the tensile moduli affected by microstructures among seven types of carbon fibers. *Compos. B Eng.* **2017**, *110*, 178–184. [CrossRef]
50. Bie, B.X.; Huang, J.Y.; Fan, D.; Sun, T.; Fezzaa, K.; Xiao, X.H.; Qi, M.L.; Luo, S.N. Orientation-dependent tensile deformation and damage of a T700 carbon fiber/epoxy composite: A synchrotron-based study. *Carbon* **2017**, *121*, 127–133. [CrossRef]
51. Borodulina, S.; Motamedian, H.R.; Kulachenko, A. Effect of fiber and bond strength variations on the tensile stiffness and strength of fiber networks. *Int. J. Solids Struct.* **2016**, *154*, 19–32. [CrossRef]
52. Wan, Y.; Straumit, I.; Takahashi, J.; Lomov, S.V. Micro-CT analysis of internal geometry of chopped carbon fiber tapes reinforced thermoplastics. *Compos. A Appl. Sci. Manuf.* **2016**, *91*, 211–221. [CrossRef]
53. Ferguson, J.C.; Panerai, F.; Lachaud, J.; Martin, A.; Bailey, S.C.C.; Mansour, N.N. Modeling the oxidation of low-density carbon fiber material based on micro-tomography. *Carbon* **2016**, *96*, 57–65. [CrossRef]
54. Liu, J.; Li, C.; Liu, J.; Cui, G.; Yang, Z. Study on 3D spatial distribution of steel fibers in fiber reinforced cementitious composites through micro-CT technique. *Constr. Build. Mater.* **2013**, *48*, 656–661. [CrossRef]
55. Sencu, R.M.; Yang, Z.; Wang, Y.C.; Withers, P.J.; Rau, C.; Parson, A.; Soutis, C. Generation of micro-scale finite element models from synchrotron X-ray CT images for multidirectional carbon fibre reinforced composites. *Compos. A Appl. Sci. Manuf.* **2016**, *91*, 85–95. [CrossRef]
56. Cosmi, F.; Bernasconi, A. Micro-CT investigation on fatigue damage evolution in short fibre reinforced polymers. *Compos. Sci. Technol.* **2013**, *79*, 70–76. [CrossRef]
57. Ning, Z.; Liu, R.; Elhajjar, R.F.; Wang, F. Micro-modeling of thermal properties in carbon fibers reinforced polymer composites with fiber breaks or delamination. *Compos. B Eng.* **2017**, *114*, 247–255. [CrossRef]
58. Headrick, R.J.; Trafford, M.A.; Taylor, L.W.; Dewey, O.S.; Wincheski, R.A.; Pasquali, M. Electrical and acoustic vibroscopic measurements for determining carbon nanotube fiber linear density. *Carbon* **2019**, *144*, 417–422. [CrossRef]
59. Putignano, C.; Menga, N.; Afferrante, L.; Carbone, G. Viscoelasticity induces anisotropy in contacts of rough solids. *J. Mech. Phys. Solids* **2019**, *129*, 147–159. [CrossRef]
60. Xia, Y.; Cai, J.; Perfect, E.; Wei, W.; Zhang, Q.; Meng, Q. Fractal dimension, lacunarity and succolarity analyses on CT images of reservoir rocks for permeability prediction. *J. Hydrol.* **2019**, *579*, 124198. [CrossRef]
61. Liu, X.; Zhu, C.; Guo, J.; Liu, Q.; Dong, H.; Gu, Y.; Liu, R.; Zhao, N.; Zhang, Z.; Xu, J. Nanoscale dynamic mechanical imaging of the skin–core difference: From PAN precursors to carbon fibers. *Mater. Lett.* **2014**, *128*, 417–420. [CrossRef]
62. Long, D.A. Infrared and Raman characteristic group frequencies. Tables and chartsGeorge Socrates John Wiley and Sons, Ltd, Chichester, Third Edition, 2001, Price £135. *J. Raman Spectrosc.* **2004**, *35*, 905. [CrossRef]
63. Musiol, P.; Szatkowski, P.; Gubernat, M.; Weselucha-Birczynska, A.; Blazewicz, S. Comparative study of the structure and microstructure of PAN-based nano- and micro-carbon fibers. *Ceram. Int.* **2016**, *42*, 11603–11610. [CrossRef]
64. Jin, S.; Guo, C.; Lu, Y.; Zhang, R.; Wang, Z.; Jin, M. Comparison of microwave and conventional heating methods in carbonization of polyacrylonitrile-based stabilized fibers at different temperature measured by an in-situ process temperature control ring. *Polym. Degrad. Stab.* **2017**, *140*, 32–41. [CrossRef]
65. Park, O.-K.; Lee, S.; Joh, H.-I.; Kim, J.K.; Kang, P.-H.; Lee, J.H.; Ku, B.-C. Effect of functional groups of carbon nanotubes on the cyclization mechanism of polyacrylonitrile (PAN). *Polymer* **2012**, *53*, 2168–2174. [CrossRef]
66. Batchelor, B.L.; Mahmood, S.F.; Jung, M.; Shin, H.; Kulikov, O.V.; Voit, W.; Novak, B.M.; Yang, D.J. Plasticization for melt viscosity reduction of melt processable carbon fiber precursor. *Carbon* **2016**, *98*, 681–688. [CrossRef]
67. Berrueco, C.; Álvarez, P.; Díez, N.; Granda, M.; Menéndez, R.; Blanco, C.; Santamaria, R.; Millan, M. Characterisation and feasibility as carbon fibre precursors of isotropic pitches derived from anthracene oil. *Fuel* **2012**, *101*, 9–15. [CrossRef]
68. Li, X.; Zhu, X.-Q.; Okuda, K.; Zhang, Z.; Ashida, R.; Yao, H.; Miura, K. Preparation of carbon fibers from low-molecular-weight compounds obtained from low-rank coal and biomass by solvent extraction. *New Carbon Mater.* **2017**, *32*, 41–47. [CrossRef]

69. Şahin, K.; Fasanella, N.A.; Chasiotis, I.; Lyons, K.M.; Newcomb, B.A.; Kamath, M.G.; Chae, H.G.; Kumar, S. High strength micron size carbon fibers from polyacrylonitrile–carbon nanotube precursors. *Carbon* **2014**, *77*, 442–453. [CrossRef]
70. Spörl, J.M.; Ota, A.; Son, S.; Massonne, K.; Hermanutz, F.; Buchmeiser, M.R. Carbon fibers prepared from ionic liquid-derived cellulose precursors. *Mater. Today Commun.* **2016**, *7*, 1–10. [CrossRef]
71. Guo, T.; Xia, Q.; Shao, Y.; Liu, X.; Wang, Y. Direct deoxygenation of lignin model compounds into aromatic hydrocarbons through hydrogen transfer reaction. *Appl. Catal. A Gen.* **2017**, *547*, 30–36. [CrossRef]

© 2019 by the authors. Licensee MDPI, Basel, Switzerland. This article is an open access article distributed under the terms and conditions of the Creative Commons Attribution (CC BY) license (http://creativecommons.org/licenses/by/4.0/).

MDPI
St. Alban-Anlage 66
4052 Basel
Switzerland
Tel. +41 61 683 77 34
Fax +41 61 302 89 18
www.mdpi.com

Applied Sciences Editorial Office
E-mail: applsci@mdpi.com
www.mdpi.com/journal/applsci

www.ingramcontent.com/pod-product-compliance
Lightning Source LLC
LaVergne TN
LVHW070413100526
838202LV00014B/1447